CHICAGO PUBLIC LIBRARY
HAROLD WASHINGTON LIBRARY CENTER
R0019596270

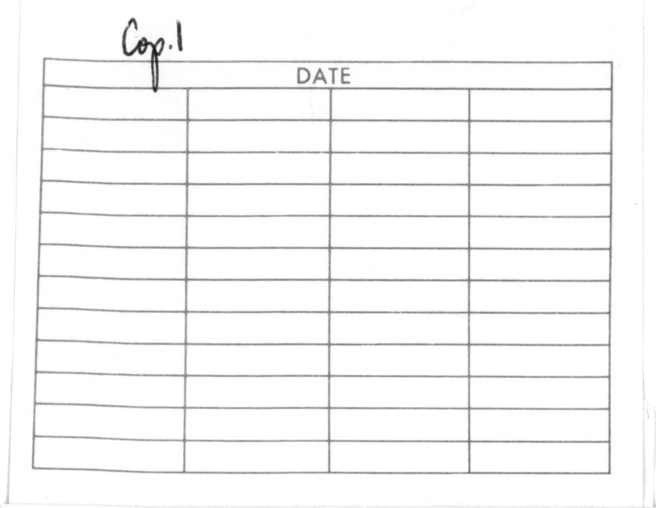
DATE

D1550388

THE SOVIET ENERGY SYSTEM:
Resource Use and Policies

SCRIPTA SERIES IN GEOGRAPHY

Series Editors

Richard E. Lonsdale • Antony R. Orme • Theodore Shabad

Shabad and Mote · Gateway to Siberian Resources, 1977

Lonsdale and Seyler · Nonmetropolitan Industrialization, 1979

Rodgers · Economic Development in Retrospect, 1979

Pyle · Applied Medical Geography, 1979

Dienes and Shabad · The Soviet Energy System, 1979

THE SOVIET ENERGY SYSTEM: Resource Use and Policies

Leslie Dienes
University of Kansas
and
Theodore Shabad
Columbia University

1979

V. H. WINSTON & SONS
Washington, D.C.

A HALSTED PRESS BOOK

JOHN WILEY & SONS
New York Toronto London Sydney

V. H. Winston & Sons, a Division of Scripta Technica, Inc., Publishers
1511 K Street, N.W., Washington, D.C. 20005

Distributed solely by Halsted Press, a Division of John Wiley & Sons, Inc.

Library of Congress Cataloging in Publication Data:

Dienes, Leslie.
The Soviet energy system.

(Scripta series in geography)
1. Power resources–Russia. 2. Energy policy–Russia. I. Shabad, Theodore, joint author. II. Title. II. Series.
TJ163.25.R9D53 333.7 78–20814
ISBN 0–470–26629–5

Composition by Marie A. Maddalena, Scripta Technica, Inc.

CONTENTS

PREFACE

The vast fuel and energy potential of the USSR has been thrust into the forefront of world attention for two countervailing reasons. On the one hand, the policy of East-West détente and the associated increase of Soviet participation in world trade have raised expectations that a growing share and rising volume of the Soviet Union's energy resources, especially petroleum and natural gas, may continue to be available to Western industrial countries. It is thought by some that Soviet resources may help alleviate what has been perceived as an energy crisis in Western Europe and Japan, but also to some extent in the United States. On the other hand, some analysts, in the Central Intelligence Agency of the United States, have predicted that Soviet production of crude oil, which has been ranking first in the world since 1974, will peak in the early 1980s. Such predictions have been prompted by two major factors: the discovery of new oil reserves has failed to keep up with the growing volume of production and the forced-draft production techniques used by the Russians tend to accelerate the depletion of existing oil fields. It has even been suggested that declining petroleum production in the 1980s may make the USSR a net importer of oil, thus adding to the pressure on world oil supplies and aggravating the energy crisis.

Soviet planners, who for 20 years have relied on hydrocarbons for most of the fuel increments, are apparently confident that they can avoid an early downturn in domestic oil output, while projecting continued rapid growth for the gas industry. But they concede that economies are needed and stress that hydrocarbons, particularly petroleum, are too valuable to be burned under boilers for the production of steam and electricity. In an effort to conserve these resources and to make more oil available for petrochemicals and export (taking advantage of high world prices), the planners have proclaimed a new energy policy that would reduce the role of oil as a power-station and industrial fuel and give more attention to the use of coal, especially that of cheap strip-mined lignite. They also appear to look to the Soviet Union's vast natural gas reserves to help reduce the domestic requirements for petroleum, permitting continued export, but increasingly also to earn valuable foreign exchange from the sale of gas as well.

The resolution of the Soviet energy problems, not least the proposed shift of emphasis toward gas and coal, has been complicated by the spatial discrepancy between the concentration of population and economy in the western (European) third of the USSR and the presence of most fuel and energy resources in the eastern (Asian) portion of the country, particularly Siberia. From prospecting through development and transport, this pronounced spatial discrepancy affects all branches of the Soviet energy industries, but in a uneven fashion. Energy policy and planning thus impinge on problems of regional development and the territorial structure of the economy as a whole.

The object of this book is to analyze the alternatives that Soviet energy planners face in meeting both the requirements of an expanding economy, the world's largest after that of the United States, and the need for exporting high-priced energy products to pay for imports of advanced industrial technology from the West. Some of the issues under consideration in the USSR, such as the proposed conversion from oil to coal, have parallels in the United States. A renewed stress on hydroelectric power and an apparently wholehearted commitment to nuclear energy to reduce the strain on fossil fuels are also considered.

In the introductory chapter, Leslie Dienes, one of the two main authors, highlights the strong interdependence and connectivity within the energy system and places the latter in the context of geographic regions and the institutional and political environment of the Soviet state. In Chapter 2, he then examines the prevailing trends in the energy economy, including the growth and structure of demand by economic sectors and technological uses, the changing resource mix and the geographic pattern of demand.

This preliminary analysis, treating energy as a more or less uniform end-product, is then followed by a detailed examination of the geography of primary energy supplies by Theodore Shabad, the other principal author. Examining the relationship between markets and producing areas and the pattern of transport flows. Shabad reviews the history of development and the geographic distribution of the fossil fuels (petroleum, natural gas and the solid fuels) and of hydroelectric power. Particular attention is given to recent development projects and to the basic areas of production on which the Soviet Union will have to rely for its future energy supplies.

Philip R. Pryde of San Diego State University has contributed a chapter on nuclear power, the other form of primary electricity that adds to aggregate energy supplies without making use of conventional fossil fuels. The geography of uranium, the nuclear fuel, on which there is virtually no published literature, is discussed by Shabad on the basis of indirect evidence.

Chapters 7 through 11 by Dienes are devoted to the analysis of the complex problems of utilization and allocation of energy and to the modeling, planning and administration of the Soviet energy system. The author examines the increasingly important role of the electric power industry, with an emphasis on thermal power, the conflict between export opportunities and domestic fuel needs, the possibilities of fuel substitution in the different technological uses, Soviet energy forecasts and long-term prospects, administrative and policy conflicts and the debate within the Soviet Union on the optimal course to follow.

This treatment of the Soviet energy system and its future prospects appears at a time when the Soviet Union, for undisclosed reasons, has greatly restricted the availability of published statistical information about the regional distribution of production and about exports and imports of all fossil fuels. The present volume thus represents an analysis of available data up to the imposition of secrecy on the systematic publication of fuel production and trade statistics starting in 1977.

Leslie Dienes wishes to express his sincere appreciation to his colleagues Professors Thomas R. Smith and John P. Augelli of the University of Kansas for their critical reading of his part of the manuscript. Their comments have been invaluable in improving style and weeding out inconsistencies in the argument. He also wants to thank Marilyn Odell for her help with library sources for several chapters and in typing the first draft. Among Soviet specialists, he is happy to acknowledge an intellectual debt to Professor Robert W. Campbell of Indiana University, whose research on Soviet energy problems has influenced him greatly. With the exception of the nuclear power map, all maps and diagrams were designed by Norman Carpenter, and executed by Gerard Blood and Mr. Carpenter, of the Cartographic Service of the Department of Geography at the University of Kansas; the two principal authors are much indebted to them. Finally, Leslie Dienes wishes to thank the Research Committee of the university's Faculty Senate, which provided financial assistance for cartographic work and for the final typing of his chapters.

Chapter 1

INTRODUCTION

Economic development involves the enhancement and use of natural resources through human skills and technology. In recent years, energy has emerged as both the focus and a symbol of global anxiety concerning the management of those natural bounties. The painfully uneven distribution of energy reserves, the massive global flows and the attendant questions of interdependence and security clearly point to the international nature of the energy problem. But large resources may have an international impact without directly entering world trade. They may ease pressure on alternative supplies elsewhere, and mere perception of them can influence policy well beyond the borders within which they are found.

Through sheer size, Soviet energy resources and the Soviet energy system have global impact. Soviet fossil fuel and waterpower potentials are easily first in the world. A role such as played by the Middle East today is at least conceivable for the USSR, if mankind's shift to nonconventional and/or renewable energy sources is long delayed. The Soviet Union is already second in the world in both the production and the consumption of total energy but has till now depleted its reserves far less than other industrial nations. The sheer magnitude of the energy it consumes (more than the nine-member European Economic Community) and the rapid growth rate of this demand mean that the way the USSR chooses, and is able, to satisfy its requirement and manage its resources is a matter of importance for the world as a whole. The Soviet Union is already the largest oil and coal producer (the latter in terms of tonnage though not calories produced) and is certain to become the largest gas producer soon after 1980. In oil exports today it holds third place, it is a significant coal exporter, and it is rapidly becoming a substantial exporter of natural gas.

The development and use of energy resources claim a high share of total investment and research efforts everywhere. Since 1928, the beginning of the five-year plans, the Soviet energy industries have received some three-tenths of all productive capital invested in industry, i.e., not counting that for supporting infrastructure.[1] If investment in energy utilization at the consumer end is included, the Soviet energy system probably claims a full half of all productive industrial investment and well over one-sixth of total investment in the national economy.[2] Given such high shares and the uneven and mismatched distribution of resources and population, investment strategies in the energy industries also have momentous consequences on the regional dimensions and structure of the economy as a whole and, in turn, are influenced by them.

In the USSR, with state ownership of the means of production and a strongly centralized system of resources allocation, planners' control over the energy sector is applied with particular force. This sector represents one of the most important "commanding heights" of a socialist economy where spontaneous market forces have never been permitted to intrude. Therefore the perception of specialists and planners concerning long-range energy prospects and the management of these crucial resources with respect to domestic needs and foreign trade reveal much about the possible course of Soviet economic strategy. Finally, the relatively straightforward linkages and substitution possibilities within this strongly interconnected sector make it possible to examine the way a centralized economy deals with allocation–substitution problems among commodities and regions, and consider the concepts and tools that are employed to optimize resource use in a goal-directed system.

THREE DIMENSIONS OF THE SOVIET ENERGY SYSTEM

Any attempt to understand the dynamics of Soviet energy supply and demand requires a three-dimensional view. The first dimension is represented by the high degree of sectoral interdependence, connectivity and substitutability within the energy economy. The second dimension is the geographical context, the third the institutional and political one.

The Interfuel and Sectoral Context

In the medium and longer run, in some cases even on short notice, primary energy sources and energy forms are substitutable to a large degree, though, of course, not without significant costs. This is particularly true if one excludes motor fuels and coke from the fuel balance and considers furnace and boiler uses only (including those of households), which in the USSR account for some four-fifths of all energy demand. Changing scarcity relationships may radically alter the value of different fuels, creating powerful pressures for large-scale substitution as witnessed in the world today.

Beyond their higher calorific content and greater transportability, petroleum and natural gas also effect large savings over solid fuels in virtually all fields of use because of greater flexibility, combustion efficiency, cleanliness and convenience. This saving is far from uniform. The advantage over coal (and lignite, peat and shale) under boilers is more modest than in the more specialized furnace uses, while most of the transportation and military sectors, other mobile demand and chemical synthesis today are technologically tied to petroleum.

Mobile and chemical uses have comprised only a fifth of gross Soviet energy demand. Therefore, one of the consequences of the post-Stalin shift to liquid and gaseous fuels was the large-scale consumption of oil and gas not only in the above uses but also in the furnace and boiler markets. With some important variations, this shift and the consequent consumption pattern paralleled those achieved in West Europe at roughly the same time. Such a consumption pattern requires crude oil to be only lightly refined: cracking, reforming and other secondary refining processes, therefore, are relatively poorly developed and close to half of refinery output consists of heavy products, such as residual fuel oil and gas oil. Though the shift to oil and gas in the United States and Canada has been even more marked, the North American pattern has been distinguished by a much smaller role of oil in boiler and furnace uses. Until very recently, the lucrative and large gasoline market in the United States has kept down the share of petroleum as a boiler fuel, while price regulation and, lately, antipollution laws encouraged the extensive penetration of natural gas, at least until the early 1970s. North American refineries refine in greater depth, maximizing the yield of light products, especially gasoline.

Over the next decade and a half, Soviet experts project a significant growth in the share of boiler use (for electricity, steam and hot water) in gross energy consumption from about 45 to almost 60 percent (Table 3). A key question of Soviet energy policy is the choice of fuel for the rapidly expanding boiler market, particularly for power plants, which account for three-fourths of all boiler use. In the early 1970s about one-quarter of all petroleum products and over two-fifths of natural gas were burned for the generation of electricity, steam and hot water,[3] i.e., in uses where the technological and economic potential of these quality fuels is not maximized. The mounting anxiety among Western powers over the adequacy of future hydrocarbon supplies has made a deep impression on Soviet leaders. The USSR's own oil and gas industries must also cope with significant problems, including a rapid rise in costs. In addition, proved reserves of oil do not appear entirely satisfactory, and prospects for this pivotal energy source seem more worrisome than for other fuels. Petroleum and natural gas are also crucial hard-currency earners and are desperately needed by the USSR's East European partners as well. Soviet officials have become more concerned with husbanding their hydrocarbon resources and, like their United States counterparts, have begun to advocate greater reliance on coal and a rapid expansion of nuclear electricity.

The Geographic Context

The Soviet energy economy also operates in a geographic space that embraces one-sixth of the earth's surface, displays huge variations in resource endowment, extraction possibilities and accessibility and suffers from a crushing areal discordance between energy demands and supplies. To a large extent, this geographic space also comprises the six East European countries of the Soviet-led Council for Mutual Economic Assistance (CMEA), which are politically but increasingly also economically integrated into the Soviet system and are critically dependent on Soviet natural resources.

Soviet planners must match demand and supply over an area larger than North America (north of Panama) with a population of 260 million. With the European CMEA countries, this realm is three times the size of the continental United States and has 368 million consumers. The lack of suitable, long-distance waterways for bulk transport enhances dependence on extra long pipelines and railway hauls. The burden of distance is awesome. Petroleum and gas from West Siberian and Central Asian fields are now piped 1,500 to 2,000 miles to major Soviet consumer points; crude oil already moves 1,000 miles farther west to European cities, and the huge pipelines under construction for the export of gas will be equally long. Large quantities of Kuznetsk Basin coal are railed 1,500 to 2,000 miles to European Russia and exported to Japan through Pacific ports (3,600 miles).

From the perspective of the energy economy, this world breaks into three different realms, though each, of course, contains several dissimilar regions.

Soviet Asia, twice the size of the continental United States, contains nine-tenths of the country's potential energy reserves and vast stores of most other minerals. Sixty-seven million people inhabit that land, but only half of them live in Siberia and adjoining northern Kazakhstan, where most of these resources are found.[4] The greater part of potential reserves are inaccessible with present-day technology and most of the coal and waterpower potential may always remain so. Still, Soviet Asia, and particularly Siberia, represents the biggest storehouse of usable, untapped energy reserves in the Soviet realm (possibly the world as well), and the only part of this realm with a huge energy surplus.

Soviet Asia is a harsh land. This is particularly true of Siberia north of the Trans-Siberian Railway, a forbidding region of either primeval swamp or continuous permafrost, with mean January-to-July temperature variations of 70–90°F, where technology and construction experience from the southern provinces do not suffice. Unfortunately, virtually all the petroleum and gas reserves of Siberia are found in these parts. As in the American Arctic, the environment, the sparse population and the absence of transportation and communication facilities present formidable obstacles to resource exploitation. The deserts and semideserts of Kazakhstan and Central Asia are only a little less harsh; their infrastructure, while better developed than that of northern Siberia, is still inadequate. The large manpower reserves found fairly close at hand in the oases of these desert regions comprise unskilled, untrained and linguistically isolated labor for the most part.

Pacific Siberia, i.e., the Trans-Baikal territories of the Asian USSR, must be considered separately. Conceivably the richest part of the Soviet Union in mineral resource potential, its remoteness from the industrial heartland of the country has retarded detailed prospecting and actual development. Resource exploitation, so far, depends on certain precious and rare nonferrous metals, lumber from the littoral and energy industries serving this local market. Little of the vast fuel potential has been proved up to commercial levels, and prohibitive transport costs will prevent these reserves from having a significant impact on the domestic energy economy. Export stimulus is expected to play a key role in large-scale energy development in Pacific Siberia as it is beginning to do already to some degree.

The European USSR, including the Urals and the Caucasus, still contains three-fourths of the Soviet population and industry, despite decades of efforts to promote development in the East. Its climate, while rigorous, presents no real hindrance to resource development and the region enjoys a transportation and communication network well developed by Soviet standards. The European USSR itself is larger than the rest of Europe and has a considerable range and large reserves of energy resources as well as other minerals. However, these resources have been intensively exploited for a century and, at current levels of technology, are able to provide only modest increases in fuel output in the coming years. Worsening geological conditions have also raised both average and marginal fuel costs. More importantly, the rapid growth in energy demand, coupled with increasing resource depletion, has led to a huge energy shortfall in the European USSR–a deficit that today amounts to about half that of Japan and 30% that of the European Community.[5] By 1980 the European USSR is expected to satisfy only about 60% of its energy requirement from its own resources and by 1990 only 40%.[6] The region is looking to Soviet Asia for an ever larger portion of its supplies.

East Europe, more precisely the six CMEA nations, makes up a compact area but contains 107 million people. The three northern states (East Germany, Poland, and Czechoslovakia) are already highly developed, with large metallurgical and heavy chemical industries that are particularly energy-intensive. The region's fuel reserves consist overwhelmingly of coal and lignite; only Rumania has substantial, but diminishing, resources of petroleum and gas. The hydraulic potential of the six countries is limited and can make only an insignificant contribution to the energy supply. Czechoslovakia and East Germany have reached per capita levels of energy demand that exceed those of West Germany.[7] This fact gives little joy to planners, for in large part it is a reflection of an energy structure that is still dominated by poor quality fuels, with consequent low rates of heat capture, high handling cost and staggering environmental damage.

East Europe imports more than one-sixth of its total energy requirement, and this share is destined to rise significantly in the future.[8] Domestic energy sources and nuclear power cannot cover incremental demand despite the renewed and costly investment effort. Besides, the contribution of oil and gas to the fuel (and chemical raw material) mix is by far the smallest among all industrial regions of the world. Some increase in the share of hydrocarbons is thus also necessary to improve the efficiency of these economies. Until the middle 1970s, these countries, except Rumania, have received all but a small fraction of their petroleum import from the USSR, which also began piping appreciable quantities of natural gas.

Whether the USSR will maintain or increase petroleum deliveries to its CMEA partners in light of its own growing domestic needs and opportunities for enlarged exports to hard-currency markets is thus a significant factor in the future patterns of Soviet energy policy. Although the USSR has extracted heavy investment commitments for Soviet fuel developments from East Europe and has now raised the prices of oil charged to its partners to near world level, it must clearly forgo real economic benefits from any petroleum supplied to the CMEA rather than to the West.

The Institutional and Political Context

The energy sector represents a most vital underpinning of a country's economic and military strength. The high technology, huge investment requirements and long payoff periods result in oligopolistic institutional arrangements, strong government involvement or outright ownership. In the USSR, government ownership and control express themselves through the institutions of strong energy ministries. Their dozens of research institutes do most of the applied research, generate the technical-economic data and, together with centers of the Academy of Sciences, draw up the alternative strategies for their respective fuel-energy industries. The selection and coordination of alternatives into a more or less comprehensive energy plan by Gosplan, the economic planning agency, which itself includes high ranking members of the ministries, is therefore a process of intense bargaining, lobbying and dispute.

While the rationality of long-range energy decisions clearly embodies the economic reality of relative costs and prices (as far as they can be projected), without doubt it also involves the long-term goals of national security, full control over supplies, the political stability of the whole Soviet bloc and the bureaucratic interests of the responsible energy agencies. Decisions concerning the fuel-energy sector are closely intertwined with questions involving the spatial dimensions and structure of the economy as a whole. But the full costs of the various spatial alternatives of industrial expansion, resource transfer and substitution over different time horizons have never been and probably cannot be assessed. At any rate, the hundreds of models for limited optima, prepared by dozens of institutes, admittedly employ many inconsistent assumptions, parameters and ranges of goals. The uncertainties leave a relatively wide field for sectoral and regional interest groups to push their pet projects and developmental schemes. To a degree, the assumptions regarding long-run resource availability (particularly in the case of hydrocarbons) and technology developments and the methodology of forecasts themselves become instruments in the struggle for investment funds.

In the short run, the management and routine operation of the energy system is mostly free from political interference, employs economic concepts as guideposts and, subject to their soundness, follows principles of economic optimization. Cost-price considerations, the savings from substitution of petroleum fuels in the domestic economy versus the value of hard currency earned from sales abroad, for example, seem to have been the prime consideration behind Soviet oil exports (to the West) in the past. Even the so-called "dumpings" of the early 1970s were of a short duration for the purpose of breaking into the market. The Russians apparently were selling well above cost, though–combined with the actions of the independent oil companies–their sales contributed to a series of price cuttings, to destabilization of the world oil market and, probably unexpectedly, to the price-raising actions of the chief oil-producing nations. Internally, fuel allocation among sectors and regions is aided by linear programming and other algorithms; rent payments and finding costs have been introduced in the mineral industries; production and transport costs are computed with interest charges, and prices tend to reflect marginal costs, at least among large regions.

Economics and politics are more closely intertwined in Soviet policy toward East Europe's energy needs. Its dependence on Soviet oil, and increasingly on gas, nuclear reactors and enriched uranium, clearly constitutes an immense political leverage constantly perceived whether applied or not. The USSR has a long-term interest both in the economic viability and political stability of the East European states. However, it is equally aware of the opportunity costs of committing large and highly valuable fuel resources, particularly oil and gas, to East Europe in place of hard-currency markets. The USSR appears reluctant to increase petroleum deliveries to its partners any further and would like to reduce the amount supplied but cannot embark on such a course lightly. The resolution of this question will form a significant part of Soviet energy policy in the future.

NOTES

[1] *Narodnoye khozyaystvo SSSR* (henceforth *Nar. khoz. SSSR*), various issues.

[2] N. Feytel'man, "Timely issues in the development of the fuels and energy complex," *Ekonomicheskiye Nauki*, 1976, No. 4, p. 32; Ye. A. Nitskevich, "Problems of perfecting the fuels and energy balance of industry," *Promyshlennaya Energetika*, 1976, No. 8, pp. 30–31.

[3] U.S. Central Intelligence Agency, *Soviet Long-Range Energy Forecasts*, A(ER)75-71, September 1975.

[4] *Nar. khoz SSSR za 60 let* (Moscow: Statistika, 1977), pp. 49–54.

[5] Consumption of state-supplied fuel, i.e., excluding fuel gathered by the population, was 1,100 million tons of standard fuel (SF) in 1976 in the European USSR, including the Urals. Production in 1975 was 918 million tons, with negligible growth in 1976, making a deficit of about 190 million tons. (One ton of SF is 7 million kilocalories, roughly equivalent to the heat content of a metric ton of very good hard coal.) The amounts of fuel shipped westward from the Asian USSR to the European USSR are greater than the deficit in the European USSR because they include exports passing through western ports or border points. Total westward fuel shipments from the Asian USSR were 360 million tons of SF in 1975. Data from *Energetika SSSR v 1976–1980 godakh* (Moscow: Energetika, 1977), pp. 148–149; Ya. Mazover, "Location of the fuel-extracting industry," *Planovoye Khozyaystvo*, 1977, No. 11, p. 139; U.S. Central Intelligence Agency, *Handbook of Economic Statistics, 1977* (Washington, September 1977, ER 77-10537), pp. 75–76.

[6]G. V. Yermakov et al., "Trends in the development of the nuclear power industry," *Ninth World Energy Conference. Transactions* (Detroit, 1974), Vol. 5, p. 279.

[7]U.S. Central Intelligence Agency, *Handbook*, op. cit., pp. 44 and 76; United Nations, Department of Economic and Social Affairs, *World Energy Supplies, 1971-75* (New York, 1976).

[8]John R. Haberstroh, "Eastern Europe: Growing energy problems," in U.S. Congress, Joint Economic Committee, *East European Economies: Post Helsinki* (Washington: Government Printing Office, 1977), p. 383.

Chapter 2

TRENDS IN THE SOVIET ENERGY ECONOMY

Long-run correlations between economic development and the growth of aggregate energy use have been high. Over shorter periods such a systematic relationship is harder to detect. The general experience has been that in the period of early industrialization, when countries build up their basic industrial infrastructure and capital stock, the growth rate of energy consumption surges ahead of the growth rate for national income and the energy intensiveness of the economy rises. In the later stages of development, the opposite situation tends to prevail, and energy intensiveness declines. This means a greater economic punch from a unit of energy input and, other things being equal, is a sign of increased efficiency.

The change in the energy–GNP relationship over time in countries with different economic structure, physical endowment and social customs is a complex and controversial matter, still largely unresearched. The perception of that relationship by decision-makers is critical for the formulation of public policy. It is obviously crucial for command economies, such as that of the USSR, which are guided and managed by long-range plans.

To understand the trends and processes affecting the energy economy of any state, a look at gross consumption alone will not suffice. One must also examine the sectoral and geographic configurations of primary energy demand and supply, the mix of resources and the transformation–utilization processes through which primary energy passes to the final consumers. Not only do these factors have a fundamental influence both on the quantities and the economics of total energy use, but they represent crucial ingredients of comprehensive energy planning.

Whether by reliance on the market mechanism or on mathematical programming methods, efficiency rules require sectoral and regional allocations that maximize economic returns from the use of given resources for the whole economic system.

THE GROWTH OF ENERGY DEMAND

The beginning of the Soviet industrialization drive was associated with high annual increases in aggregate energy consumption and, as in most other aspects, the Soviet industrial buildup was unique in its energy intensity (Table 1). Since World War II, the rapid decline in that rate and the relationship of energy demand to economic growth suggest that the USSR has become more careful and effective in energy use. In fact, by the second half of the 1960s, the increase in Soviet demand was comparable to growth rates in America and West Europe, while Soviet GNP continued to rise faster than United States national income. A brief look at this problem of the energy–GNP relationship is necessary here to gauge future trends and assess Soviet energy needs.

If the Soviet definition of national income and its growth rates are accepted, the energy intensiveness of the Soviet economy declined significantly during the 1950s and less rapidly but noticeably in the subsequent decade. According to Soviet figures, a unit of energy input into the economy has been creating greater increments of national income, enhancing Soviet strength and raising the standard of living. Using the lower Western estimates for the growth of Soviet GNP, one finds a rise in energy intensiveness during the 1950s, followed by a small decline.[1] The literature of the late 1960s and early 1970s shows that Soviet experts at that time expected a continued decrease (i.e., improvement) of energy–GNP ratios until 1990, albeit at a lower rate.[2] More recently they have been silent on the subject and closer inspection throws doubt on that attractive notion.

Changes in energy–GNP ratios based on aggregate energy input do not measure the development of "true" energy intensity of an economy. More suggestive is the change in net (useful) energy consumption relative to national income and the development of gross consumption weighted by productivity coefficients for energy sources in principal uses. Such an assessment for the USSR was made recently by the author in a research report with striking results. Computed first with net (useful) instead of gross consumption, then with productivity weights for different fuels in stationary and mobile uses, the energy intensiveness of the Soviet economy was shown to rise substantially rather than decline throughout the post-World War II period.[3] So far, at least, there seems to be no evidence of any systemic shift in the USSR toward a less energy-demanding economic structure as the economy becomes more developed and complex. Consequently, we cannot from past experience deduce that expansion of the national income in the future can be accomplished by lower rates of useful energy input, and to project such a trend for the future is, in my opinion, a serious mistake. And if past experience suggests no decrease in net

Table 1. Annual Rates of Increase in Gross Energy and Electricity Consumption, Gross National Product and Industry Output in the USSR (in percent)

Period	Gross energy consumption	Gross electricity consumption	Gross national product		Industry	
			Soviet data	Western estimate	Soviet data	Western estimate
1929-32	17.2	28.2				
1933-40	12.3	17.2				
1933-37	7.3	21.8				
1938-40	21.2	10.1				
1951-55	8.3	13.3	11.3	6.0	13.1	11.3
1956-60	6.3	11.4	9.1	5.8	10.4	8.7
1961-65	5.8	11.6	6.5	5.0	8.6	7.0
1966-70	4.5	7.8	7.7	5.5	8.4	6.8
1971-75	5.1	6.9	5.7	3.8	7.4	6.0

Notes and sources:

Column 1: Prewar growth rates computed from data on apparent consumption in Robert W. Campbell, *The Economics of Soviet Oil and Gas.* Baltimore: Johns Hopkins, 1968, p. 4. Postwar growth rates from *Nar. khoz. SSSR. 1922-1972*, p. 61 and *Nar. khoz. SSSR v 1975 g.*, p. 112. Campbell's series includes additions to stock; the *Nar. khoz.* series only energy actually used. In addition, methodological differences make the prewar and postwar periods not strictly comparable.

Column 2: Computed from official data in *Nar. khoz. SSSR. 1922-1972*, pp. 155 and 1958 and *Nar. khoz. SSSR v 1975 g.*, p. 235 and in *Vneshnyaya torgovlya SSSR za 1971 god*, p. 27, and *Vneshnyaya torgovlya SSSR v 1975 g.*, p. 25.

Columns 3 and 5: Promyshlennost' SSSR. 1964, p. 35 and *Nar. khoz. SSSR v 1975 g.*, pp. 49, 51 and 53.

Columns 4 and 6: from Rush V. Greenslade, "The real gross national product of the USSR, 1950-1975," in U.S. Congress, Joint Economic Committee, *Soviet Economy in a New Perspective.* Washington, 1976, p. 273.

National income and industry data have been omitted for prewar years because of unreliability.

energy use per ruble of national income, aggregate energy requirement (i.e., the demand for raw fuels, hydro and nuclear electricity) per unit of GNP can continue to decline only if utilization efficiency goes on improving at a significant pace.

In the previous decades, the decline in aggregate energy–GNP ratios was brought about entirely by the rapid shift to more efficient hydrocarbon fuels in industry and transport and by long-run improvements in heat capture and other energy utilization processes. As in other countries, electrification greatly enhanced the flexibility and efficiency of factory, and much later, agricultural operations and was a strong factor in the improvement in the energy–GNP relationship despite the very large, though declining conversion losses in the generation of electric power. It is obvious that most of these beneficial changes, e.g., the shift to oil and gas-fired plants, diesel and electric railways, etc., are unrepeatable once completed; other technological improvements are asymptotic in nature and are rapidly approaching their respective limits.

In contrast to the dramatic rise in energy efficiency during the 1950s and, still more, during the 1960s, progress in this field in the forthcoming decade is expected to be only marginal. From 1950 to 1960 the utilized portion of gross energy demand, the portion not squandered as waste heat and lost during transport and transmission, advanced from a little less than 28 to 31%, then rose rapidly to about 40% in 1975. This represents a 44% improvement in thermal efficiency, most of which came after 1960, coinciding with the upsurge of oil and gas output. For the forthcoming decade, Soviet experts expect a mere 1 to 2% rise in this thermal efficiency ratio (net/gross consumption) to somewhere between 40 and 42%.[4]

Most of the readily available technological steps in improving energy efficiency have already been taken, while continuing electrification of the economy (i.e., the still rising share of energy consumed via electric power) must now proceed with only marginal improvement in the heat rates of generating stations. At the same time, the mounting dependence on poorer quality and/or less accessible resources requires massive energy inputs, the energy costs of environmental protection are growing and the gradual expansion of motorization and income-elastic household demand should also push the economy towards greater energy intensity, however slightly. Largely as a consequence of worsening accessibility, the energy industries have become large direct consumers of their own output,[5] reducing substantially the portion of net energy available. Self-use, with internal losses, amounts to 12–13% of gross energy output and has been rising since 1970 after a period of relative decline.[6] If one adds indirect use of energy through structures, machinery and materials, the energy industries themselves may consume from 25 to 30% of their own output.[7]

A crucial trend in global energy consumption to which virtually all countries, including the USSR, seem to conform is the increasing portion of energy consumed as electric power. Since 1928, electricity consumption in the Soviet Union, too, has been expanding at a greater average annual rate than total energy demand and considerably faster than national income or industrial output. The only exception was the immediate prewar period when intense war preparation distorted the trend (Table 1). In 1928 a mere 2.2% of Soviet aggregate energy supply was converted to electricity, by the present decade more than one-fifth (Table 3). The versatility of electric power and its technological advantage in a widening field of production processes make the continued spread of electrification a crucial vector in the modernization of the economy. In the long run, the inexorable pressure on the accessible resources of fossil fuels and the growing contribution of nonconventional energy sources (which may be harnessed on a large scale only via electricity) should also help raise the share of electric power in the energy balance. The contribution of primary electricity (power not generated from conventional fossil sources) to aggregate energy resources is still small–less than 4% both in the Soviet Union and the United States. However, the growth of nuclear energy and the beginning of solar and geothermal power are destined to raise this share to one-fifth or more of total energy supply by the end of the century.[8]

Table 1 also reveals that in the USSR the growth rate of electricity consumption has fallen faster than that of total energy demand–a situation not found in Western industrialized countries, at least until today. Certain Soviet specialists consider such a trend ominous in the technological competition with the West. Because of its unique, nonstorable nature and resulting problems, its increasing importance as a refined form of energy and its contribution to primary energy supplies as well, the electric power industry will be analyzed in separate chapters.

THE STRUCTURE OF ENERGY CONSUMPTION

In a modern economy, primary resources pass through complex stages of interindustry processing and transaction to satisfy final demand. Little of total resource inputs today passes to consumers in an unprocessed form. Energy is no exception. In the course of development, primary energy resources in the USSR, too, are increasingly refined and transformed. The primary energy branches provide the flows of crude oil, natural gas, coal and other solid fuels, hydroelectric power and uranium ore to the two processing segments: to fuel-refining factories and to electric power and boiler plants. The former refine and upgrade raw fuels into petroleum products, coke and briquettes; the latter transform both the raw fuels and some of the refined fuels into electric power, steam and hot water (Fig. 1). Some raw fuels still flow directly to the rest of the economy and they also comprise most energy exports. By 1975, about four-fifths of all primary energy resources were refined or transformed into more usable and/or valuable forms compared with two-thirds in 1960.[9]

To trace the flow of aggregate energy input to the major sectors of the economy and, still more, to attempt cross-country comparisons of such distribution through time is treacherous. Information on the USSR is available mostly in the form of percentages and the methodology behind them is seldom revealed and seems full of inconsistencies. Striving for greater consistency by constructing his own balances, Robert Campbell provides the following sectoral pattern (Table 2).

Self-consumption by the energy industries, losses and nonenergy uses are factored out and power stations are made a separate sector, since their output will itself have to be distributed. As in the United States, the broad sectoral breakdown in the USSR appears remarkably stable through time, but a number of important differences from the United States pattern stand out. In the USSR, losses and direct use by the energy industries themselves account for one-eighth of gross energy input into the economy, a share twice as high as in the United States. This points to a relatively wasteful and inefficient technology level in the Soviet Union, although the significant import share and more favorable

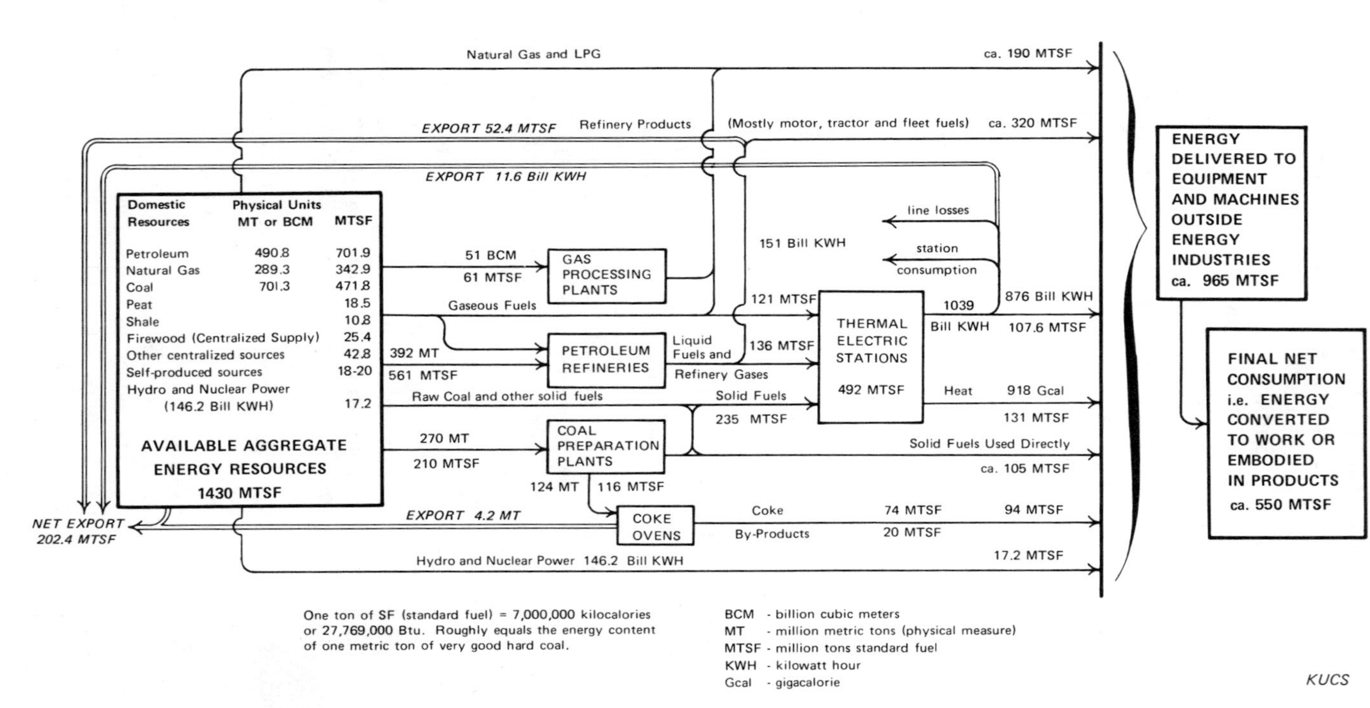

Fig. 1. Flow chart of energy production and consumption in the USSR, 1975.

Table 2. Structure of Gross Domestic Energy Consumption by Consuming Sector (in percent)

Sector	Area	1960	1965	1970	1975
Own use and losses	USSR	12.8	11.7	10.8	12.6
	West Europe	8.2	7.4	6.3	5.9
	USA	5.5	5.7	4.1	6.0
Nonenergy	USSR	2.8	3.1	3.9	4.4
	West Europe	2.2	3.8	6.1	3.3[a]
	USA	3.7	3.2	4.1	2.3[a]
Electric power	USSR	25.5	28.3	30.6	32.8
	West Europe	26.9	27.4	26.9	28.5
	USA	21.2	23.5	26.2	29.6
Industry	USSR	30.1	29.3	28.5	28.5
	West Europe	27.2	25.8	24.9	24.4
	USA	23.3	23.2	21.7	21.2
Transportation	USSR	11.8	9.9	7.7	5.5
	West Europe	12.4	12.8	13.2	14.7
	USA	23.1	22.6	22.2	24.2
Households, municipal, and other	USSR	17.0	17.7	18.5	16.2
	West Europe	23.1	22.8	22.6	23.2
	USA	23.2	21.8	21.7	16.7

[a]These declines are misleading—for earlier years all naphtha was treated as nonfuel use, but in 1975 it is allocated to industry.

Source: Robert W. Campbell, *Soviet Fuel and Energy Balances*. Santa Monica, CA.: Rand Corporation, Research Report R-2257, 1978, Table 7.

geographic conditions probably mitigate self-consumption by the energy industries in the United States. (Geological conditions in the United States are better for coal but worse for oil and gas and these factors may roughly balance themselves out.) The Soviet pattern also differs from the American by the smaller and sharply declining importance of the transport sector, which claimed half the United States share in aggregate energy consumption in 1960 and less than a quarter in 1975. The sharp decline is largely explained by the shift in railroad haulage from steam to diesel and electric traction (though the latter is not assigned to transport in this table), which is far more efficient in energy use. And the embryonic development of private motoring, of course, results in far smaller energy demand by this sector compared to the United States or even Western Europe. Industry, on the other hand, is an appreciably greater energy consumer in the Soviet Union, which indicates the high priority assigned to investment and defense compared with consumption.

Finally, the relatively high share of household and municipal consumption in

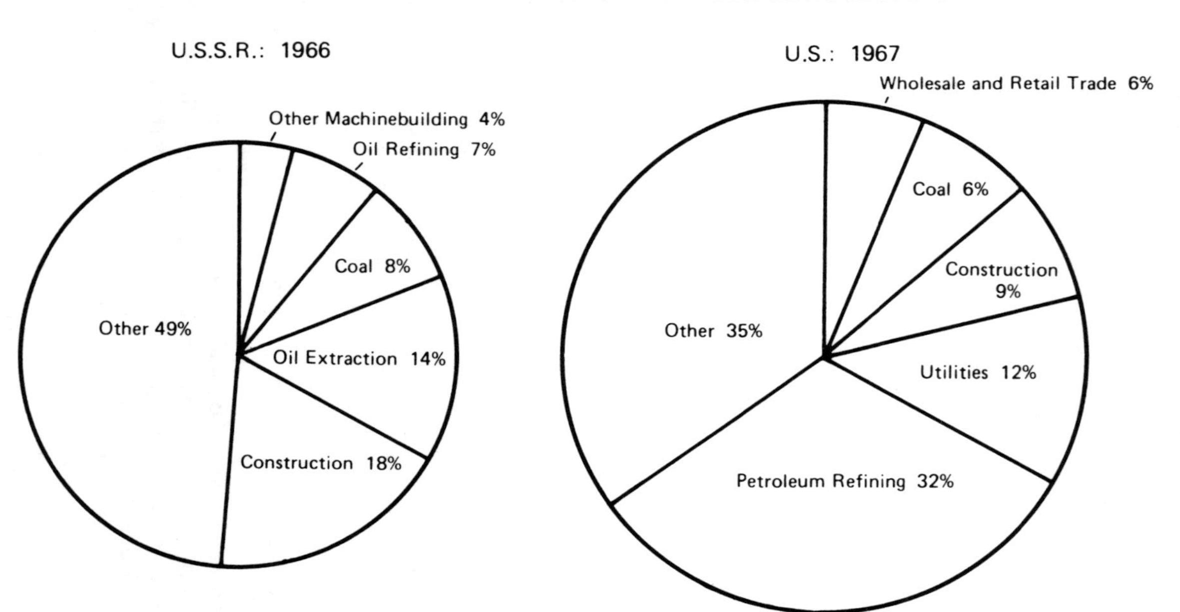

Fig. 2. Final shares of fuel consumption in the United States and the Soviet Union.

the USSR, not far from that in the United States, is deceptive on two counts. First, well over one-fourth of all Soviet energy supply to the household and municipal sector is firewood and a full fifth is made up of wood and peat gathered and cut by consumers themselves and not furnished by the state.[10] This, of course, results in a low efficiency of energy use, economic waste and a huge burden borne by the population, particularly in the countryside. Second, a smaller share of the energy distributed via electric power goes to the household and municipal sector in the Soviet Union than in the United States (14% versus some 58%).[11] And while Soviet households also receive significant quantities of captured waste heat from electric stations, most of that heat, almost four-fifths of the total, is also, like that of electricity, sent to industrial consumers.[12] Therefore, when the energy provided through power stations is so distributed, the dominance of industry is further enhanced, the share of the transport sector increases somewhat and that of households is further reduced. According to Campbell, industry in the USSR consumes a larger absolute amount of energy than in the United States, though Soviet statisticians themselves concede that their country's industrial output still falls some 20% below that of the United States.[13]

Most heavy energy-using industries produce inputs to other intermediate sectors rather than for final demand, i.e., public and private consumption, defense, fixed investment, net export and net addition to inventories. Yet it is final demand, which, through the sum of independent consumer decisions and/or the planning of government agencies, is the chief determinant of the structure of an economy. Based on Soviet and United States input-output tables, a careful Western study traces direct and indirect fuel consumption through the intermediate (producer goods) sectors to final uses in 1966–67 (Fig. 2). The author shows that while 32% of aggregate American fuel input "went directly or indirectly to the provision of refined oil products for final demand, largely to quench the thirst of a private automobile, in the USSR refined oil products accounted for a comparatively paltry 7% of fuel needs." Private utilities accounted for 12% of United States fuel requirement for final demand but only 3% of corresponding Soviet needs; wholesale and retail trade respectively 6 and 2%. Reflecting planners' preferences and the emphasis on investment and growth, construction was the largest fuel-using sector in the USSR, accounting directly or indirectly for 18% of fuel requirement. With much greater emphasis on private housing, construction was also a significant fuel using sector (9% of the total) in the United States.

While it underlines the priorities and developmental strategy of Soviet planners, such a broad distribution still does not reveal very much. Far more significant for an analysis of the energy system are the trends in the technological processes and forms in which energy is consumed. Chapter 8 examines this topic in detail, but an initial idea may be gained from Table 3. (The table excludes nonenergy uses, e.g., chemical raw materials.)

The most striking trend is the rapidly declining portion of primary energy inputs used directly in consumer installations and the corresponding growth in the share of such inputs used to generate electricity, steam and hot water. This is

Table 3. Consumption of Primary Energy Resources by Functional Categories (excludes nonenergy uses)

Category	1930		1950		1960		1971		Projected (1990–95)
	In million gigacalories	(%)[a]	In million gigacalories	(%)[a]	In million gigacalories	(%)[a]	In million gigacalories	(%)[a]	(%)[a]
Generation of electricity	67	5.6	319	11.5	745	15.7	1655	20.1	29
Generation of steam and hot water	112	9.3	400	14.4	845	17.9	1840	22.4	30
Direct use of primary energy:	980	81.7	1984	71.2	3000[b]	63.3	4440	53.9	37
(a) for high temperature industrial processes (in furnaces, kilns, ovens, and related equipment)	172	14.3	618	22.2	1080[c]	22.8	1815	22.1	14.5
(b) for medium and low temperature processes (space and water heating, cooking, etc.)	608	50.7	966	34.7	960[c]	20.3	950	11.5	6.5
(c) for mobile machines and power tools	190	15.8	395	14.2	960[d]	20.3	1675	20.4	16.5
(d) for lighting	10	0.9	5	0.2	—	—	—	—	
Losses in transport	40	3.3	82	2.9	150	3.2	295	3.6	3.5
TOTAL	1199	100.0	2785	100.0	4740	100.0	8230	100.0	100.0

Notes and sources:

[a]Percentages may not add up because of rounding.

[b]Estimated by using the share given by M. A. Vilenskiy, *Ekonomicheskiye problemy elektrifikatsii SSSR*. Moscow: Nauka, 1975, p. 14.

[c]Estimated from combined high temperature and medium-low temperature total. The slight rise in the share of high-temperature processes between 1950 and 1960 and slight decline between 1960 and 1971 is plausible in view of the relatively heavier emphasis on metallurgy during the 1950s.

[d]Estimated by using the share given by Vilenskiy, op. cit., p. 17 for all fuel-burning transport equipment and other mobile machines with internal combustion engines. The rise in that share between 1950 and 1960 and its stabilization during the 1960s is, again, plausible. The rapid growth in demand by construction and agricultural machinery and heavy trucks coincided with the continued dominance of the inefficient steam locomotive in railway haulage. During the 1960s, the shift to diesel (and electric) traction helped to counteract the swift rise in fuel consumption by trucks, agricultural and construction machinery.

All estimated figures rounded.

1930–1971 from A. A. Beschinskiy and Yu. M. Kogan, *Ekonomicheskiye problemy elektrifikatsii.* Moscow: Energiya, 1976, pp. 413–15. Projected breakdown from ibid., p. 23. The percentages given were recomputed to exclude projected nonenergy uses and exports but to include a 3.5% loss in transport.

particularly true if one excepts energy that runs mobile units, where direct consumption of fuels via the internal combustion engine is still dominant. Specialists expect this trend to continue, with the stated policy to further increase the centralization of heat supply and to drastically reduce the need for small heating devices, such as furnaces for individual apartments, commercial buildings and small industrial plants.

The generation of electricity, steam and hot water comprise boiler demand, where fuels burned in a furnace apply their heat to a steam-raising device, the boiler. By contrast, in the more specialized equipment of industrial furnaces, ovens and kilns, heat from the combustion of fuels is applied directly to materials being processed (e.g., in the smelting of ores, the cement and ceramic industries, etc.). Boiler use constitutes the most flexible part of the energy sector where direct substitution among energy sources is technically and economically at its most feasible. Under boilers, solid fuels suffer a far smaller technological disadvantage compared with natural gas and petroleum products than in the more specialized furnace uses, while the motor and naval markets, agricultural field operations and most of chemical synthesis are technologically tied to petroleum. If solid fuels are substantially cheaper on an even calorific basis than liquid and gaseous fuels, their cost advantage may more than counteract their technological drawbacks under boilers. Not surprisingly, the relationship between hydrocarbon and solid fuel reserves and prices in the world today has focused attention on the boiler market, where key policy decisions concerning an economically and strategically more rational fuel mix must be made.

Three-fourths of boiler fuel demand (about a third of aggregate fuel consumption and more in stationary uses) in the USSR is accounted for by electric power stations.[14] They form a pivotal part of the energy system and policy decisions concerning primary energy consumption and the energy mix. Soviet conventional thermal station capacities at the end of 1975 were divided between heat and power stations (TETs) and condenser electric stations (KES) in a proportion of 35 to 60%. The rest, a mere 5% and diminishing part of the total, was comprised of mobile and other small diesel units.[15] Heat and power

Table 4. Share of Urban Heat Consumption Covered by Heat and Power Stations (in percent)

Sector	1965	1970	1975	1980 (plan)
Industry	47.4	48.9	49.0	50.0
Residential	21.3	26.4	27.0	33.0
All urban consumption	38.4	41.4	42.0	44.6

Source: V. P. Korytnikov ed., *Rabota TETs v ob"yedinennykh energosistemakh.* M. Energiya, 1976, p. 21.

Table 5. Regional Fuel Consumption in the USSR (1965 and 1970)

Region	All fuels		Boiler and furnace fuels					
	(%)	Index of per capita demand (USSR = 100)	In million tons of standard fuel		(%)		Index of per capita demand (USSR = 100)	
	1965	1965	1965	1970	1965	1970	1965	1970
Group 1								
Northwest	6.1	121	42	50	5.95	5.53	118	110
Central Russia	10.5	91	74	96	10.48	10.62	91	93
Donets-Dnieper[a]	14.8	176	131	140	18.55	15.49	221	187
Urals	12.7	194	100	121	14.16	13.38	216	215
Group 2								
Volga-Vyatka	2.7	76	19	23	2.69	2.54	75	74
Central Chernozem	2.9	84	21	26	2.97	2.88	86	88
Volga	8.4	110	59	78	8.36	8.63	110	114
North Caucasus	4.3	74	27	36	3.82	3.98	66	67
Southwest[a]	4.3	49	24	40	3.40	4.42	39	52
South[a]	1.6	63	7	15	1.00	1.66	40	63
Baltic	2.7	87	18	23	2.55	2.54	82	81
Belorussia	2.3	62	15	21	2.12	2.32	57	62
Moldavia	0.6	41	4	6	0.57	0.66	39	45
Transcaucasia	3.0	61	22	30	3.12	3.32	63	65
Group 3								
West Siberia	7.0	134	49	65	6.94	7.19	132	145
East Siberia	5.3	169	30	39	4.25	4.31	136	140
Far East	3.4	142	20	24	2.83	2.66	118	110
Group 4								
Kazakhstan	3.0	57	26	42	3.68	4.65	70	87
Central Asia	4.4	58	18	29	2.55	3.21	33	38
USSR	100.0	100	706	904[b]	100.00	100.00	100	100

Notes:

[a]The 1970 figures for the Southwest and South are slightly overestimated and those for the Donets-Dnieper underestimated. The Ukrainian total for boiler and furnace fuels was broken down to the three component regions shown according to estimated total fuel consumption. These shares for the Southwest and South are known to be somewhat higher than their shares in boiler and furnace fuel consumption.

[b]Apparently excludes blast furnace gases, which explains the discrepancy between this table and Table 9.

Sources: For all fuels (1965), A. E. Probst et al., *Razvitiye toplivnoy bazy rayonov SSSR.* M. Nedra, 1968, p. 45. For boiler and furnace fuels (1965 and 1970), VINITI, *Razrabotka neftyanykh i gazovykh mestorozhdeniy*, Tom. M. 1972, pp. 44–45. Ukrainian total broken down according to percentages in Akademiya nauk SSSR, Inst. ekon., *Zakonomernosti i faktory razvitiya ekonomicheskikh rayonov SSSR.* M. Nauk, 1965, p. 146; Institut Kompleksnykh Transportnykh Problem, *Perevozki gruzov.* M. Transport, 1972, p. 78 and V. A. Shelest, *Regional'nyye energoekonomicheskiye problemy SSSR.* M. Nauka, 1975, p. 212. Shelest supplies percentages on total fuel consumption in 1970 for all regions, but his figures do not add up to 100%, do not seem compatible with those of Probst for 1965 and thus could not be used. Regional per capita indices computed according to official population data of Jan. 1 for the following year from appropriate issues of *Nar. khoz. SSSR*.

stations supply both power and heat to industry and/or urban neighborhoods via special bleeding turbines; most condenser stations generate only electricity and, as a rule, are of greater size and are located outside urban centers. Heat and power stations play a vital and increasing role in the heat supply of cities both for the heating of structures and for industrial processes. These stations furnish half of all the heat consumed by industry in the form of steam and hot water, 27% of all heat consumed by the urban residential sector and 42% of all heat consumed by cities in total (Table 4). In some parts of the country the role of heat and power stations in the heat supply of cities is even greater, reaching 56% in the Volga region and 63% in East Siberia.[16]

THE GEOGRAPHIC PATTERN OF DEMAND

As made clear in the Introduction, the geographic context is one of the crucial dimensions of the Soviet energy system. Demand must be analyzed as it varies through the vast Soviet ecumene; trends in the spatial pattern of consumption and the way it is satisfied must be examined. The priorities of central planners in the sectoral allocation of energy find an immediate geographic expression, since heavy industries, the most conspicuous users, are areally highly concentrated. Regional per capita variations in energy demand, particularly in stationary uses, tend to follow the distribution of heavy industry. The Spearman rank correlation coefficient between per capita consumption of all boiler and furnace fuels, virtually synonymous with stationary energy demand, and the per capita value of industrial fixed assets over the major Soviet economic regions indicates a close relationship for the years 1965 and 1970 (R_s = 0.93 and 0.92).[17] Variations in per capita consumption among these areal units ranged from 33 to 221% of the Soviet average in 1965 and from 38 to 215% in 1970 (Table 5 and Fig. 3). With a finer regional mesh, such differences, of course, would be much greater. Because Soviet regional statistics are frequently inaccurate and sometimes inconsistent, Table 5 should not be taken too precisely, but a number of useful conclusions from it can be made.

Not surprisingly, the Donets-Dnieper and Ural regions, the two iron and steel bastions are the leading per capita fuel consumers followed by the provinces of Siberia and the Volga (with ferrous and nonferrous metal smelting, woodpulp, etc. dominating in the former, petroleum refining and petrochemicals in the latter). The Northwest, where iron-steel and pulp-paper industries are important, is also high on the list. All ethnic non-Russian provinces, with the exception of the eastern Ukraine (Donets-Dnieper) rank well below the Soviet mean. The poorly industrialized but agriculturally significant regions (North Caucasus, Southwest, South, Belorussia, Central Asia, Moldavia) and some coastal provinces, particularly the Far East, show appreciably higher shares in total fuel consumption than in boiler and furnace fuel demand. This speaks of the relatively greater importance of mobile uses, requiring tractor and fleet fuel, in their consumption mix than in that of the country as a whole. The comparatively low per capita fuel demand in the Central Russian and Baltic

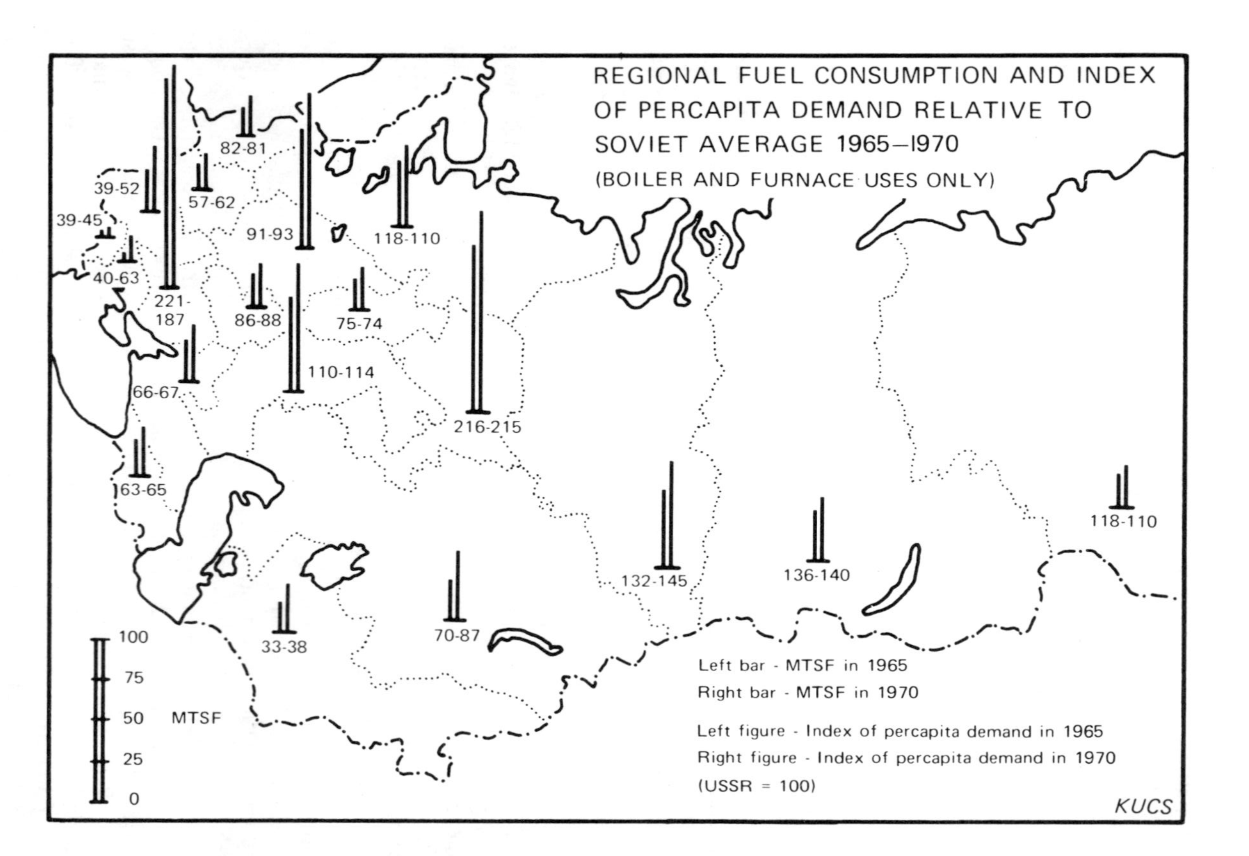

Fig. 3. Regional fuel consumption in the USSR. (MTSF = million tons of standard fuel.)

regions is not due to industrial backwardness, but to the great emphasis on precision and light manufacturing in their industrial structure.

The first four regions in Table 5 represent the traditional, pre-World War II concentrations of industry, the industrial foundation of the first three having been laid well before the Bolshevik Revolution and that of the Urals, too, having important Czarist roots. These traditional concentrations still accounted for nearly one half of all stationary (boiler and furnace) fuel demand in 1965 and 45% in 1970. Their share in total fuel consumption was not much smaller. Excepting the Kuzbas, the industrialization of Siberia and that of the Volga area is essentially a post-World War II, and primarily post-1955, development, with emphasis on heavy energy-using processes. While the share of the traditional manufacturing regions in fuel consumption has been gradually declining (though the Central Russian Region, the urban-industrial agglomeration around Moscow, managed to increase its share between 1965 and 1970), those of Siberia grew steadily, but not spectacularly through the 1960s. There are indications that this trend has accelerated in the present decade and that the relative drop in the Far East may also have been reversed. The percentage and per capita indices of fuel use also understate the share of East Siberia in total energy consumption because of the greater contribution of hydroelectricity to aggregate primary energy demand than in the country as a whole (around 15% versus less than 4% in the USSR).[18]

Since the late 1950s, the spread of manufacturing from the old established centers has gradually been affecting formerly backward and heavily agricultural provinces as well, both in the European and Asian parts of the Soviet Union. The growing labor shortage and the over-loading of the supporting infrastructure have resulted in declining and lower-than-average marginal productivity in established industrial centers and in relatively greater economic (not to mention social) advantages from expansion in the rural periphery.[19] Such a spread of manufacturing has led to a faster-than-average growth of both total and per capita fuel-energy consumption in once backward regions. This is evident even for the short span of 1965-70 shown in Table 5. Nearly all regions in Groups 2 and 4 improved their share in boiler and furnace fuel use and all but the Baltic (which was already highly developed economically by the mid-1960s) bettered their position on a per capita basis. In Central Asia and Transcaucasia, the burgeoning population growth makes per capita improvements difficult to achieve. The rapid growth of fuel demand along the Black Sea littoral and in the western half of the Ukraine (South and Southwest) appears especially noteworthy. So it is in Kazakhstan, where general industrial-agricultural expansion and energy-intensive development were both responsible for the upsurge.

Since the early 1960s momentous changes have been wrought in the supply-demand relationships of Soviet provinces in the energy field. In 1960, the European USSR as a whole was still an energy-surplus area: it fully satisfied its own consumption needs and supplied most of Soviet fuel exports as well. Large fuel deficits in the central part of the Russian Plain, the western periphery and the Urals were more than counterbalanced by big surpluses in the eastern

Ukraine, Volga Region, North Caucasus and Transcaucasia and the Komi ASSR.[20] Energy between the two sets of provinces also began to move by wire through the high-voltage power grid under development. Only the Ural region received appreciable quantities of fuel from Soviet Asia and most of it was coking coal, but Siberia was dependent on the Volga-Ural fields for most of its oil supplies. West Siberia had a big fuel surplus in the form of Kuzbas coal, but most of the Asian provinces suffered some fuel deficit, made good from the Kuzbas and the Volga oil pools (Fig. 3).

Through the 1960s the alignment of fuel-poor and fuel-rich regions in the European USSR remained unchanged. However, while the deficit of the former increased sharply, the surplus of the latter diminished rapidly[21] and that of Transcaucasia disappeared by the close of the decade.[22] The first half of the 1970s saw the Ukraine turn into a net fuel importer, as the surplus in the Donets-Dnieper area could no longer counterbalance the shortfall in the rest of the republic.[23] By the end of the 1970s net outshipment from the North Caucasus region will also come to an end[24] and only the Volga region will enter the 1980s with a substantial fuel surplus. Already by the mid-1960s the European Soviet Union was unable to provide both for its own needs and shoulder fuel exports as well; in 1970 its own consumption alone outran its resources by a substantial margin.[25] These provinces have continued to be unable to furnish more than a fraction of their incremental needs. Through the first half of the 1970s they could augment their energy output by a mere 7.4% (1.4% per year) and only a 3% total growth was forecast between 1975 and 1980.[26] By 1980, the European USSR and the Urals were expected to produce at best only 60% of their energy requirements.[27]

The deepening energy shortfall in the European areas resulted in a massive and ever-increasing flow of fuels from the Asian USSR westward. This flow more than doubled from 1970 to 1975, and was to double again by 1980 (Table 6). The cis-Ural provinces are no longer able to satisfy most of Soviet export

Table 6. Movement of Fuels from the Asian USSR to the Urals and Westward[a]

Fuels	1970	1975	1980 plan
Oil (million tons)	15.0	113	242
Natural gas (billion m^3)	44.8	104	224
Coal (million tons)	65.8	96	120
Total (million tons of SF)	130.0	361	708

[a]Apparently includes exports through western border points and ports.

Source: Energetika SSSR v 1976–1980 godakh. Moscow, 1977, p. 148.

commitments and an increasing share also originates in Soviet Asia. The oil and gas fields east of the Urals and Caspian Sea have been connected with the Friendship (Druzhba) oil pipeline and the Brotherhood (Bratstvo) gas pipeline, carrying Soviet exports westward. The development of Far Eastern hydrocarbon and coal resources to be shipped though Pacific ports will further enhance the dominance of the Asian USSR in fuel supplies. The five-year plan 1976–80 called for Soviet Asia to furnish over nine-tenths of the increment in Soviet coal production, all increment in gas production and more than 100% of the increment in petroleum output to compensate for declines west of the Urals.[28]

Since the early 1960s, the pattern of energy demand and supply also altered among the regions of Soviet Asia, with the most dramatic changes taking place in Central Asia, Kazakhstan and West Siberia. In 1960, gross energy supply about equaled demand in the four union republics of Central Asia, while Kazakhstan was a net importer of fuel,[29] its surplus in coal more than counterbalanced by big deficits in petroleum products. The intervening years saw both regions enlarge their petroleum industries and, more importantly, to become massive suppliers of natural gas and power station coal respectively to the Russian Republic. By 1976, the Central Asia–Central Russia and Central Asia–Urals pipeline systems accounted for three-quarters of all natural gas transported from the Asian to the European parts of the country,[30] and Kazakhstan accounted for perhaps two-fifths of Asian coal shipped to the West.

These Moslem republics enjoy far less plentiful proved resources of both gas and coal than Siberia. However, the absence of permafrost and a somewhat more tolerable environment with respect to Central Asian gas and closer location to major industrial centers with respect to Kazakhstan coals spurred their exploitation after the mid-1960s at a faster pace than that of Siberian gas and coal. By 1975–76, combined fuel production in Central Asia and Kazakhstan doubled consumption and in Central Asia alone exceeded it by 2.7–2.8 times.[31] Outshipment of fuel from these Moslem republics (some 120–125 million tons of standard fuel) roughly equals that from the Volga region, by far the largest fuel-surplus region throughout the 1960s. Such massive exports of energy, particularly of natural gas, combined with the still low level of fuel use in Central Asia, has generated resentment in these ethnic non-Russian areas and raised the specter of too rapid depletion of valuable resources. The burgeoning population growth and an awareness of the ethnic distinctness and geographic isolation of these republics lend urgency to the problem. The problems and prospects of such massive fuel exports and the region's long-term energy prospects are now discussed in scholarly publications with some degree of openness.[32]

In Siberia, all economic regions expanded their output and demand of energy greatly after 1960. Yet despite vast reserves and still greater undiscovered potentials. East Siberia and the Far East have remained regions of net fuel deficit because of the large quantities of petroleum they must bring in. However, with the exploitation of huge crude oil and natural gas deposits in the Ob' Basin, West Siberia has blossomed into by far the greatest fuel producer and supplier in the country's history—a position it is destined to retain for at least another decade and a half. In 1977, West Siberia produced 520 million tons in standard fuel equivalent, almost 60% of which was oil and an additional 10% natural gas.[33] By contrast, the province consumed less than one-fourth as much and shipped some 400 million tons of standard fuel to other areas and abroad.[34] The belated, but now rapid, development of the world's largest known accumulation of gas reserves in the northern Ob' Basin and the further growth of crude oil output are bound to increase outshipment of fuel from West Siberia in the future.

The examination of the geographic pattern of demand and supply since World War II (as through the entire Soviet period) shows that supplies have shifted in space more radically than demand. In particular, the dominance of the European USSR in energy consumption, as in total economic activity, is continuing. About three-quarters of all Soviet energy is still consumed in the European USSR, including the Urals and the Caucasus, and close to 65% is consumed west of the Urals.[35] By contrast, the cis-Ural regions in 1975 produced only a half and the Urals a mere 6% of all primary energy resources and their combined share is destined to fall by 1980 to some 45% of all primary energy output.[36]

Current planning policies impose severe restrictions on the further location of energy-intensive industries in the European provinces. Expansion of such industries is to take place mostly in south-central Siberia as well as central and northeast Kazakhstan, where vast supplies of cheap but low-quality coals and hydroelectricity can provide heat and power at uncommonly low costs. However, beyond a narrow range of industries, further accelerated development in these areas would involve heavy commitments of manpower on a scale impossible to sustain, given the precipitous decline of Slavic birth rates and the net outflow of people from Siberia since the late 1950s.[37] Per capita consumption of fuel in the region between the Urals and Lake Baikal already exceeds the Soviet average by almost one half, while per capita electricity use is more than 70% above the Soviet mean and will be difficult to raise much further.[38] Meanwhile the broad base of the Soviet economy will continue to remain in the west. Here ongoing mechanization and electrification, the further chemicalization of agriculture and the spread of industry to the rural periphery should continue to increase per capita energy demand. Therefore, even strongly pro-Siberian planners expect the European provinces to account for at least 70% of all fuel-energy consumption even by 1990.[39] The heart of Soviet energy policy will remain the problem of furnishing these regions with fossil fuels, whose incremental supplies must now come entirely from Soviet Asia.

THE ENERGY MIX

The Soviet energy mix today is thoroughly dominated by conventional fossil fuels. They account for over 95 percent of both consumption and production and are expected to shoulder up to four-fifths of all primary energy demand even at the turn of the century. Despite a long-standing ideological penchant for big waterpower projects, the share of hydroelectricity has been declining since the early 1960s and will continue to diminish steadily in the future. The contribution of nuclear power has so far been negligible, with less than one percent of total primary energy. Its rapid growth, however, is one of the cornerstones of Soviet energy policy (Chapter 6). It comprises the most swiftly expanding portion of the energy balance and, in the European USSR, it should make a substantial contribution to incremental supplies in the forthcoming years. Be that as it may, the focus of this section must remain firmly on fossil fuels, the predominant part of the energy structure.

During the Stalin years, the Soviet fuel mix was characterized by structural conservatism, a slowness to shift toward more economic sources of energy. For this period Shimkin reported poor correlation between output trends of major fuels and their cost and profitability.[40] Excluding firewood destined for the domestic sector of the fuel budget, the USSR experienced a mere 1% increase in the calorific content per ton of fuel produced between 1928 and 1955. The increase for the world as a whole was 20%.[41] The emphasis on low calorific energy sources adversely affected productivity in extraction and refining. Output of raw mineral fuels per extraction worker did not even double between 1928 and 1950, and output of refined products (coke, coal concentrates, petroleum fractions etc.) per refining worker actually declined. It also burdened the Soviet economy with excessive transport and handling costs and deprived it of a wide range of by-products.[42]

The stress on solid fuels, whose relative contribution actually rose to reach a peak of almost four-fifths of both total fuel consumption and production in the early 1950s, may have been partly rational, given Soviet emphasis on heavy industries and inadequate reserve and cost information concerning alternatives. Still, it is clear that Soviet planners were slow in appreciating the economic advantages of petroleum and, particularly, natural gas; in that perception bureaucratic rigidities, excessive fear of investment risks and the personal preferences of Stalin also played parts.

When the decision to accelerate the development of the oil and gas industries was finally taken, the change was swift and profound. The excellent geological and gravimetric studies of the previous decades were followed up by a vast increase in exploratory activities and were rewarded by a spectacular expansion of hydrocarbon reserves. Output grew apace. By the mid-1960s, petroleum and natural gas dominated the Soviet fuel balance, a decade later their contribution passed three-fifths of all consumption and approached two-thirds of all production. Between 1955 and 1975, 84.2% of the 1,110 million standard tons of increment in aggregate fuel production on a calorific basis was accounted for by these two fuels (Tables 7 and 8).

Such a structural shift in the Soviet fuel mix was unquestionably rational. It would have been physically impossible to achieve such an increment (a 3.3-fold growth) primarily from solid fuels. Had the share of oil and gas remained constant for these two decades, an increment of roughly 1,200 million physical tons in coal production, a 4.1-fold increase, would have been necessary, or an average of 60 million tons a year. Combined with replacement of exhausted mines, such an expansion would have required the construction of perhaps 1,500

Table 7. Soviet Fuel and Energy Production and Net Export

Year	Petro-leum	Natural gas	Coal and lignite	Peat	Oil shale	Fire-wood[b]	All fuels	Net export
				In million tons of SF[a]				
1940	44.5	4.4	140.5	13.6	0.7	34.2	237.9	-2.0
1950	54.2	7.3	205.7	14.8	1.3	27.9	311.2	-7.8
1955	101.2	11.4	310.8	20.8	3.3	32.4	479.9	(2.1)[c]
1960	211.4	54.4	373.1	20.4	4.8	28.7	692.8	49.1
1965	346.4	149.8	412.5	17.0	7.4	33.5	966.6	107.6
1970	502.5	233.5	432.7	17.7	8.8	26.6	1221.8	152.9
1975	701.9	342.9	471.8	18.5	10.8	25.4	1571.3	202.4
1976	743.1	380.3	479.0	11.3	11.0	24.6	1649.3	239.1
1977	780.5	410.0	486.0	14.0	11.4	24.6	1726.5	264.9
1980 plan	915	520	ca. 550	ca. 19	ca. 12	ca. 18	ca. 2035	
				(%)				
1940	18.7	1.8	59.1	5.7	0.3	14.4	100	
1950	17.4	2.3	66.1	4.8	0.4	9.0	100	
1955	21.1	2.4	64.8	4.3	0.7	6.7	100	0.3
1960	30.5	7.9	53.9	2.9	0.7	4.1	100	7.1
1965	35.8	15.5	42.7	1.7	0.8	3.5	100	11.1
1970	41.1	19.1	35.4	1.5	0.7	2.2	100	12.5
1975	44.7	21.8	30.0	1.2	0.7	1.6	100	12.9
1976	45.0	23.1	29.0	0.7	0.7	1.5	100	14.5
1977	45.2	23.7	28.2	0.8	0.7	1.4	100	15.3
1980 plan	45.0	25.6	27.0	0.9	0.6	0.9	100	

Hydro and nuclear electricity

Year	Hydroelectricity		Nuclear electricity	
	At heat value of power (1kwh = 123 grams SF)	At fuel rates for central thermal stations	At heat value of power (1kwh = 123 grams SF)	At fuel rates for central thermal stations
1940	0.6	3.3		
1950	1.6	7.5		
1955	2.8	12.3		
1960	6.3	23.8	negl.	negl.
1965	10.0	33.8	negl.	negl.
1970	15.3	45.6	0.4	1.3
1975	15.5	42.8	2.5	6.9
1976	16.7	45.7	3.1	8.4
1977	18.1	48.8	n.a.	n.a.
1980 plan	24.2	64.4	9.8	26.0

Notes and *Sources* to Table 7 on page 33.

Notes:

[a]Standard (or conventional) fuel: 1 ton = 7 million kilocalories or 27.8 million Btu's. The 1975 conversion ratio for 1000 cu. meters of gas was 1.195 ton of SF. Earlier ratios were slightly different. Soviet hard coals per physical ton range from 0.57 to 0.93 ton of SF and lignites from 0.29 to 0.57 ton of SF. The average heat content of Soviet coal mined in 1975 was 0.70 ton of SF.

In the official gross export and import figures of fuels in tons of SF, Soviet statistical sources apparently include coal and coke on the basis of one natural ton equaling one standard ton. No attempt was made to adjust these totals.

[b]Does not include firewood gathered by population for own use.

[c]Official totals in SF equivalent for this year are unavailable and net export was computed by converting the physical quantities traded to calorific equivalent.

Sources: All figures from 1940–1976 inclusive are taken directly or derived by simple computation from *Nar. khoz. SSSR 1922-1972*, p. 61 and *Nar. khoz. SSSR v 1976 g.*, pp. 83 and 204. Data for 1977 computed from physical outputs converted to calorific equivalents. *Ekonomicheskaya gazeta*, No. 6, 1978, p. 5; *Elektricheskiye stantsii,* No. 1, 1978, pp. 1–2; T. Shabad, "News Notes," *Soviet Geography: Review and Translation*, June 1978; *Nar. khoz. SSSR v 1976 g.*, pp. 206–207.

---

million tons of new capacity over the 20 years, more than 70% of it underground.[43] Since large, modern deep mines have production capacities ranging from 6 to 10 million tons a year and take a decade to construct, such an expansion appears beyond the realm of the possible even today.

Increments of capital and labor inputs in the fuel industries also would have had to be several times greater, and the cost of an average calorie to the Soviet economy would have been considerably higher. Not only has there been a large difference between the average cost of coal (as that of oil shale and peat) and the average cost of oil and gas, but this difference seems to have increased until the late 1960s. (Production costs of oil and gas were falling sharply until the middle of the 1960s while those coal were rising steadily.)[44] Between 1955 and 1970, the growing share of petroleum and gas reduced the mean production cost of one ton of standard fuel by some 3 rubles (from 10 to 7 rubles) and yielded large economies through more efficient heat capture, locomotive power, reduced handling charges and far greater flexibility in chemical synthesis. Delivered costs were much higher at both dates, but savings due to the structural shift appear to have been even larger than indicated by production costs alone.[45] Since the mid-1960s both the production and delivered costs of oil and gas have been rising rather rapidly, apparently faster than those for coal.[46] Incremental capital requirements per unit of new capacity in the oil and gas industries are increasing particularly sharply.[47] Still, the gap between the cost of solid fuels and that of hydrocarbons remains substantial on a calorific basis both on the average and at the margin.

The upsurge of oil and, later gas, output also generated huge benefits to the USSR through international trade. Petroleum and lately also natural gas have

Table 8. Soviet Energy Mix (1965, 1975 and 1976)

Type of energy	Production		Apparent gross consumption		Net export	
	Mill. tons of SF[a]	(%)	Mill. tons of SF	(%)	Mill. tons. of SF	(%) of production
1965						
Petroleum	346.4	34.6	258.4	28.9	88.0[b]	25.4
Natural gas	149.8	15.0	149.3	16.7	0.5	0.3
Coal and lignite	412.5	41.2	393.8	44.1	18.7	4.5
Peat, oil shale and firewood	57.9	5.8	57.9	6.5	–	
Hydroelectricity[d]	33.8	3.4	33.8	3.8	–	
Nuclear power[d]	–	–	–	–		
Total	1000.4	100.0	893.2	100.0	107.2[c]	10.7
1975						
Petroleum	701.8	43.3	526.2	37.0	175.6[b]	25.0
Natural gas	342.9	21.2	334.6	23.6	8.3	2.4
Coal and lignite	471.8	29.1	455.3	32.1	16.5	3.5
Peat, oil shale and firewood	54.7	3.4	54.7	3.8	–	
Hydroelectricity[d]	42.8	2.6	42.8	3.0	–	
Nuclear power[d]	6.9	0.4	6.9	0.5	–	
Total	1620.9	100.0	1420.5	100.0	200.4[c]	12.4
1976						
Petroleum	743.1	43.6	541.1	36.9	202.0[b]	27.2
Natural gas	380.3	22.3	363.6	24.8	16.7	4.4
Coal and lignite	479.0	28.1	461.3	31.4	17.7	3.7
Peat, oil shale and firewood	46.9	2.8	46.9	3.2	–	
Hydroelectricity[d]	45.7	2.7	45.9	3.1	–	
Nuclear power[d]	8.4	0.5	8.4	0.6	–	
Total	1703.4	100.0	1467.2	100.0	236.4[c]	13.9

Notes:

[a]1 ton of SF (standard fuel) = 7 million kilocalories. Roughly equivalent to the heat content of one ton of very good hard coal.

[b]Includes crude oil and refined products.

[c]To these should be added some 0.5 million ton of SF equivalent of coal and gas in 1965 and 2–3 million tons in 1975 and 1976 exported via electricity. This explains the discrepancy in total export between Tables 7 and 8.

[d]Hydroelectricity and nuclear power added to the fuel mix according to the heat rate at Soviet thermal stations for the respective years, i.e., the amount of fuel needed to generate an equivalent quantity of thermal power.

Sources: Fuel production from *Nar. khoz SSSR za 60 let*, p. 204. Exports and imports in physical units from respective issues of *Vneshnyaya torgovlya SSSR* and *The Oil and Gas Journal*, Aug. 15, 1977, pp. 28–29.

proved to be the most sellable commodities on hard-currency markets, without the need for an elaborate component and service network and the quality control presented by manufacturing exports. The dramatic world price changes since late 1973, in particular, enable the Soviet Union to finance a large share of its Western imports with energy exports, petroleum providing from one-third to over two-fifths of all Soviet hard-currency earnings in recent years.[48] In 1976, in response to a disastrous harvest, resulting grain imports and a huge trade deficit of the previous year, the USSR allocated two-thirds of the increment in petroleum output for export, shipping abroad close to 30% of its output.[49] Finally, Soviet oil has been vital to modernization and economic growth in Eastern Europe, with gas increasingly playing a similar role.[50]

Table 8 shows that for many years most of net Soviet fuel export has consisted of crude oil and petroleum products. Since the early 1960s, in fact, 25 to 30% of all petroleum produced was shipped abroad.[51] In the future, natural gas exports are expected to rise dramatically, but oil is expected to dominate Soviet fuel exports for some time to come. Because of this position in fuel exports, petroleum plays a more modest role in the Soviet consumption mix than in the production mix. It was only in 1975 that, on an even calorific basis, the gross consumption of oil finally equaled that of solid fuels (Table 8). With the drastic changes in the world price of petroleum and growing demand in East Europe, the relative export value of liquid fuels versus their domestic utility increased still further, a problem that will be taken up later.

As in North America and Western Europe, solid fuels remain more important in boiler and furnace uses than in the fuel balance as a whole; coal (with coke) continues to be the dominant fuel to this day, while the contribution of oil shale has actually increased since 1960 (Table 9). Yet, in the generation of electricity, the Soviet Union has decreased its reliance on coal to a greater extent than did the United States (Table 10). Until recently, the lucrative gasoline market in the United States has kept down the share of oil as a power-station fuel, while price regulations encouraged an earlier and more extensive penetration of natural gas. By 1965, however, natural gas accounted for a full quarter of the fuel supply to Soviet thermal power stations, a share it has kept ever since. At the same time, the contribution of petroleum as a power-station fuel grew rapidly throughout the 1960s and the first half of the 1970s. By the early 1970s, liquid and gaseous fuels came to dominate the fuel balance of Soviet power stations, a position they never achieved in the United States. These shares also translate into large absolute quantities, comprising about 180 million tons of oil equivalent (95 million tons of petroleum and 100 billion m^3 of gas) in 1975,[52] or well over the total energy export of the Soviet Union.

Since power stations consume about a third of all aggregate energy input to the Soviet economy[53] and represent a use where the advantages of oil and gas are not maximized, it is crucial to understand the causes of this trend. Through most of the 1960s, a lag in development of the urban gas distribution network, the difficulties of gas storage and the consequent problem of regular utilization of long distance gas-pipeline capacity resulted in large amounts of surplus gas to power stations, particularly on an interruptable basis. This surplus was mostly a

Table 9. Soviet Fuel Structure in Boiler and Furnace Uses, 1960-1980 (in percent)

Fuel	1960	1970		1975		1980 (forecast)	
	(%)	Mill. tons of SF	(%)	Mill. tons of SF	(%)	Mill. tons. of SF	(%)
Coal	58.4	354.2	37.9	387.7	32.9	436.2	29.5
Coke and coke-oven gas	10.3	88.3	9.5	97.6	8.3	110.5	7.5
Blast furnace gas	not. incl.	22.8	2.5	25.9	2.2	28.7	1.9
Peat	4.1	16.2	1.7	13.9	1.2	17.6	1.2
Oil shale	0.4	6.1	0.7	8.4	0.7	8.6	0.6
Wood	4.3	18.4	2.0	16.9	1.4	13.7	0.9
Natural gas (including liquified gases)	10.1	242.7	26.2	360.3	30.6	530.0	35.9
Fuel oil	12.4	143.8	15.5	215.2	18.3	257.1	17.4
Others (mostly other petroleum products)[a]	12.4	57.0	6.1	52.1	4.4	74.1	5.0
Total[b]	100.0	926.7	100.0	1178.0	100.0	1476.5	100.0

Notes:

[a]Calculated as a residue. Assumed to be gas and diesel oils for the most part.

[b]Total may not add up to 100% because of rounding.

Sources: For 1960, N. V. Mel'nikov, *Mineral'noye toplivo.* M. Nedra, 1966, p. 164. For 1975-80, *Energetika SSSR v 1976-1980 gg.* (1977), p. 149. Mel'nikov's percentages were corrected to eliminate the share of synthetic gases, which he gives separately, but which are made from other fuels.

temporary phenomenon and began to disappear by the latter part of the decade. Thus the proportion of gas in the fuel supply of power plants doubled between 1960 and 1965, then quickly stabilized, though similar shares since then, of course, still mean larger quantities so used each year. The tremendous growth in the share of petroleum fuels has been primarily a function of the mounting fuel deficit west of the Urals, aggravated by the increasing problem of covering the peak load, for which coal-fired plants are by nature unsuited. As in the United States, an additional, but significant stimulus for the enhanced use of both oil and gas in electric stations was provided by the growing environmental concern.

In the European USSR, growth in the extraction of coal has lagged far behind the growth in energy demand. By 1970, west of the Urals power stations alone required from 50% more fuel input on a calorific basis than all the noncoking coal furnished by the miners of the cis-Ural provinces; by 1975, this requirement was about 60% greater.[54] Even with peat and oil shale production added, the gap remained large indeed, especially since some steam coal must be utilized for

Table 10. Soviet and United States Fuel Structures in Thermal Electric Stations (in percent)

Fuels	Soviet fuel structure					U.S. fuel structure[a]		
	1960	1965	1970	1975	1980[b]	1960	1970	1975
Natural gas	12.3	25.6	26.0	25.7	25.1	26.0	30.0	20.8
Liquid fuels	7.5	12.8	22.5	28.8	28.0	7.6	14.7	20.0
Coal	70.9	54.6	46.1	41.3	42.5	66.4	55.4	59.3
Peat	7.0	4.5	3.1	2.0	2.6	0.0	0.0	0.0
Oil shale	1.0	1.5	1.7	1.7	1.4	0.0	0.0	0.0
Others	1.3	1.0	0.6	0.5	0.4	negl.	negl.	negl.
Total	100.0	100.0	100.0	100.0	100.0	100.0	100.0	100.0

Notes:

[a]U.S. percentages given for all stations, including nuclear and hydro. Percentages recomputed to include conventional thermal stations only. Total may not add up to 100% because of rounding.

[b]Forecast.

Sources: For Soviet Union *Energetika SSSR v 1976–1980 gg.* (1977), p. 151, For U.S., *Statistical Abstract of the United States*, 1976, p. 553.

other purposes. Stations in Siberia, the Far East and Kazakhstan overwhelmingly depend on coal, but in other regions solid fuels are less significant for the generation of electricity and in a few provinces their role is small (Table 11). As in most countries, reducing or at least stabilizing the demand for oil and gas in power stations has become an urgent policy question in the USSR as well. The directives of the 10th Five-Year Plan (1976–80) stated that the future growth of electricity generation would proceed chiefly on the basis of coal, primarily strip-mined coal, and the expansion of nuclear and hydroelectic capacities.[55] Because of geological conditions, the strip mining of coal is almost entirely restricted to the Asian USSR, and the large-scale transport of that coal (or of the power generated from it) is still far from solved. This spatial dilemma will be treated fully in later chapters, but Tables 11 and 12 show that only a stabilization of relative proportions could be expected during the 1976–80 plan period. The actual quantities of hydrocarbons burned in power stations were to increase significantly between 1975 and 1980, by over 20 million tons for oil products and some 24 billion m^3 for natural gas.[56] West of the Volga, even the share of petroleum in the fuel structure of thermal stations had to be expanded slightly (Table 11).

In the European USSR, natural gas and oil also dominate in other stationary uses. They cover two-thirds of fuel demand under industrial boilers and, except for coking coal and peat and firewood cut by rural households, virtually all furnace fuel demand.[57] Because of the huge technological advantage of these quality fuels in these more specialized uses and the concern for clean air, there

Table 11. Fuel Structure of Soviet Electric Stations in Major Economic Regions in 1969–80 (in percent)[a]

Region	1969–70			1975			1980 (forecast)		
	Petroleum products	Natural gas	Solid fuels	Petroleum products	Natural gas	Solid fuels	Petroleum products	Natural gas	Solid fuels
Cis-Volga USSR and European North of which:	20.5	26.7	52.8	35.0	17.0	48.0	36.1	16.0	47.9
Ukraine and Moldavia	7.9	34.7	57.4						
Belorussia	20.5	16.0	63.5						
Baltic Republics	28.5	14.5	57.0						
Transcaucasia	72.4	25.0	12.6	42.8	55.7	1.5	43.0	54.0	3.0
Volga Region	59.4	23.2	17.4	55.6	34.9	9.5	47.5	40.5	12.0
Urals	11.2	28.5	60.3	16.4	28.6	55.0	7.1	29.2	63.7
Kazakhstan	12.1	11.5	76.4	} 14.0	37.9	48.1	17.9	31.8	50.3
Central Asia	16.0	61.8	22.2						
West Siberia	11.4	1.1	84.5	10.7	8.1	81.2	8.9	16.8	74.3
East Siberia	0.7	0.0	99.3	1.0	0.0	99.0	1.2	0.0	98.8
Far East	8.7	1.8	89.5	8.4	2.6	89.0	9.3	2.3	88.4
USSR	22.5	26.0	51.5	28.8	25.7	45.5	28.0	25.1	46.9

Sources: Energetika SSSR v 1976–1980 gg. (1977), p. 153 and *Energetika SSSR v 1971–1975 gg.* (1972), pp. 172–73.

[a]Refers only to electric stations under the jurisdiction of the Ministry of Electric Power. These stations generate 94% of all Soviet electricity. All percentages add up to 100.

Table 12. Share of Natural Gas in Boiler and Furnace Fuel Demand of Soviet Regions (in percent)

Region	1962	1965	1970
USSR	14.9	21.3	27.9
European provinces:	22.0	26.4	32.2
Northwest	9.2	16.9	20.9
Baltic	3.4	11.0	10.7
Belorussia	15.8	16.6	18.2
Central Region	25.7	32.3	37.2
Central Chernozem	14.1	19.1	24.3
Volga-Vyatka	15.5	21.8	22.2
North Caucasus	33.0	47.0	46.5
Volga Region	26.9	33.9	29.8
Transcaucasia	46.6	41.3	38.6
Ukraine	21.3	29.2	35.0
Urals	1.7	15.6	37.8
Asian provinces:	2.6	5.5	12.1
West Siberia	0.0	0.0	0.7
East Siberia	0.0	0.0	1.4
Far East	2.1	2.4	4.7
Kazakhstan	2.3	2.8	12.6
Central Asia	18.0	37.7	57.0

Source: A. Ye. Probst and Ya. A. Mazover (eds.), *Razvitiye i razmeshcheniye toplivnoy promyshlennosti.* Moscow: Nedra, 1975.

are no plans to substitute, or resubstitute, coal for hydrocarbons or even to decrease the relative share of oil and gas in the future.[58] In fact, the continued modernization of rural households, the spread of propane burners and oil stoves, is bound to augment somewhat the role of gas and oil in furnace uses. Rural households in 1975 were to account for 37% of total heat consumption by the domestic sector, but their needs were still mostly met by solid fuels and agricultural wastes.[59]

During most of the 1960s natural gas increased its share in the fuel balance of nearly all economic regions (Table 12), but at the expense of solid fuels, not oil. The gradual peaking of most European gas reservoirs and the delays and formidable transport difficulties in making big supplies available from Soviet Asia slowed down the growth in gas consumption by the latter part of the 1960s (yearly increments in output declined substantially between 1965 and 1973). With the belated arrival of Siberian supplies, the contribution of gas is growing significantly again in many regions of the European USSR.[60] Because natural gas is considered a more abundant fuel than petroleum, some substitution of gas for refinery products may also be expected in the future in stationary, nonpower plant, uses. Such substitution may be effected with a minimum of cost and is already thought to have taken place to a degree.[61]

NOTES

[1]U.S. Central Intelligence Agency, *Soviet Long-Range Energy Forecasts*, A(ER)75-71, September 1975, p. 26; Herbert Block, "Energy syndrome, Soviet version," *Annual Review of Energy*, 1977, No. 2, pp. 461-66.

[2]C.I.A., *Soviet Long-Range Energy Forecasts*, op. cit., p. 25, Note 3.

[3]Leslie Dienes, "Soviet energy policy and the hydrocarbons," in: Association of American Geographers, Discussion Paper No. 2. Project on *Soviet Natural Resources in the World Economy* (Syracuse University, 1978), pp. 4-14.

[4]A. A. Beschinskiy and Yu. M. Kogan, *Ekonomicheskiye problemy elektrifikatsii* (Moscow: Energiya, 1976), pp. 23-24 and 200-202.

[5]For example, without losses, internal demand by the Soviet coal industry is 3.5% of its production; the oil industry directly consumes 14% of its output in field operations and refining and uses some 100 KWH of electricity per ton of oil produced. Gas pipelines in the Soviet Union consumed 4.2% of gas output as fuel in 1975 (in addition to large losses); this share is expected to rise to more than 6% by 1980. In the electric power industry, line losses and station use account for almost 15% of gross electricity consumption in the USSR. L. Yakubovich et al., *Pryamyye khozyaystvennyye svyazi po postavkam uglya i slantsa* [Direct Economic Linkages in Coal and Oil Shale Deliveries] (Moscow: Nedra, 1974), p. 28; *Neftyanik*, 1975, No. 6, p. 7; M. B. Ravich, *Toplivo i effektivnost' yego ispol'zovaniya* [Fuel and the Cost-Effectiveness of Its Use] (Moscow: Nauka, 1971), p. 203; *Energetika i transport*, 1975, No. 1, p. 122; L. Samsonov, *Rol' elektrifikatsii v povyshenii effektivnosti obshchestvennogo proizvodstva* [The Role of Electrification in Raising the Cost-Effectiveness of Social Production] (Moscow University, 1974), p. 142; *Energetika SSSR v 1976-1980 godakh* (Moscow: Energiya, 1977), pp. 13, 15, 151 and 181-182; United Nations, Economic Commission for Europe, *Annual Bulletin of Gas Statistics for Europe 1975* (Geneva, 1976), pp. 56-57.

[6]Robert W. Campbell, *Soviet Fuel and Energy Balances*, Research Report R.2257 (Santa Monica, CA: RAND Corporation, 1978).

[7]The Soviet fuels and energy industries in the middle 1970s consumed almost two-thirds of all steel pipe, up to one-fifth of other iron and steel products, 15% of copper and aluminum and 15% of the machinery output. A. Vigdorchik et al., "Methods of optimization . . ," *Planovoye Khozyaystvo*, 1975, No. 2, p. 29.

[8]The only serious attempt to compute total (direct and indirect) self-consumption of energy by the energy industries appears to have been made for Britain. The self-consumption share was found to be more than 30%. In view of the greater geographic and transport burdens in the USSR, Soviet self-consumption share can be assumed to be greater. P. F. Chapman et al., "The energy cost of fuels," *Energy Policy*, September 1974, pp. 231-244.

[9]P. S. Neporozhnyy et al., "Fuel and power economy of the Soviet Union at the current stage . . . ," *Ninth World Energy Conference. Transactions* (Detroit, 1974), Vol. II, p. 149.

[10]Campbell, op. cit., 1978.

[11]*Energetika SSSR v 1976-1980 godakh*, op. cit., p. 51; U.S. Bureau of Mines, *U.S. Minerals Yearbook 1974* (Washington: Government Printing Office, 1976), pp. 40-41.

[12]Data are available only for 1970. In that year, common-use and industrial heat and power stations combined distributed 22 percent of their heat supply to households and 78%

to industry. *Energetika SSSR v 1971-1975 godakh* (Moscow: Energiya, 1972), p. 81.

[13]Campbell, op. cit., 1978.

[14]*Energetika SSSR v 1976-1980 godakh*, op. cit., p. 151; V. Kudinov and S. Litvak, "Concerning the fuels and energy balance," *Vestnik Statistiki*, 1972, No. 9, p. 41.

[15]A. M. Nekrasov and V. Kh. Khokhlov, "Electric power in the 9th five-year plan," *Izvestiya VUZ, Energetika*, 1976, No. 9, p. 6.

[16]V. P. Korytnikov, ed., *Rabota TETs v ob"yedinennykh energosistemakh* [The Operation of Heat and Power Stations Within Interconnected Power Systems] (Moscow: Energiya, 1976), p. 9.

[17]Regional indices of per capita fuel consumption are from Table 5. Regional indices for per capita industrial fixed assets are available only for 1965 in A. I. Vedishchev. "Commensuration of economic development levels of the economic regions of the USSR," in Akademiya Nauk SSSR and Gosplan SSSR, SOPS, *Ekonomicheskiye problemy razmeshcheniya proizvoditel'nykh sil SSSR* [Economic Problems of Location of the Productive Forces of the USSR] (Moscow: Nauka, 1969), p. 63. They may also be computed for 1968 without Moldavia. L. Dienes, "Investment priorities in Soviet regions," Association of American Geographers, *Annals*, September 1972, p. 440.

[18]For the share of hydroelectricity in total power output within economic regions, see *Energetika SSSR v 1971-1975 godakh*, op. cit., p. 144, and V. A. Ryl'skiy et al., *Elektroenergeticheskaya baza ekonomicheskikh rayonov SSSR* [The Electric Power Base of Economic Regions of the USSR] (Moscow: Nauka, 1974), p. 178.

[19]Dienes, op. cit. (1972), pp. 440-445.

[20]R. W. Campbell, *The Economics of Soviet Oil and Gas* (Baltimore: Johns Hopkins, 1968), p. 13.

[21]From data on production and consumption shares in calorific equivalents. A. Ye. Probst, ed., *Razvitiye toplivnoy bazy rayonov SSSR* [Development of the Fuels Base of Regions in the USSR] (Moscow: Nedra, 1968), pp. 45 and 48.

[22]Total fuel demand in 1970 was estimated from boiler and furnace consumption as given in Table 5. Production of different fuels in physical units is from *Nar. khoz. SSSR v 1975 godu*, pp. 240-243. They have been converted to tons of standard fuel.

[23]Consumption for 1970 in the Ukraine is from *Nardne hospodarstvo Ukrains'kol RSR v 1971 rotsi* [Economy of the Ukrainian SSR in 1971], in Ukrainian, p. 66. Reasonable rates of growth, somewhat below national growth rates, were applied from 1970 to 1975. Production data in physical units, from *Soviet Geography*, April 1977, pp. 264-269, have been converted to tons of standard fuel.

[24]Consumption was projected from 1970 boiler and furnace fuel demand (Table 5) by applying reasonable growth rates, somewhat below national rates. Gas production in the North Caucasus has been declining sharply since the late 1960s and oil production from the early 1970s. *Soviet Geography*, April 1977, p. 270, and May 1977, pp. 348-250; Economist Intelligence Unit, *QER Special*, No. 14, June 1973, pp. 7-8; N. P. Akopyan, "Main tasks in gas extraction in the North Caucasus," *Gazovaya Promyshlennost'*, 1976, No. 9, pp. 2-3. Physical units of production were converted to standard fuel.

[25]Data from Tables 5 and 7, from Probst, op. cit. (1968), pp. 45 and 48, and *Soviet Geography*, April 1978, pp. 273-285. Physical units were converted to standard fuel. Table 5 shows boiler and furnace fuel consumption only, which comprises four-fifths of total fuel demand.

[26]Nekrasov and Khokhlov, op. cit., p. 4; *Energetika SSSR v 1976-1980 godakh*, op. cit., p. 148.

[27]G. V. Yermakov et al., "Trends in the development of the nuclear power industry," *Ninth World Energy Conference. Transactions* (Detroit, 1974), Vol. 5, p. 279.

[28]*Energetika SSSR v 1976-1980 godakh*, op. cit., p. 147; *Neftyanik*, 1976, No. 5, pp. 1-4; *Soviet Geography*, April 1978, pp. 273-285.

[29]Campbell, op. cit. (1968), p. 13.

[30]Total capacity of the Central Asia–Central Russia pipeline system is 68 billion m³ a year; that of the older Bukhara–Urals system, 21 billion m³. S. A. Orudzhev, *Gazovaya promyshlennost' po puti progressa* [The Gas Industry on the Road of Progress]. (Moscow: Nedra, 1976), p. 44; K. N. Bedrintsev et al., eds., *Sredneaziatskiy ekonomicheskiy rayon* [Central Asian Economic Region] (Moscow: Nauka, 1972), pp. 60-61.

[31]Production of fuels in 1975 and 1976, from *Soviet Geography*, April 1977, pp. 264-269, has been converted to tons of standard fuel. Consumption figures for 1970 appear in S. Kh. Khusainova and Yu. S. Gudkov, *Gazifikatsiya i gazosnabzheniye Kazakhstana* [Gasification and Gas Supply of Kazakhstan] (Moscow: Nedra, 1975), p. 90, and in L. N. Serovoy, "Problems of the fuels and energy balance. Central Asia," Diploma thesis (Moscow University Geography Faculty, 1975), p. 15, quoting official data from the Central Statistical Administration. Reasonable growth rates, somewhat above the national average, were applied to estimate 1975-76 consumption.

[32]See, for example, P. K. Savchenko and A. R. Khodzhayev, *Toplivno-energeticheskiy kompleks Sredneaziatskogo ekonomicheskogo rayona* [The Fuels and Energy Complex of the Central Asian Economic Region] (Tashkent: Uzbekistan, 1974); S. Chokin, *Energetika i vodnoye khozyaystvo Kazakhstana* [Electric Power and Water Management in Kazakhstan] (Alma-Ata: Kazakhstan, 1975), p. 77; K. M. Kim, *Sovershenstvovaniye struktury toplivno-energeticheskogo balansa Sredney Azii* [Perfecting the Structure of the Fuels and Energy Balance of Central Asia] (Tashkent: FAN, 1973), pp. 197-198 and 208-209. Half of the huge reserves of Gazli were exhausted by the middle 1970s (Brents et al., *Ekonomika gazodobyvayushchey promyshlennosti* [Economics of the Gas-Extraction Industry] (Moscow, 1975), p. 172, and gas from the Turkmenian fields is being transmitted out of Central Asia at an equally rapid rate. At the same time, gasification within Central Asia is spreading slowly, and the region is importing increasing amounts of Siberian and Kazakhstan coals and must rely even on noncommercial fuel sources. Three-tenths of the fuel consumed in Central Asia has worse economic characteristics than the gas and fuel oil exported by the region. The Central Asian republics, of course, are not independent and, in view of low labor productivity within the region, such large fuel exports to the RSFSR may well maximize economic benefits for the Soviet economy as a whole, though the present author has yet to see a model attempting to demonstrate that point. An unpublished student thesis (Serovoy, op. cit) seeks to show that the relatively low gas consumption within Central Asia also has favorable consequences by stimulating the exploitation of new coal deposits within the region.

[33]Output in physical units, from *Soviet Geography*, April 1978, pp. 274-282, has been converted to tons of standard fuel.

[34]Boiler and furnace fuel demand for 1970 is available from Table 5. Total fuel consumption is about 25% larger. The 1976 demand is estimated by applying reasonable growth rates, somewhat above the national average.

[35]See Table 5. Also N. V. Mel'nikov et al., *Toplivno-energticheskiye resursy SSSR* [Fuels and Energy Resources of the USSR] (Moscow: Nedra, 1971), p. 510; V. A. Shelest, *Regional'nyye energoekonomicheskiye problemy SSSR* [Regional Energy-Economic Problems of the USSR] (Moscow: Nauka, 1975), p. 212, and other sources.

[36]*Energetika SSSR v 1976-1980 godakh*, op. cit., p. 149.

[37]In both the Russian Republic and the Ukraine, the net reproduction rates have been below replacement level for over a decade, while in Belorussia this level was reached in 1970. B. S. Khorev and V. M. Moiseyenko, *Sdvigi v razmeshchenii naseleniya SSSR* [Shifts in the Distribution of Population in the USSR] (Moscow: Statistika, 1976), pp. 30-35; B. S.

Khorev and V. M. Moiseyenko, eds., *Migratsionnaya podvizhnost' naseleniya v SSSR* [Migratory Mobility of Population in the USSR] (Moscow: Statistika, 1974), p. 80. The outflow continued in the first half of the 1970s at a reduced rate. Since 1959, the birth rate in Siberia has also declined much faster than in the USSR or RSFSR as a whole. Ye. D. Malinin and A. K. Ushakov, *Naseleniye Sibiri* [The Population of Siberia] (Moscow: Statistika, 1976), pp. 25 and 42–45. Reports on idleness of machinery because of labor shortages are common.

[38]Table 5 and Ryl'skiy, op. cit., pp. 34–35.

[39]See, for example, Probst, op. cit., pp. 218–219, and Yermakov, op. cit., p. 279.

[40]D. B. Shimkin, *The Soviet Mineral Fuel Industries, 1928–1958* (Washington: Bureau of the Census, 1961), pp. 43–44.

[41]S. D. Fel'd, *Yedinyy energeticheskiy balans narodnogo khozyaystva* [The Combined Energy Balance of the Soviet Economy] (Moscow: Ekonomika, 1964), pp. 48 and 51.

[42]Shimkin, op. cit., p. 80. Shimkin calculates that over a 15-year period, after 1939–40, Soviet fuel production rose from 24 percent to 32% of the United States level, but Soviet labor productivity fell from 23 to 14% of the American level; ibid, p. 53.

[43]Computed from the average calorific value of the different fossil fuels, as given in *Nar. khoz. SSSR v 1975 godu*, pp. 239–243.

[44]The average production cost of oil, as calculated by Soviet sources, declined from nearly 5 rubles in the first half of the 1950s to less than 3 rubles by 1965; the average production cost of natural gas dropped by 70%. A. A. Konstantinova, "The cost-effectiveness of technical progress in the extractive industries," unpublished dissertation (Moscow: Institute of Economics, 1975), p. 55, and M. Brenner, *Ekonomika neftyanoy i gazovoy promyshlennosti SSSR* [Economics of the Oil and Gas Industry of the USSR] (Moscow: Nedra, 1960, p. 254.

[45]In the USSR, the cost of transporting solid fuels over a distance of 1,000 km or more exceeds the cost of transporting not only crude oil but even natural gas in terms of equal calorific content. The savings derived from an increased role of oil and gas would thus have to be larger on the basis of delivered cost than that of production cost alone. A. Probst, "Ways of developing the fuel economy of the USSR," *Voprosy Ekonomiki*, June 1971, pp. 51–62. Also L. Dienes, "Geographical problems of allocation in the Soviet fuel supply," *Energy Policy*, June 1973, pp. 4–5, and L. Dienes, *Locational Factors and Locational Developments in the Soviet Chemical Industry* (University of Chicago, Dept. of Geography Research Paper 119, 1969), especially Chapters 2 and 7.

[46]For crude oil, production cost reverted almost to the 1950–55 level of 5 rubles by 1973. Because of the rapid growth of surface mining and its increase in total coal output, the average production cost of coal per physical ton remained unchanged from 1969 to 1974; there was a small increase in costs in terms of standard fuel. Konstantinova, op. cit., p. 55, and A. D. Ayushiyev, *Finansy proizvoditel'nykh ob"yedinenii ugol'noy promyshlennosti* [Finances of the Corporations in the Coal Industry] (Moscow: Nedra, 1976), p. 92.

[47]The incremental capital requirements per unit of new capacity in the oil and gas industries increased as follows: In the oil industry, unit capital requirements per metric ton were, in 1966–70, 99.2 rubles for production and 10.3 for transportation; in 1971–75, 116.6 and 24 rubles, respectively; in the gas industry, unit capital requirements per 1,000 m^3 were, in 1966–70, 77.5 rubles for production and 43.8 for transportation; in 1971–75, 96 and 97.6 rubles, respectively. A. Lalayants, "Two societies and two policies in the development of the fuels and energy complex," *Planovoye Khozyaystvo*, 1975, No. 4, pp. 20–31.

[48]U.S. Central Intelligence Agency, *Recent Developments in Soviet Hard-Currency Trade*, ER 76-10015 (January 1976), p. 21; E. E. Jack et al., "Outlook for Soviet energy," in U.S. Congress, Joint Economic Committee, *Soviet Economy in a New Perspective* (Washington:

Government Printing Office, 1976), pp. 475–477; T. Shabad, *The New York Times*, June 11, 1977, pp. 1 and 27.

[49] Shabad, ibid.

[50] L. Dienes, "Energy demand in Eastern Europe from 1960 to the present," *Energy Policy*, June 1976, pp. 119–129.

[51] *Nar. khoz. SSSR* and *Vneshnyaya torgovlya SSSR* (Foreign Trade of the USSR], statistical yearbooks, various issues.

[52] Computed from *Energetika SSSR v 1976–1980 godakh*, op. cit., pp. 149 and 151.

[53] Two-fifths of boiler and furnace fuel consumption, and thus 33.3% of total fuel use. Ibid, pp. 149 and 151, and *Nar. khoz SSSR v 1975 godu*, p. 112. This, of course, includes fuel used for the production of both electricity and steam heat.

[54] Production in natural tons, from *Soviet Geography*, April 1977, p. 264, has been converted to tons of standard fuel. Fuel consumption in power stations has been computed from *Energetika SSSR v 1976–1980 godakh*, op. cit., pp. 149 and 151, which also gives production of coal in standard fuel equivalent for 1975.

[55] *Materialy XXV s"yezda KPSS* [Proceedings of the 25th Congress of the Soviet Communist Party], 1976, p. 140. Kosygin's speech, March 1, 1976.

[56] Computed from *Energetika SSSR v 1976–1980 godakh*, op. cit., pp. 149 and 151.

[57] Neporozhnyy, et al., op. cit. (1974), p. 160.

[58] For example, an authoritative source states that hydrocarbon fuels comprise almost 80% of the fuel supply of industrial boilers and heat and power stations. During the period 1976–1980 alone, these consumers will need an increment of natural gas and petroleum amounting to 80 million tons of standard fuel (which equals 56 million tons of oil equivalent). N. A. Dolezhal' and L. A. Melent'yev, "The role of the nuclear energy system in the fuels and energy complex of the USSR," *Vestnik Akademii Nauk SSSR*, 1977, No. 1, p. 89.

[59] *Energetika SSSR v 1976–1980 godakh*, op. cit., p. 77.

[60] Orudzhev, op. cit., pp. 31–33.

[61] The substitution of natural gas for high-sulfur fuel oil produced by the refineries of Bashkiria is reported in V. I. Manayev, "Bashkir ASSR, its complex today and tomorrow," *EKO (Ekonomika i organizatsiya promyshlennogo proizvodstva)*, 1977, No. 2, pp. 46–59. These refineries should therefore increase the share of lighter distillates in their output mix, a task that has traditionally been hampered by delays in the construction of facilities for deeper refining.

Chapter 3

ENERGY SUPPLIES: PETROLEUM AND NATURAL GAS

Having examined the overall trends in the Soviet energy economy, including the geographic pattern of demand and the energy mix, we will now consider the development and location of the primary energy sources, including the hydrocarbons (petroleum and natural gas), the solid fuels (mainly coal, but also such secondary fuels as oil shale and peat), and the two forms of primary electricity (which does not involve the combustion of fossil fuels), namely hydroelectricity and nuclear power. The present chapter is concerned with the hydrocarbons; subsequent chapters will deal with the soild fuels as well as both hydroelectric and nuclear power.

The discussion of each type of primary energy is hampered to some extent by the well-known Soviet reluctance to publish detailed statistics on economic activities, in particular, the regional location of production. In the case of petroleum, reserve figures have been considered a state secret and regional production figures have been published most consistently and in the most easily accessible form only for the 15 union republics, the basic constituent states of the Soviet Union. A republic-wide figure for the vast Russian SFSR, the Soviet state extending from the central European USSR through the Urals and across Siberia to the Pacific coast, is of limited utility, and there has been little information in absolute figures for the distribution of production within the RSFSR. Most of the detailed production figures used in the discussion that follows and shown in Table 13 have been computed and estimated from a wide range of data assembled from scattered sources, including regional Soviet newspapers, and published from time to time in the monthly news notes of the translation journal *Soviet Geography: Review and Translation.* In a further

Table 13. Regional Crude Oil Production in USSR[a]
(million metric tons)

Region	1940	1945	1950	1955	1960	1965	1970	1975	1976	1977	1978 est.	1980 plan
USSR	31.1	19.4	37.9	70.8	148	243	353	491	520	546	572	640
RSFSR	7.0	5.7	18.2	49.3	119	200	285	411	445	478	509	554
EUROPEAN USSR	28.9	16.2	32.1	63.1	134	213	265	259	254	245	235	230
incl. Urals	*29.1*	*16.7*	*32.6*	*64.2*	*138*	*225*	*289*	*299*	*296*	*289*	*280*	*276*
European RSFSR	6.3	4.4	17.0	47.2	114	184	227	221	219	213	207	192
incl. Urals	*6.5*	*4.9*	*17.5*	*48.3*	*117*	*197*	*251*	*261*	*261*	*257*	252	*237*
Komi ASSR	0.070	0.17	0.52	0.55	0.81	2.2	7.6	11	13	15	18	25
North Caucasus	4.6	1.9	6.0	6.5	12	21	35	24	23	22	21	20
Krasnodar Kray	2.2	0.76	3.2	4.1	7.0	6.2	5.9	6.2	6	6		
Stavropol' Kray	–	–	–	–	1.6	4.5	6.4	7	7	7		
Chechen-Ingush ASSR	2.2	0.87	2.3	2.1	3.3	9.0	20	9	8	7		
Dagestan ASSR	0.15	0.29	0.51	0.3	0.23	1.0	2.2	2	2	2		
Volga	1.7	2.3	10.5	40.1	101	161.2	184.4	186	183	176	169	147
Tatar ASSR	–	0.01	0.95	14.4	46.3	79.7	101.9	103.7	100	97		
Bashkir ASSR	1.5	1.3	5.5	14.8	25.3	40.7	39.2	39.1	40	40		
Kuybyshev Oblast	0.2	1.0	3.5	7.3	22	33.4	35.0	34.8	33	31		
Others	–	–	0.5	3.6	7	7.4	8.3	8	8	8		
Urals	0.18	0.5	0.53	1.1	3.5	12.4	24.0	39.6	42	44	44	45
Perm' Oblast	0.15	0.2	0.30	0.6	2.3	9.8	16.1	22.3	23.5	24		28–30
Orenburg Oblast	0.03	0.27	0.23	0.48	1.2	2.6	7.4	13.9	14	14		
Udmurt ASSR	–	–	–	–	–	–	0.48	3.4	4.4	5.5		8
Volga-Urals	*1.9*	*2.8*	*11.0*	*41.2*	*104.3*	*173.6*	*208.4*	*226*	*225*	*220*	*213*	*192*

Table 13. (cont'd)

Region	1940	1945	1950	1955	1960	1965	1970	1975	1976	1977	1978 est.	1980 plan
EUROPEAN USSR (cont'd)												
Belorussian SSR	–	–	–	–	–	0.04	4.2	8.0	6.2	5	4	8
Ukrainian SSR	0.35	0.25	0.29	0.53	2.2	7.6	13.9	12.8	11.6	10.5	9.5	8.6
Georgian SSR	0.04	0.04	0.04	0.04	0.03	0.03	0.02	0.26	0.88	1	1	3.0
Azerbaijan SSR	22.2	11.5	14.8	15.3	17.8	21.5	20.2	17.2	16.5	15.7	14	19.7
ASIAN USSR (excl. Urals)	2.0	2.7	5.2	6.6	10	17	64	192	224	257	292	364
Siberia	0.51	0.8	0.62	0.95	1.6	3.4	34	151	184	220	257	317
West Siberia	–	–	–	–	–	1.0	31.4	148	181.7	218	254	315
Tyumen' Oblast	–	–	–	–	–	1.0	28.0	143	176	211	246	305
Tomsk Oblast	–	–	–	–	–	–	3.4	5	6	7.1	8	10
Sakhalin	0.51	0.8	0.62	0.95	1.6	2.4	2.5	2.4	2.4	2.4	3	2.5
Kazakhstan	0.70	0.79	1.1	1.4	1.6	2.0	13.1	23.9	23.3	21	20	26.9
Central Asia	0.76	1.1	3.4	4.2	7.3	11.8	16.8	17.5	16.6	16	15	
Kirghizia	0.024	0.019	0.047	0.12	0.46	0.31	0.30	0.23	0.23	0.2		0.2
Uzbekistan	0.12	0.48	1.3	1.0	1.6	1.8	1.8	1.4	1.3	1.2		1.0
Tadzhikistan	0.030	0.020	0.020	0.017	0.017	0.047	0.18	0.27	0.31	0.34		0.4
Turkmenia	0.59	0.63	2.0	3.1	5.3	9.6	14.5	15.6	14.8	14	13	

[a]Includes gas condensate (see Table 23, p. 98).

Sources: Production by union republics was available in official statistical yearbooks until 1977, when publication of regional data ceased. Production for autonomous republics within the RSFSR appeared in the national statistical yearbooks from 1964 through 1974, covering the years 1940, 1960, 1965, and 1970. Production of the Asian regions, including Siberia, was published in the yearbooks through 1973. For other areas and other years, output data was assembled by the author on the basis of absolute and percentage figures published in a wide range of scattered sources, including regional newspapers.

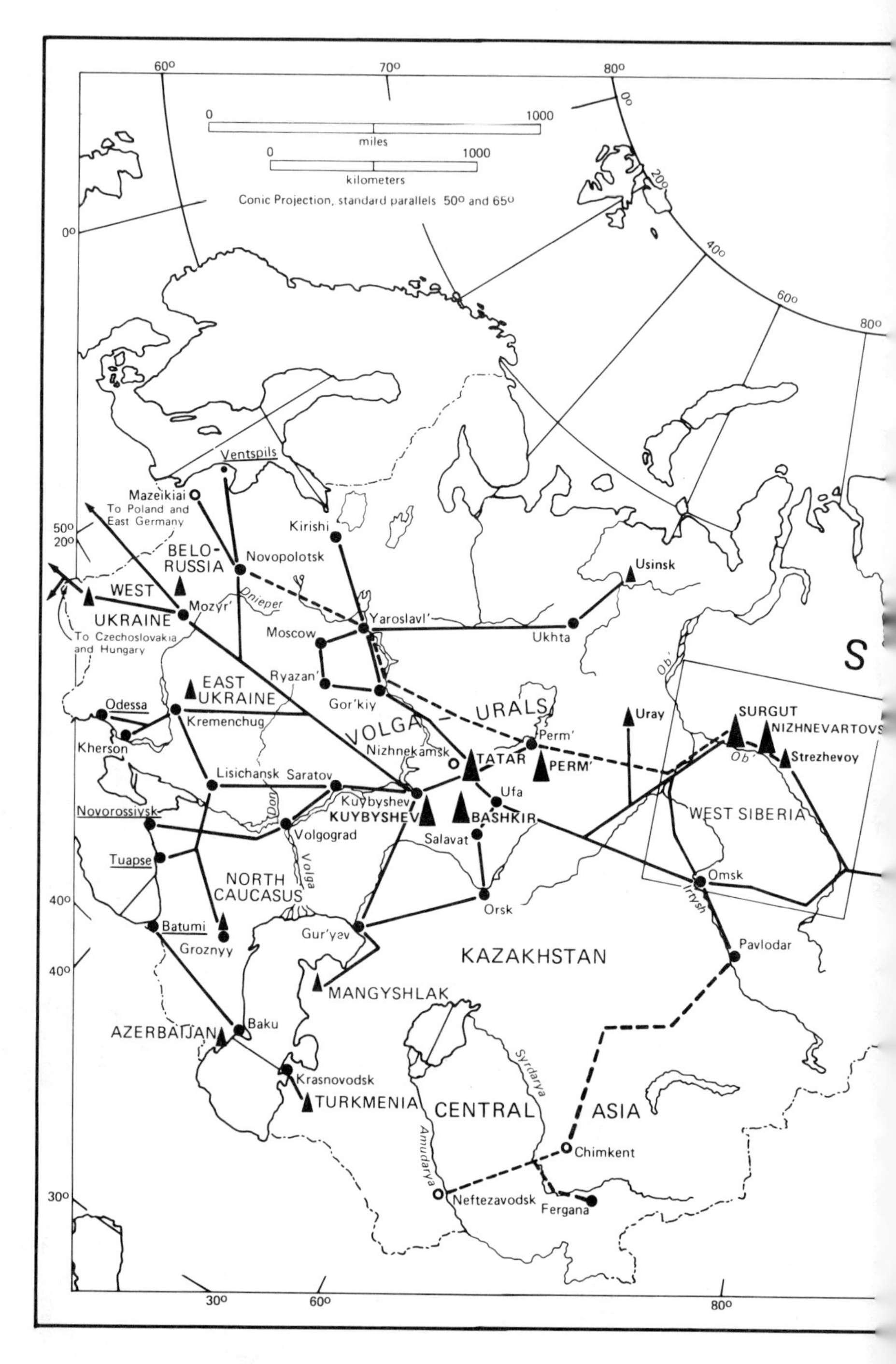

Fig. 4. Major oil fields, pipelines and refineries.

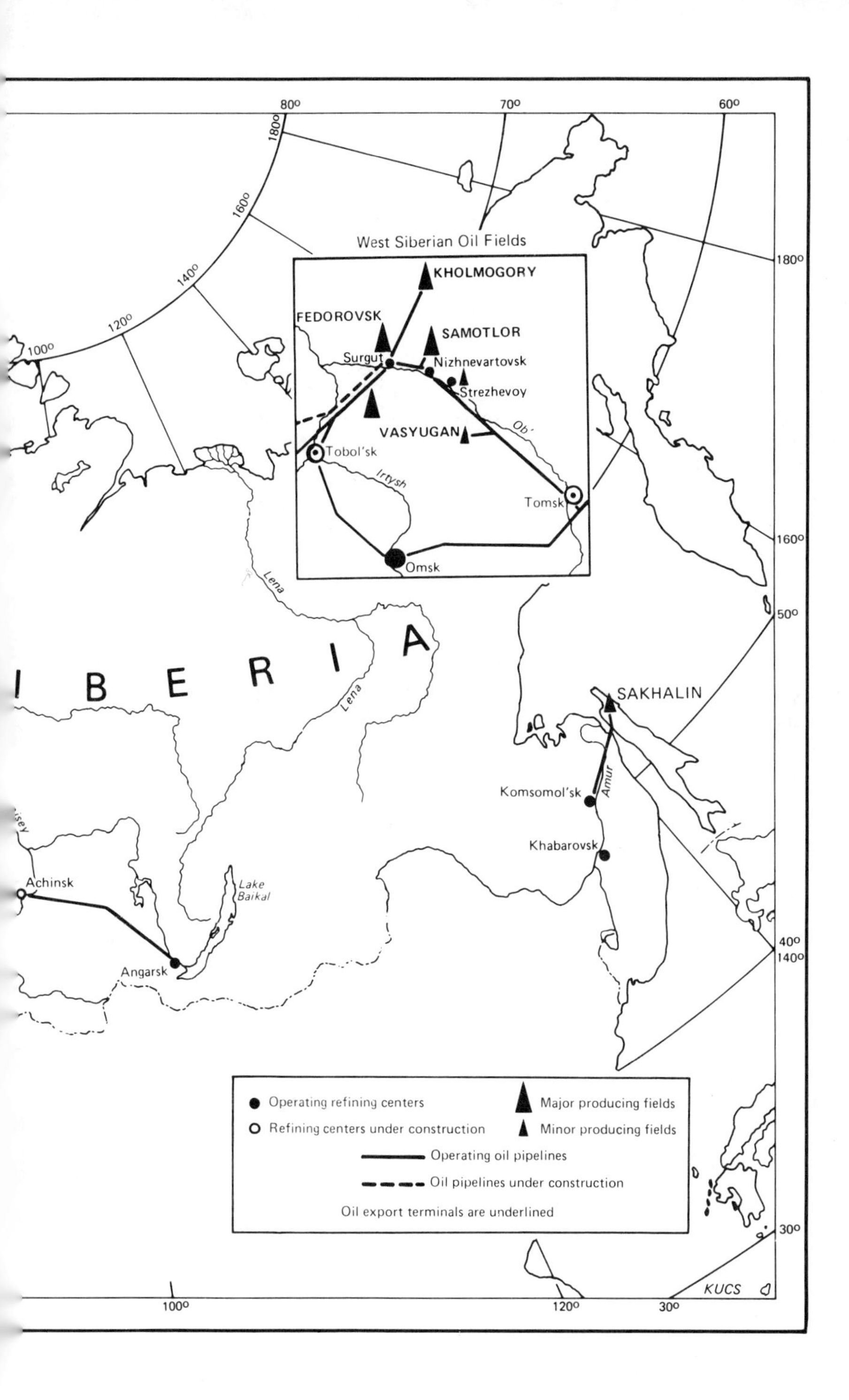
West Siberian Oil Fields
KHOLMOGORY
FEDOROVSK
SAMOTLOR
Surgut
Nizhnevartovsk
Strezhevoy
VASYUGAN
Ob'
Tobol'sk
Irtysh
Tomsk
Omsk
Lena
S I B E R I A
SAKHALIN
Komsomol'sk
Amur
Khabarovsk
Achinsk
Lake Baikal
Angarsk
Operating refining centers
Refining centers under construction
Major producing fields
Minor producing fields
Operating oil pipelines
Oil pipelines under construction
Oil export terminals are underlined
KUCS

statistical restriction, imposed as of 1977, the Soviet Union ceased the publication of data on foreign oil trade in physical units and domestic output by union republics, excepting thus far only the RSFSR, the Ukraine and oddly a wholly negligible oil producer, the Tadzhik SSR of Central Asia.

In the case of natural gas, statistical restrictions had been originally somewhat more relaxed, perhaps because natural gas became a significant factor in the Soviet energy system only in the mid-1950s, at a time when some of the secrecy of the second half of the Stalin era was being lifted. Information on Soviet reserves of natural gas has been published over the years and regional production data have been more easily available than for petroleum, not only at the level of union republics, but also at a more detailed level within republics. Most of the regional figures for natural gas production (Table 18) have thus been derived from absolute data. However, the statistical secrecy imposed in 1977 on foreign trade and production data of fossil fuels also affected natural gas. Except for the overall production of the Soviet Union, systematic output data for natural gas is now being published only for the RSFSR as a whole.

PETROLEUM

Although the Soviet Union became the world's leading producer of crude oil in 1974 and ranks third in oil exports (behind Saudi Arabia and Iran), the prospects for petroleum appear to be by far the most uncertain among the Soviet primary energy sources. Despite the absence of official reserve data, it has become evident in recent years that new discoveries have not kept up with continued growth of production, that the reserves-to-production ratio, a key indicator of future prospects, has been deteriorating, and only about 10 years worth of output at the current high levels may be at the disposal of Soviet planners. (For a more detailed discussion of the long-term outlook, see Chapter 9.) There is certainly no doubt that annual increments of crude oil production peaked in 1975, at 31.8 million metric tons, and have been declining since at a steady rate, contrary to the projections that went into the current (10th) five-year plan. The plan, which runs from 1976 through 1980, envisaged average annual growth of 29.8 million tons, up from an average rate of 27.6 million tons during the preceding five-year period. However, the increments dropped from 29 million in 1976 to 26 million in 1977, where they appear to have leveled off in 1978. Growth in West Siberian output, which has been slowing down, was being eroded by production declines in older European fields (Table 13).

Early History of Oil Production

Before World War II, about 80% of Soviet oil production was concentrated in the Caucasus, where commercial output dated from the 1870s at Baku (Azerbaijan SSR) and from the 1890s at Groznyy (Chechen-Ingush ASSR) (Fig. 4). Out of a national production of 31.1 million tons in 1940, Baku yielded 22.2 million, a record level that was not to be reached again after World War II (a postwar peak of 21.7 million tons was attained in 1966). On the eve of the war, Groznyy contributed 2.2 million tons, having dropped from a peak of 8.1 million tons in 1931.[1] In contrast to Baku, which never recovered its prewar output levels, Groznyy did pass through a revival in the postwar period following the discovery in the late 1950s of new reserves in deeper Cretaceous rocks below the previously productive Miocene beds. However, the upsurge proved to be shortlived. Having steadily risen through the 1960s, Groznyy output peaked at 21.6 million tons in 1972[2] and then declined precipitously, to about 7 million tons in 1977.[3] Soviet projections had expected the Cretaceous discoveries to maintain high output levels, and after the successful performance of the 1960s, the 9th five-year plan had envisaged a 1975 goal of 25 million tons;[4] the actual 1975 production was about 9 million tons.

There had been relatively little exploration for oil before World War II as national fuel policy still favored a coal-oriented economy. The Baku and Groznyy fields, though threatened by the Germans' military drive into the Caucasus, were not seized during the war, but the lag of exploratory drilling and wartime disruptions nevertheless led to a decline of production. In the Baku district, after a final spurt to 23.5 million tons in 1941, output dropped to 15.7 million in 1942, 12.7 in 1943, 11.8 million in 1944, and 11.5 million in 1945,[5] or less than one-half the prewar level. In an effort to restore Baku production to its earlier magnitude, Soviet producers began the development of Caspian offshore fields.

The first major offshore field, which began production in 1951-52, was at Neftyanyye Kamni (Oil Rocks), a reef area 25 miles east of Baku's Apsheron Peninsula and linked to the mainland by a causeway. This field, which peaked at 7 million tons in 1970,[6] was replaced in the early 1970s as the leading Baku producing area by a second offshore area extending from the Cape Sangachaly, on the mainland southwest of the Apsheron Peninsula, over a distance of 20 miles into the Caspian to Duvannyy Island and Bulla Island. In 1975, two-thirds of Azerbaijan oil production was derived from offshore fields, including 4.5 million tons from the Duvannyy–Bulla field[7] and 3.8 million from Neftyanyye Kamni.[8] However, offshore development in the Baku region was unable to compensate for declining production onshore, and Azerbaijan output continued to drop. The poor long-term prospects were reflected in a drastic downward revision of the 10th five-year plan. The original 1980 goal of 22.15 million tons, announced at the Azerbaijan party congress in January 1976, was reduced on re-evaluation to 19.69 million tons the following November at the republic's Supreme Soviet,[9] including 13.7 million offshore and 6 million onshore. Even the reduced five-year plan goal appeared to be beyond the grasp of the Baku region.

Production continued to drop in the late 1970s (Table 13), mainly because offshore fields were unable to meet their goals. It was hoped that the development in 1977 of a new inland field, at Muradkhanly, 10 miles northwest of Imishli, would help stabilize and perhaps even slightly raise Azerbaijan production.[10] The discovery of the Muradkhanly field, in the arid Mil Plain between the Kura and Araks river, has stimulated an oil-exploration program in central Azerbaijan. However, with now only about 3% of Soviet oil production, it is doubtful that the Baku region, which had been the world's main producer at the turn of the 20th century, would even again play more than a marginal role.

Rise of the Volga-Urals

The Baku-dominated phase of Soviet oil development had actually come to an end in the mid-1950s, when the Soviet industry began to make both a quantitative leap and a geographic leap from the early, modest production levels of Baku and the rest of the Caucasus to the "Second Baku," as the new Volga-Urals province was originally dubbed. Under the pressure of the emergency of World War II, greater attention had been given to this vast new potential oil-bearing province, identified in the 1930s between the Volga River and the Urals. Beginning with the first, accidental oil strike (in the process of exploring for potash) in April 1929 at Verkhne-Chusovskiye Gorodki, east of Perm', commercial production began successively in 1932 at Ishimbay (Bashkir ASSR), in 1936 at Krasnokamsk (Perm' Oblast), in 1937 near Syzran' (Kuybyshev Oblast) and Tuymazy (Bashkir ASSR), and in 1938 at Buguruslan (Orenburg Oblast), and Stavropol' (the present Togliatti; in Kuybyshev Oblast). The early Volga-Urals finds, in relatively shallow Permian and Carboniferous rocks (to 4,000 feet deep), were of limited productivity, and their combined output on the eve of the war was 1.85 million tons, or 6% of the Soviet total, with Ishimbay alone contributing 1.4 million.

The intensified wartime exploration of the Volga-Urals, continuing after the war, uncovered the full potentialities of the region, whose most productive fields were associated with deeper Carboniferous and especially Devonian strata. The giant Romashkino field, at Al'met'yevsk in Tatar ASSR, was discovered in 1948. Additional large reserves were identified in the Bashkir ASSR, where the center of production shifted progressively from the Tuymazy–Oktyabr'skiy area in the late 1950s to the Belebey–Aksakovo district and, in the early 1960s to the Arlan area of northwest Bashkiria, where the oil town of Neftekamsk arose. In Kuybyshev Oblast, the third major Volga-Urals producer, major gains were made with the discovery of the Mukhanovo field at Otradnyy in 1955 and the Kuleshovka field at Neftegorsk in 1958.

By 1949, expanding Volga-Urals production had compensated for the wartime decline of the old Caucasus fields, and in 1956–58 the three leading producing regions of the Volga-Urals–Tatar ASSR, Bashkir ASSR and Kuybyshev Oblast–each passed the output level of Baku, becoming respectively the first, second and third producers of the Soviet Union. Volga-Urals growth raised the Soviet Union to second place among world oil producers in 1961, ahead of Venezuela and behind the United States. By the mid-1960s, the region was accounting for 72–73% of total Soviet production. The start of oil operations in West Siberia, together with the shortlived increases in a number of other areas (Groznyy, Ukraine, Kazakhstan, Turkmenia), reduced the percentage share of the Volga-Urals after 1965, although absolute production in the region did not peak until 1975, when the Tatar ASSR reached its record level of 104 million tons[11] (the Bashkir ASSR had peaked in 1967 at 45.3 million[12] and Kuybyshev Oblast in 1971-72 at 35.6 million[13]). The downward trend in the Volga portion of the region appeared to be slowed somewhat by an upward trend in the Urals portion, where small production increases were being recorded in the mid-1970s in Perm' and Orenburg oblasts and in Udmurt ASSR. However, there was no doubt that the Volga-Urals phase of the history of Soviet oil development, which had succeeded the Baku phase in the mid-1950s, was about to be replaced by the West Siberian chapter.

Lesser Oil Producing Areas

The 1970s also saw the peaking of oil output in a number of other Soviet producing areas where the long-term potential had been less significant than in the Volga-Urals province. These secondary producers included the Belorussian and Ukrainian republics in the European USSR; Kazakhstan, and the Central Asian republic of Turkmenia.

In the *Ukraine*, there had been negligible production in the old Austro-Hungarian fields of Borislav and Drogobych, which in 1909 produced 2 million tons, or 5 percent of world output. By the time the fields passed from Poland to the Soviet Union in 1939, production was down to 350,000 tons. After World War II, West Ukrainian oil output was eclipsed in the early 1950s by more productive fields in the Dolina area of Ivano-Frankovsk Oblast, also in the West Ukraine, and, after 1960, by new discoveries on the left bank of the Dnieper River in Chernigov Oblast (Priluki), in Poltava Oblast (Mirgorod) and in Sumy Oblast (Akhtyrka). By the time Ukrainian oil production peaked, at 14.5 million tons in 1972, the left-bank fields accounted for about 85% of the total. The Ukrainian 10th five-year plan, reflecting the depletion of reserves, scheduled a decline of production from 12.8 million tons in 1975 to 8.6 million in 1980; the actual decline appears to have been more rapid, with an output level of 9.5 million tons expected in 1978.[14] The Ukraine never contributed more than 4% of Soviet oil production; the value of the Ukrainian crude oil lay in its high quality (low tar, low paraffin and high yield of light distillates) and in its favorable location close to the markets of the European USSR.

The *Belorussian* oil development occurred in the mid-1960s in the Pripet sedimentary basin, northwest of the Dnieper oil-bearing province. Like the Dnieper oil, the Belorussian product was a light, low-sulfur crude, favored moreover by its geographical location. Commercial production began promptly after the first discovery in 1964 near Rechitsa, in southeast Belorussia. Production reached 4 million tons, as planned, by 1970, but longer-term projections proved overoptimistic. Instead of a 1975 goal of 10 million tons, 8 million was produced, and turned out to be the peak. Although the 10th five-year plan envisaged the maintenance of the 8-million level through 1980,[15] output dropped to 6.2 million tons in 1976[16] and is estimated to have been 4 million tons in 1978. The Belorussian oil administration, incidentally, also operates small oil fields in Kaliningrad Oblast of the RSFSR (former East Prussia), where limited reserves were identified in the late 1960s and early 1970s, and commercial oil first moved to a refinery (Novopolotsk) in 1975; about 290,000 tons was produced in that year.[17]

In *Turkmenia*, production on the southeast shore of the Caspian Sea has also had a long history, with the earliest output dating from the decade before the Bolshevik Revolution, when a peak of 208,000 tons was obtained in 1911-12 on the Cheleken Peninsula, south of Krasnovodsk. During the Soviet period, production began in 1932 at Nebit-Dag, farther inland, and by World War II the western Turkmen fields yielded 587,000 tons. A refinery opened in the area in 1943 at Krasnovodsk. Production continued to rise gradually after the war with discoveries southeast of Nebit-Dag (at Kum-Dag and Kyzyl-Kum) and again on Cheleken during the period 1948-51. But the real spurt in Turkmenia came after the discovery in 1956 of the Kotur-Tepe field, between Nebit-Dag and Cheleken, followed in 1962 by the Barsa-Gel'mes field, 10 miles south of Kotur-Tepe. Turkmen production nearly tripled during the 1960s, from 5.3 million tons in 1960 to 14.9 million in 1970; Soviet planners projected a further increase to 22 million by 1975, and there was talk about laying a pipeline eastward across the desert Kara Kum[18] to a projected second Turkmen refinery to be built at Neftezavodsk, northwest of Chardzhou. The calculations proved overoptimistic. Production faltered during the five-year period 1971-75, peaking in 1973 at 16.2 million tons (with about 60% from Kotur-Tepe and 30% from Barsa-Gel'mes), and then declining to 15.6 million by 1975 (instead of the planned 22 million). The 10th five-year plan reflected the uncertain Turkmen prospects by altogether omitting a 1980 production goal for crude oil, and the decline continued to about 13 million tons in 1978.[19] Offshore drilling in Caspian Sea structures trending westward from Cheleken (Zhdanov, LAM, Gubkin and Livanov banks) yielded additional supplies in the early 1970s, and some of the wells were connected by pipelines to the mainland.[20] But the offshore discoveries were not substantial enough to affect the overall deterioration of production. In any event, Turkmen offshore wells are operated by the Caspian Oil and Gas Administration, with headquarters in Baku, and their output is credited to Azerbaijan production.[21]

In northwest *Kazakhstan*, too, oil production dates from before the 1917 Revolution, with early output coming from the Dossor and Makat fields in the Emba district on the northeast shore of the Caspian Sea. Although production rose during the Soviet period and the Emba district received a pipeline outlet to a new refinery at Orsk in the southern Urals in 1935, output levels remained modest, reaching a peak of 979,000 tons in the wartime year 1943. In that year construction began on a local refinery, at Gur'yev, and it went into operation in 1945. In the postwar period, after a temporary decline, production expanded again after additional discoveries and reached a level of 1.6 million tons in the early 1960s, when the large Mangyshlak deposits were discovered to the south of the Emba district.

At the time of their discovery, the Mangyshlak finds, notably the Uzen' and Zhetybay fields, were considered of the same order of magnitude as some of the giant fields of the Volga-Urals and also of West Siberia, which was then beginning to reveal its potential. Early forecasts projected a Kazakhstan output of 15 million tons (including 83% from Mangyshlak) under the five-year plan ending in 1970, and 30 million tons (with nearly 90% from Mangyshlak) under the five-year plan ending in 1975. However, the crude oil at Mangyshlak, with a high paraffin content (20 to 30%) and high viscosity, causing it to congeal at temperatures of 30-32°C, posed serious problems both in extraction and in transportation. Hot water had to be injected into the fields to ease the flow of oil to the wellhead. The first installation for desalting and heating water brought in from the Caspian Sea through a 90-mile aqueduct went into operation in 1970[22] and a second aqueduct was completed in 1974.[23] In addition, a "hot" pipeline, with heaters along the way keeping the oil at 60-65°C, was completed in 1969 to the Gur'yev refinery and the following year to the refining complex at Kuybyshev, 1,000 miles away. At Kuybyshev, the Mangyshlak crude oil is being mixed with other oils both for refining and for onward transportation by pipeline. A second "hot" pipeline opened from Uzen' to Gur'yev in early 1975.[24] However, technical problems kept Mangyshlak production below expectations. Starting from 335,000 tons in 1965, the first operating year, Mangyshlak production rose to 10.4 million tons in 1970[25] and to 20.1 million in 1975,[26] considerably below the 26.5 million level envisaged for that year. About 16 million tons, or 80% of Mangyshlak oil, came from the Uzen' field, and 4 million from Zhetybay. There are indications that 1975 was the peak production year for Mangyshlak, with output slipping to 19.3 million tons in 1976, 17 million in 1977 and 15 million in 1978. The drop in Kazakhstan as a whole was slowed by a slight increase in the output of the Emba district, where additional reserves just west of the lower Ural River (at Martyshi and Kamyshitovoye) began to be developed in 1968 and accounted for half of Emba production, running at about 4 to 5 million tons in the late 1970s.[27]

About the only older producing region that appears to have at least modest prospects for output increases over the short term is the *Komi ASSR* of northern European Russia, where an exploration program has been under way near the coast of the Barents Sea, and oil production was rising in the late 1970s at the rate of 3 million tons a year. The first commercial oil was obtained in 1930 in the small Chib'yu field, in the Ukhta area, followed two years later by the unusual Yarega deposit of extra-heavy oil, which is so viscous that it could actually be extracted in a shaft mine. Both the heavy crude oil, which yielded low-temperature lubricants and asphalt, and the lighter oil were processed at a small refinery built in Ukhta in 1947–49. However, Komi production remained negligible until the discovery of larger reserves of light oil about 40 miles east of Ukhta, where three fields were developed–West Tebuk (1962), Dzh'yer (1967) and Pashnya (1970). Oil production rose sevenfold, from 806,000 tons in 1960 to 5.6 million in 1970 (Komi liquid hydrocarbon statistics include a substantial gas condensate component, which was 2 million tons in 1970, for a combined oil and condensate output of 7.6 million tons; see Tables 13 and 23).

The main Komi development came more than 200 miles northeast of Ukhta, between the Pechora River and its tributary, the Usa. There the large Usinsk field was discovered in 1963, causing the exploration effort to shift from the Ukhta district to the northern prospects. The Vozey field was discovered 35 miles north of Usinsk, and the Upper Grubeshor field followed 100 miles northwest of Usinsk. The northern Komi crude oil also proved to be high in paraffin, but more manageable than the Mangyshlak oil. At any rate, the prospects appeared promising enough to justify the construction of a 28-inch, 1,100-mile diameter pipeline from Usinsk through Ukhta to the Central Russian refinery of Yaroslavl', completed in 1974; a 250-mile, 220 KV power transmission line to Usinsk from Ukhta via Pechora in 1975 to provide a power supply to the fields;[28] and an access rail spur from the Pechora mainline at Synya.[29] In addition, a gas plant nearing completion at Usinsk in 1978 will extract natural gas liquids from gas associated with Usinsk oil production, transmitting the residual gas to a gas-burning thermal electric station under construction at Pechora.[30] Initial technical problems in the Usinsk-Vozey area, including an early loss of reservoir pressure requiring water injection for pressure maintenance,[31] delayed the start of full production until the mid-1970s. Although the 9th five-year plan (1971–75) had projected a crude oil output of 13–15 million tons by 1975, it rose only slightly, from 5.6 million in 1970 to 7.1 million in 1975.[32] But once oil production got into swing (even though marred by an unusually large water cut), it increased to 8.7 million tons in 1976, 11 million in 1977 and 14 million in 1978, aiming at a five-year plan goal of 22 million in 1980, not counting 3 million tons of natural gas liquids.[33]

Despite the current Komi growth, there was concern about the longer term. Beyond the major fields now entering production, there appeared to be no major prospects as exploration continued to be pressed to the north toward the coast of the Barents Sea.[34] Because of problems of access from the landward side, geological prospecting parties, based at Dresvyanka and Varandey on the coast of the Pechora Sea (an arm of the Barents Sea), have been supplied in winter by ice-class cargo vessels using the fast ice along the shore as unloading platforms.[35] This exploration program is taking place within the Nenets Autonomous Okrug, an ethnic subdivision of Arkhangel'sk Oblast based on the Nentsy (Samoyeds), tundra-dwelling reindeer herders, separating the Komi ASSR from the coast. The oil and gas exploration activity in the area has instilled new life in the okrug's administrative center, Nar'yan-Mar, a northern settlement of 20,000 people. On the assumption that the search for hydrocarbons would succeed, there was talk of building a new town at Nar'yan-Mar, with a projected population of 60,000.[36]

The Development of West Siberia

As long as the Soviet petroleum industry was dominated by the old oil fields in the Caucasus and on the Caspian Sea and, after the mid-1950s, by the Volga-Urals, production centers were situated in reasonable proximity to the main markets in the European USSR. The European part of the country accounted for 93% of national production in 1940; 86% in 1945 and 1950 after the wartime disruption; and again for 93% in the early 1960s. Even the small share represented by the Asian USSR consisted largely of the West Turkmen and Emba fields on the eastern shore of the Caspian Sea within easy reach of the European markets.

The discovery of a vast new oil-bearing province of West Siberia caused a fundamental eastward shift in the centers of production. The new fields, including some of the largest ever developed in the Soviet Union, were about 1,500 pipeline miles farther east than the main Volga-Urals producing areas. In contrast to the Volga-Urals, a settled area with transportation and infrastructure in place, the West Siberian fields were situated in the isolated swampy northern forest of the middle Ob' River region, about 300 to 400 miles north of the settled and economically developed zone along the Trans-Siberian Railroad. The rapid growth of West Siberian production, especially after 1970, combined with the gradual decline of the Volga-Urals, reduced the share of the European USSR in oil output. In 1977, West Siberia, at 220 million tons, equaled the combined production of the Volga-Urals, and in 1978, the entire Asian USSR roughly matched the petroleum output of the European part.

The penetration of West Siberia began in the early 1960s with the Shaim district of the Konda River valley, were oil was discovered in 1960. In the course of development of this district, which turned out to be the westernmost producing area in West Siberia, the oil town of Uray was founded in 1965, and oil began to flow in that year through a 250-mile, 30-inch pipeline to Tyumen' city. However, development in this area did not meet expectations. The population of Uray, originally projected at 30,000, leveled off at 17,000 in the 1970s, and production, having risen to 4.2 million by 1970, remained at around 5 million tons a year.

From this modest beginning, the focus of development shifted to the middle reaches of the Ob' River valley, 300 to 400 miles farther east, where more substantial discoveries were being made. The initial finds occurred in 1961 at Megion, west of Nizhnevartovsk, and in the Ust'-Balyk field, southwest of Surgut. The substantial Ust'-Balyk field gave rise to the oil town of Nefteyugansk, which was founded in 1967 and reached a population of 20,000 in 1970 and 28,000 in 1974. The first oil from these Middle Ob' fields was produced in 1964, when about 210,000 tons moved by tanker barges down the Ob' and up the Irtysh River to the refinery at Omsk. This refinery, the first of any significance in Siberia, was opened originally in 1955 to process crude oil brought in by pipeline from the Volga-Urals fields. The Trans-Siberian pipeline was later extended halfway across Siberia, and in 1964 reached Angarsk (near Irkutsk), where a second refinery had begun operations in 1961.

Being dependent on barge shipments during the brief summer shipping season, the development of the Middle Ob' proceeded slowly at first, reaching 5.8 million tons in 1967. The rate of production was stepped up when the completion of a 650-mile, 40-inch pipeline, with a transmission capacity of more than 40 million tons, provided a year-round outlet to the Omsk refinery in late 1967. West Siberian crude oil represented 17.5 percent of the throughput of the Omsk refinery, while the balance was still being met by Volga-Urals crude.[37] But by 1970, West Siberian production had risen to a level–31.4 million tons of crude oil–adequate to meet regional refinery needs and the flow in the Trans-Siberian pipeline system began to be reversed so that West Siberian crude oil could flow to refineries in the Volga-Urals region.[38]

By this time the huge potential of West Siberia had become evident. In the nine-year period 1961–69, an intensive exploration effort had identified 59 petroleum fields in the Middle Ob' region, including a supergiant, defined as having recoverable reserves of more than 500 million tons; nine giant fields (of 100 million to 500 million tons each), and nine large fields (of 50 million to 100 million tons each).[39] The supergiant field, called Samotlor for a small lake north of Nizhnevartovsk, was discovered in 1969 and, in an unusual Soviet disclosure, was said to have reserves of "about 2 billion recoverable tons."[40] It was the largest oil field discovered in the Soviet Union, of the same order of magnitude as the Romashkino field in the Tatar ASSR, long the mainstay of Volga-Urals production. The giant fields discovered in the 1960s were clustered mainly around Surgut and included Ust'-Balyk (placed in operation in 1964), West Surgut, Pravdinsk (1968) and Mamontov (1970).

During the five-year plan 1971–75, the development of the West Siberian fields was dominated by the upsurge of production in the supergiant Samotlor. Confronted with a vast amount of oil in one place and eager to raise production as rapidly as possible, Soviet planners found it efficient to concentrate their development effort in this single site. With a start of production in 1969, output at Samotlor jumped from 10 million tons in 1970 to 86.5 million in 1975 and then leveled off at 130 to 140 million in 1977–78. The incremental output at Samotlor during the 9th five-year plan 1971–75 represented 66% of the production increase of West Siberia, and 55% of the entire Soviet growth.

However, there was only one Samotlor, and after development of the field had been raised to its full capacity, Soviet planners had to bring in many of the smaller fields to maintain continued incremental output. Up to this point, development had been concentrated in the immediate vicinity of the Ob' River, which served as the principal transport route during the icefree summer shipping season (May–October), and around the two main oil towns–Surgut and Nizhnevartovsk–which during the first decade of production had grown from small, isolated Siberian villages to urban centers of 60,000–70,000 population. In its first 10 years, the crash program to develop Siberian oil resources had proceeded without the benefit of a year-round overland access route; a 400-mile railroad from Tyumen', on the Trans-Siberian mainline, did not reach Surgut until 1975, and Nizhnevartovsk the following year.

In the second half of the 1970s, the development effort shifted away from the Ob' valley as more remote fields, deep in the swampy woodlands of the West Siberian plain, were brought into production. Now that the planners no longer had the luxury of a favorably situated supergiant field, the rate of development was inevitably slowed as transport routes and power supplies had to be provided to a large number of less accessible production sites. As of the beginning of 1976, 18 fields were in production in Tyumen' Oblast; the five-year plan 1976–80 called for the addition of 22 fields if output growth was to be maintained. In 1976, seven fields were brought into production, as planned,[41] but in 1977 only three fields were added, with eight more under development in 1978. Of the 18 new fields opened up in the three-year period 1976–78, only four were accessible by road and only seven had been connected with the regional power grid, based on the gas-burning Surgut thermal power station with a designed capacity of 2,400 MW (1,800 MW installed by the end of 1978).

The expansion of the West Siberian oil-producing region was proceeding in the late 1970s in two basic directions. One was northward from the Middle Ob' valley, both in the Surgut district and particularly in the Nizhnevartovsk district. The other part of the development was focused on the Vasyugan Swamp of southwest Tomsk Oblast, where production had reached only a fraction (8.5 million tons in 1978) of the 245 million ton level of Tyumen' Oblast.

The advance from Surgut is being aided to some extent by the high-priority construction of a 400-mile railroad from the Surgut area north toward the supergiant gas field of Urengoy. Work on the line began in 1976 and has been remarkably close to schedule despite the difficult construction conditions in the remote northern swamps and the need for bringing in ballast and other roadbed materials from distant quarries. In mid-1978, the rail line reached Noyabr'skaya station (at Mile 125) serving the Kholmogory oil field, brought into production in 1976 as one of the largest producers of the post-Samotlor phase of development. A 32-inch pipeline from Kholmogory to Surgut, capable of handling 20-25 million tons of crude oil, was completed in July 1976,[42] followed by a power transmission line in 1978. The 1980 production plan for Kholmogory was set at 8 million tons.[43] Other giant fields in the Surgut district likely to play a significant role are Fedorovsk, which is situated a few miles north of Surgut, and Lyantor, 40 miles northwest of Surgut.

North of Nizhnevartovsk, no railroad was available to aid the advance, and transportation was based on barge traffic along the Agan River (a tributary of the Ob') during the brief summer shipping season, and onward truck transport on temporary winter roads during the cold season. The transfer point at the head of navigation on the Agan River, Novoagansk, was constituted in 1978 as an urban settlement, the first to be established in the region outside of the Ob' valley, Novoagansk serves as a support base for the development of nearby oil fields (Var"yegan and North Var"yegan) and of the Vyngapur gas field farther north.[44] Because of the difficult waterlogged terrain and high costs of construction, road-building has been kept to a minimum. In the first decade of development, only about 150 miles of surfaced road was built, usually in heavily traveled short segments of up to 25 miles, like the road linking the town of Nizhnevartovsk with the Samotlor field. Elsewhere, corduroy roads have been constructed by laying logs side by side transversely across the swampy ground. The cost of one mile of surfaced road in the area has been put at 500,000 to 1,000,000 rubles ($750,000 to 1.5 million compared with 100,000 to 150,000 rubles in the Volga-Urals.[45]

Because of the high costs of construction and development, Soviet planners have kept the building of permanent settlements to a minimum. Wide use has been made of the so-called work-shift method, in which oil production workers are ferried by helicopter from a base city like Surgut or Nizhnevartovsk to isolated drilling sites, to be accommodated for one or two weeks in makeshift dormitory settlements before returning for an off period to their permanent places of residence. The system appears to have worked as long as fields were developed within reasonably close distance from base cities. Now, as development is being pushed farther out, there is again talk of building a few permanent urban settlements at major production sites such as the Kholmogory field.

Although the urban population of Khanty-Mansi Autonomous Okrug, the ethnic political subdivision in which the Middle Ob' oil fields are situated, more than tripled since 1965 (to 385,000 in 1978), the supply of labor, especially of skilled labor such as oil-well drillers, remains tight. In-migration continues to be intensive, but unattractive living conditions result in a high rate of turnover as migrants leave after relatively short stays. In view of the growing need for oil drillers in West Siberia, both for exploration and for development, and the gradual decline of activity in the Volga-Urals, the Soviet authorities began flying in drilling teams from the Volga-Urals for extended tours of duty in West Siberia. This so-called "expeditionary" method of meeting the labor requirements in the new producing areas began in 1977 as workers were flown to West Siberia from such Volga-Urals regions as Tatar ASSR, Bashkir ASSR and Kuybyshev Oblast. In 1978, out of a combined exploratory and development drilling program of 5 million meters in West Siberia (35% of the Soviet Union's total planned oil drilling plan), about 700,000 meters, or 14%, was assigned to the expeditionary teams from the Volga-Urals.[46]

While oil development was being pressed northward in Tyumen' Oblast, renewed activity became evident in adjacent Tomsk Oblast, to the southeast. Tomsk Oblast was one of the original producing areas of West Siberia, with operations centered in a giant field discovered in 1962 and variously known as Sosnino-Sovetskoye or Vartovsk-Sosnino. The oil town of Strezhevoy arose in the field, situated southeast of Nizhnevartovsk, just across the Tyumen'-Tomsk oblast border. Production, though projected initially at 17–20 million tons by the mid-1970s, stabilized at around 5 million tons, and the town of Strezhevoy did not grow as anticipated while attention was focused on the nearby Samotlor field. The situation in Tomsk Oblast began to revive in 1976, when the Vakh field was put into production 60 miles east of Strezhevoy, and especially after 1977, when the advance into the Vasyugan swamps got under way. A two-pronged development drive appears to be envisaged.[47] One prong, based on Strezhevoy, which was raised to the higher urban status of oblast city in April 1978, is penetrating about 120 miles SSW to a cluster of fields on the Tomsk-Tyumen' oblast border (Olen'ye, Katyl'ga, Pervomayskoye). The northernmost field, Olen'ye, started producing in mid-1978 after having been reached by a branch pipeline from the main oil-transmission trunk that runs southeast along the Ob' River to the Trans-Siberian; a power line was scheduled to link the new producing area to the West Siberian electricity grid in 1979. The other prong, based on Tomsk city, is planned to penetrate 300 miles west to the Nyurol sedimentary basin on the Tomsk-Novosibirsk oblast line, where a major oil discovery was made in 1974. The newly discovered field, on the Malaya Icha River (a right tributary of the Tara), lies in the deep Middle Paleozoic beds associated with a large anticlinal structure of about 20 by 3–5 miles; most of the West Siberian reserves previously identified had been of Cretaceous and Jurassic age.[48] The Nyurol basin may also be developed by pressing directly northward from the Trans-Siberian Railroad, which is only about 120 miles away; plans have been announced for a pipeline running north from the rail town of Barabinsk in Novosibirsk Oblast.[49] At any rate, this development would bring into production the southernmost portion of the vast West Siberian hydrocarbon-bearing province, extending a thousand miles north to the Arctic Ocean.

Finally, an oil-producing area separated from the mainstream of the Soviet Union's economic life is the Far Eastern island of Sakhalin, off the Pacific coast of Siberia. Commercial production began in Sakhalin in 1921 under Japanese military occupation during the Russian civil war and was carried on under a Japanese commercial concession from 1925 to 1928, when 90,000 tons was produced. Under Soviet control, Okha on the northeast coast became the center of oil production, and a brief rail spur to the west coast harbor of Moskal'vo, opened in 1930, improved transportation to the mainland. In World War II, when Sakhalin output rose from 510,000 tons in 1940 to about 800,000 in 1945, a pipeline was laid in 1942 across the narrow strait to the mainland to supply a newly opened refinery at Komsomol'sk (it did not actually reach the refinery until 1952).[50] A second line was completed in 1974,[51] having also been under construction for nearly a decade. Although a number of new fields were put into production in Sakhalin, onshore reserves appeared to be limited. Annual output stabilized around 2.5 million tons after 1965 (with a peak of 2.65 million in 1968), and Sakhalin met an increasingly declining portion of the growing petroleum needs of the Soviet Far East. In the late 1960s, regional production was said to represent 40% of requirements, estimated at 6 to 7 million tons;[52] by the mid-1970s, Sakhalin production satisfied only 20 percent of regional demand, which by then had increased to around 12 million tons.[53] Indications of a larger petroleum potential offshore developed in 1971 with the discovery of the Odoptu structure, 25 miles SSE of Okha. The western limb of the anticline, which measured about 18 by 5 miles, was on shore but the axis about 2.5 miles from the coast. Inclined drilling of the northwest portion, closest to shore, yielded a strong oil flow from Middle Miocene beds. However, the Soviet Union lacked the offshore exploration technology to test the Odoptu structure, and concluded a joint development agreement with Japan in January 1975. The first test drilling indicated the existence of a relatively large field. Exploration, which is limited to the icefree months (July-October), may be completed by 1980. Under the Soviet-Japanese agreement, exploratory drilling is also under way off southwest Sakhalin. Meanwhile, onshore, production rose to 3 million tons in 1978 as a result of the development of the Mongi-Dagi field, 100 miles south of Okha.

Geography of Oil Transport and Refining

The huge growth of petroleum output in the last two decades has resulted in corresponding increases of oil transportation and refining, and the successive shifts in the centers of production from the Caucasus to the Volga-Urals and then to West Siberia have affected the pattern of transportation and refining. As long as Caucasus oil predominated, railroads, historically the principal means of

Table 14. Oil Transportation

Carrier	1940	1945	1950	1955	1960	1965	1970	1974	1975	1976	1977	1980 plan
	Length of Oil Pipelines (thou. of km)											
Total	4.1	4.4	5.4	10.5	17.3	28.6	37.4	53.0	56.2	58.6	61.9	
Crude oil	3.2		3.9	7.4	13.0	22.1	30.7	44.1	46.1	48.4		63–65
Products	0.9		1.5	3.1	4.3	6.5	6.7	8.3	10.1	10.2		
	Shipments of Crude Oil and Products (mill. of tons)											
Pipelines	7.9	5.6	15.3	51.7	130	226	340	457	498	532	559	650
Crude oil	6.8	4.0	12.6	45.3	115	205	315	424	458		514	600
Products	1.1	1.6	2.7	6.4	14.5	20.4	25.3	32.9	39.6		45	50
Railroads	29.5	21.2	43.2	77.6	151	222	302	379	388	394	406	475
Crude oil	5.3		14.1	15.6	36	50	59.6		58.3			50
Products	24.2		29.1	62	115	172	242		329			425
River Tankers	9.6	5.4	11.9	14.4	18.4	25.0	33.5	37.8	39.0	38.1	37.4	
Crude oil	0.6			2.7	2.1	2.2						
Products	9.0			11.7	16.3	23						
Sea-going tankers (domestic trade)	19.5	11.3	15.8	20.2	24.1	26.4	32.8	(35)	(36)	(36)	(36)	(38)
Crude oil	3.0			6.8	9.4	11.6						
Products	16.5			13.4	14.7	14.8						
Sea-going tankers (foreign trade)	0.1	–	–	2.8	8.4	27.1	42.3	(51)	(55)	(65)	(68)	(74)
Total Soviet sea-going tankers	19.6	11.3	15.8	23.0	32.5	53.5	75.1	85.9	91.4	100.9	104.4	

Sources: V. G. Dubinskiy, *Ekonomika razvitiya i razmeshcheniya nefteprovodnogo transporta v SSSR* [Economics of the Development and Location of Oil-Pipeline Transport in the USSR] (Moscow: Nedra, 1977), pp. 11, 13; *Perevozki gruzov* [Freight Transport] (Moscow: Transport, 1972), p. 108, and *Soviet Geography*, November 1977, pp. 701–703.

transport in the Soviet Union, handled almost three times as much traffic in all petroleum products combined (crude oil and refined products) as the rudimentary pipeline system. In 1950, railroad tank cars even carried more crude oil than was transported by pipeline (Table 14). As the pipeline system developed, expanding tenfold during the period 1950–75, an increasingly large portion of the crude oil moved from producing field to refinery by pipeline. At the present time, 85% of all Soviet crude oil moves by pipeline, 10% by rail, 5% by coastal tanker (in the Caspian Sea) and a negligible amount by river-going tanker (in the Volga basin). The proportion of pipeline to rail movements is reversed in the case of refined products, which tend to be shipped mainly by rail both because of the scattered distribution of destinations and because the construction of product pipelines has been lagging even in areas of bulk movements. In 1976, only 17% of the total length of Soviet petroleum pipelines carried refined products, and 8% of the tonnage transported by pipeline consisted of products.

In an effort to accommodate the growing traffic of crude oil from field to refinery, there has been a steady increase in pipeline diameters since the upsurge of the Soviet petroleum industry in the mid-1950s (Table 15). Previously the largest diameters in the rudimentary pipeline system were of the order of 12 inches, with an annual carrying capacity of 1.5 million tons of crude oil. The size of pipelines began to increase in the mid-1950s with the construction of the first 20-inch lines as refineries began to be located increasingly in market areas at greater distance from producing fields; the pipeline size rose to 40 inches in the mid-1960s and to 48 inches in the mid-1970s as the need for transporting growing amounts of crude oil from West Siberian fields to distant refineries increased. As of 1975, one-quarter of the total crude-oil pipeline length consisted of 40-inch and 48-inch diameters–11,300 out of 46,100 km; products

Table 15. Diameters and Capacity of Oil Pipelines

Diameter (inches)	Annual capacity (mill. tons)	Length of pipelines (thou. of km)						
		1940	1950	1955	1960	1965	1970	1975
<20		4.1	5.4	7.5	9.2	9.5	10.8	12.1
20	8	–	–	3.0	6.3	9.9	10.3	15.9
28	17	–	–	–	1.7	6.1	9.5	11.0
32	25	–	–	–	0.05	1.8	2.9	5.9
40	45	–	–	–	–	1.3	3.9	6.4
48	75	–	–	–	–	–	–	4.9
Total length of pipelines		4.1	5.4	10.5	17.3	28.6	37.4	56.2

Sources: Dubinskiy, op. cit., p. 15; *Ekonomika stroitel'stva magistral'nykh truboprovodov* [Economics of Construction of Trunk Pipelines] (Moscow: Nedra, 1977), pp. 10–11.

pipelines tended to be of smaller diameters. Although the 40-inch lines were designed to carry 45 million tons and the 48-inch lines as much as 75 million tons of crude oil a year, these designed capacities were usually slow to be reached because of delays in installing the required number of intermediate pumping stations.

In the early phase of the Soviet oil industry, when the Caucasus was dominant, more than 80% of the refinery capacity was concentrated in the two major producing areas of Baku and Groznyy as well as at Batumi, the Black Sea port in Soviet Georgia, which was connected by pipeline with the Baku fields. At that time, crude oil from the Caucasus also moved typically from the Caspian Sea up the Volga River to refinieries situated at rail crossings (Saratov, Gor'kiy, and Konstantinovskiy near Yaroslavl') for distribution of products by rail. Other refineries were situated at rail-to-sea transfer points, such as the Caspian coastal refinery of Makhachkala and the Black Sea refineries of Tuapse and Batumi, or at sea-to-rail transfer points, such as the Ukrainian refineries at Odessa and Kherson. During this phase of limited oil production, refineries tended to be small, with an average throughput capacity of less than 2 million tons.

The situation changed dramatically as the center of crude-oil production shifted to the Volga-Urals region during the 1950s. The rapid increase of output, which rose nearly tenfold during the 1950s, first led to an expansion of refining capacity in the producing region itself. Large refinery complexes developed at Ufa and in the Ishimbay-Salavat area of the Bashkir ASSR, in the Kuybyshev area and at Perm'. Of the major producing areas in the Volga-Urals, only the Tatar ASSR had no refinery of its own. But as crude-oil production continued to rise, it was found more efficient to build additional refineries in market areas and to develop a pipeline system capable of handling the growing flow of oil from producing field to refinery. With the Volga-Urals region as the focus, this pipeline system evolved in three basic directions:

(1) Eastward into Siberia, where refineries had gone into operation at Omsk (in 1955) and at Angarsk near Irkutsk (in 1960). The first crude-oil pipeline from the Volga-Urals to Omsk, a 20-inch line 827 miles long, was completed in 1955; a second Trans-Siberian pipeline, of 28-inch diameter and 2,287 miles long, reached Omsk in 1960 and Angarsk in 1964.

(2) Westward to Eastern Europe, with the Druzhba (Friendship) pipeline system carrying Volga-Urals crude oil to new refineries at Plock (Poland), Schwedt (East Germany), Bratislava (Czechoslovakia) and Szazhalombatta (Hungary). The first string of the system, of 40-inch diameter, went into operation in 1964; a second string, with 48-inch pipe, was added in the mid-1970s. In addition to serving export functions, the Druzhba system also plays an important role in serving domestic refineries. A southern branch, starting at Michurinsk, reached the Ukrainian refinery of Kremenchug in 1974.[54] A northern branch, starting at Unecha, reached the Belorussian refinery at Novopolotsk in 1965, and was extended in 1968 to the Baltic oil-export terminal of Ventspils in Latvia. This northern spur will also supply crude oil to a refinery to be completed in 1980 at Mazeikiai in Lithuania. A 275-mile segment from

Novopolotsk to Mazeikiai was laid by Polish workers from 1975 to 1977, but the refinery itself has suffered delays.[55]

(3) Northwestward to a cluster of market-oriented refineries in Central and Northwest Russia. They included a second Gor'kiy refinery (at Kstovo), opened in 1958; Ryazan' (1960); expanded capacity at the old Moscow refinery, and a refinery that went on stream in 1966 at Kirishi near Leningrad.

The new market-oriented refineries, which helped raise the apparent throughput of crude oil from about 130 million tons in 1960 to 290 million in 1970, used mainly straight-run refining processes without cracking or other secondary methods with a view to maximizing the output of residual fuel oil. This fuel-oil emphasis was particularly strong in the Volga valley, the Caucasus and in some parts of the European USSR west of the Volga River, where an increasingly large share of thermal electric power was generated by oil-burning plants. In Siberia, where fuel oil played a less important role as a power-station fuel and where agricultural machinery in the newly developed Virgin Lands offered a market for diesel fuel, the fuel-oil emphasis in refinery processes was less pronounced, and a greater effort was made to derive more light distillates like gasoline and diesel fuel from crude oil. Despite these efforts to insure a regional differentiation of the refinery mix in accordance with the market, the Soviet oil-refining industry in general was not designed to maximize the output of light fractions. Because of the low level of private passenger car ownership, gasoline in particular represented a far smaller component of the refinery mix than, say, in the United States. Refinery completions moreover failed to keep in step with the rapid growth of crude-oil production, giving rise to a growing exportable surplus of crude oil. Crude-oil exports nearly quadrupled in the 1960s, from 16.6 million tons in 1960 to 63 million in 1970 as the gap between crude-oil production and refinery capacity widened (Table 16).

This trend became further accentuated in the 1970s as the growing volume of West Siberian production came on line and exports were further stimulated after 1973 by the rise in world prices. Growth in refining capacity was achieved mainly by expanding existing centers, and the completion of new refining centers lagged despite an avowed aim of further decentralizing the refining industry and bringing it closer to markets. Of the seven new refineries scheduled to be opened during the 9th five-year plan (1971-75), including four in the Asian USSR, only the western Mozyr' refinery, on the Druzhba pipeline system in Belorussia, came on line, in 1975. In the 10th five-year plan (1976-80), the second largest Ukrainian refinery, at Lisichansk, opened in 1976, and the long-planned Pavlodar refinery, in northeast Kazakhstan, in 1978. Work on others (Mazeikiai in Lithuania, Nizhnekamsk in Tatar ASSR, Achinsk in Siberia, Chimkent in southwest Kazakhstan, and Neftezavodsk in Turkmenia) was continuing.

The onrush of West Siberian oil required further expansion of the pipeline network, with increasingly large pipe diameters. In the early phase of West Siberian development, production could be accommodated by the first 40-inch line, laid in 1967 from the producing fields to the Omsk refinery and by the reversal of oil transmission in 1970 in the old Trans-Siberian pipeline system. But

Table 16. Estimate of Soviet Refinery Throughput and Products Consumption[a] (million metric tons)

Year	Crude-oil production	Net crude-oil export	Apparent refinery throughput	Net products export	Apparent domestic products consumption
1950	37.9	0	38	+1.5	39
1955	70.8	−2.3	68	−1.3	67
1960	148	−16.6	131	−12.2	119
1965	243	−43	200	−19.1	181
1970	353	−63	290	−28	262
1971	377	−70	307	−28.8	278
1972	400	−68	332	−29.5	302
1973	429	−72	357	−31.5	325
1974	459	−76	383	−34.6	348
1975	491	−87	404	−36.2	368
1976	520	−104	416	−36.9	379
1977	546	−118	428	−38	390

[a]The estimates ignore field and refinery losses and fuel use in refineries.

the growth of Siberian oil output placed increasing demand on the provision of additional outlets. These followed four basic directions:

(1) Southeastward into eastern Siberia. A 48-inch line, starting at the Samotlor field and running along the Ob' River, was completed in 1972 to a junction with the existing Trans-Siberian system at Anzhero-Sudzhensk, and was extended the following year as far east as Krasnoyarsk. It was apparently envisaged originally as the initial link of a pipeline that might carry West Siberian oil to the Pacific coast for export to Japan, but the plan was dropped. The present segment will be adequate to feed crude oil to the Achinsk refinery (near Krasnoyarsk) when it starts operation. Future oil flows across Siberia to the Pacific coast, if indeed such oil flows will be generated by the mid-1980s, are to be handled by unit tanker trains on the Baikal-Amur Mainline after it goes into operation.[56] Despite expressed doubts that the Soviet oil potential would be adequate by the mid-1980s to generate a massive flow eastward across Siberia, design planning for the BAM in the late 1970s continued to envisage the movement of heavy oil trains for 10 to 15 years after the start of service on the new railroad.[57]

(2) Southward into Kazakhstan and Central Asia. A 32-inch line, designed ultimately to handle a flow of 25 million tons a year, was completed from Omsk to Pavlodar, a distance of 283 miles, in 1977, feeding West Siberian crude oil to the first 6-million ton unit that went on stream in 1978 at the Pavlodar refinery.[58] Opening up an entirely new flow line for West Siberian oil, the

pipeline is to be extended ultimately to refineries under construction at Chimkent and Neftezavodsk. A spur branching off at Chardara near Chimkent would run southeastward to an existing refinery at Fergana in the Uzbek SSR. The Fergana refinery is now receiving West Siberian oil by rail.[59]

(3) Southwestward to the European USSR. Since 1970, when the flow in the existing Trans-Siberian pipeline system was reversed west of Omsk, two new 48-inch lines have been added to the westward transmission capacity. One, designated the Samotlor–Al'met'yevsk pipeline, was inaugurated in 1973; the other, called the Nizhnevartovsk–Kuybyshev pipeline, began transporting oil in 1976. Each of these pipelines, about 1,400 miles long, has a designed carrying capacity of 70 to 75 million tons after all pumping stations are installed. If the old Trans-Siberian system, with a capacity of 25 to 30 million tons, is added, the present pipeline capacity available to carry West Siberian crude oil to the European USSR is 170 to 180 million tons. The actual westward flow rose from a negligible amount in 1970 to 85–90 million tons in 1975 and appeared to approach the existing transmission capacity in the late 1970s.[60] The five-year plan 1976–80 envisaged a westward flow of 210 to 220 million tons, on the assumption that goals for the production of crude oil and for the installation of new Siberian refining capacity would be met.

(4) A new pipeline corridor directly westward across the Urals. In an effort to expand transmission capacity to handle the expected increases in West Siberian production, construction began in the winter of 1977–78 on a 48-inch, 2,122-mile trunk pipeline from Surgut, one of the Siberian oil towns, to the Belorussian refinery center of Novopolotsk. In contrast to previous pipeline routes, which ran along the Trans-Siberian Railroad through Kurgan and Chelyabinsk, the new transmission system is carving a more northerly route across the Urals at the latitude of Perm'. Beyond Perm' it will serve the Central Russian refinery cluster of Gor'kiy, Ryazan', Yaroslavl' and Moscow before proceeding to Novopolotsk, where it would also feed into the export pipeline to the Baltic terminal at Ventspils and into the spur leading to the Lithuanian refinery at Mazeikiai. Plans called for first-stage operation, as far as Perm', in 1979 and completion in 1981, but construction across the swampy West Siberian plain to the Urals was reported to be running behind schedule.

The increasing westward flow of Siberian oil in the mid-1970s spurred the construction of additional trunk lines within the European USSR, both to refineries and to export terminals. Most of the construction was concentrated in the southern regions. To accommodate the movement of crude oil to the Black Sea export terminal of Novorossiysk, a 32-inch, 945-mile line was completed in 1974 from Kuybyshev to the tanker port, which now loads 30 to 35 million tons of crude oil a year. The pipeline provided a direct outlet for West Siberian oil to the Black Sea.[61] The export terminal of Tuapse, southeast of Novorossiysk, was rebuilt and modernized to accommodate larger tankers.[62] In 1977 another direct pipeline, of 48-inch diameter and 676 miles long, was completed between Kuybyshev and the Ukrainian refinery at Lisichansk. It was then extended to the Kremenchug refinery, also in the Ukraine, and to the Odessa oil terminal on the Black Sea, which has been handling a growing volume of exports.[63] Work was under way in 1978 to expand the pipeline system serving the Groznyy refinery complex in the Northern Caucasus, where crude-oil production was declining after its surge of the late 1960s and early 1970s. The new pipeline will provide access for Siberian crude oil to Groznyy.[64]

The avowed Soviet policy of reducing the use of fuel oil and making more oil available for petrochemical uses and for export has resulted in a fundamental shift in the refinery mix. Greater attention than in the past is being given to secondary refining methods such as cracking and reforming to reduce the final yield of fuel oil and enhance the yield of light distillates, including high-octane gasoline for motor fuel in the expanding motor vehicle fleet, low-octane gasoline as a petrochemical feedstock and diesel fuel. Light fractions, accounting for 45–50% of the product mix in the late 1970s, were to rise to 60–65% by 1990.[64a]

NATURAL GAS

While the future of Soviet petroleum production appears clouded by uncertainty over the magnitude of proved and probable reserves, there is little controversy over the huge size of natural gas reserves. In contrast to the secrecy surrounding Soviet oil resources, official reserve figures for natural gas have been published by the Soviet Union not only for the country as a whole, but for individual regions. They show a remarkably rapid growth of explored reserves since the late 1960s, with at least 70% of all new discoveries concentrated in the giant fields of northwest Siberia's Tyumen' Oblast (Table 17). The main problem in the case of natural-gas supplies is their remote location astride the Arctic Circle, farther north than the West Siberian oil fields and even more distant from existing transport routes. Gas production in many fields of the European USSR has been peaking since the early 1970s, and the center of the industry has been shifting increasingly to West Siberia. The first giant field, Medvezh'ye, started production in 1972, followed by the second, Urengoy, in 1978. In that year, West Siberia accounted for 25% of Soviet natural gas. The five-year plan 1976–80 called for this share to rise to 35%, with further increases expected in the 1980s (Table 18).

Regional production data for natural gas has generally been more easily available than for oil output until the Soviet Union in 1977 imposed secrecy on the regional distribution of most fossil fuels production. However, in contrast to oil production, the output of natural gas in the Soviet Union has been organizationally, and therefore statistically, more complex. Soviet crude oil is extracted almost entirely by the Ministry of the Petroleum Industry, except for the small gas condensate component of liquid hydrocarbons that is contributed by the Ministry of the Gas Industry. In 1978, out of a total Soviet production of 572 million tons of liquid hydrocarbons, 9 million consisted of gas condensate, produced mainly by the gas industry, and 563 million tons of crude oil. In the case of natural gas, production originates both from the Gas Ministry and from the Oil Ministry. Out of a total 1978 natural gas production of about 370 billion cubic meters, 310 billion was produced by gas fields under the jurisdiction of the Gas Ministry and 60 billion originated in the Oil Ministry. About one-half, or 30

Table 17. Explored Reserves of Soviet Natural Gas ($A+B+C_1$ categories, in trillion m^3 at year-end)

Region	1950	1955	1960	1965	1970	1973	1975
USSR	0.17	0.69	2.34	3.57	15.8	22.4	25.8
RSFSR	0.089	0.45	1.08	1.70	12.3	18.1	21.3
European USSR	*0.17*	*0.64*	*1.59*	*1.80*	*3.35*	*4.4*	*4.2*
Komi ASSR	0.021	0.021	0.017	0.038	0.41	0.37	
Orenburg Oblast	0.004	0.005	0.021	0.025	1.13	2.11	
Krasnodar Kray	–	0.076	0.42	0.47	0.29	0.26	
Stavropol' Kray	0.027	0.23	0.28	0.23	0.20	0.17	
Ukrainian SSR	0.070	0.23	0.54	0.66	0.81	0.87	
Azerbaijan SSR	0.009	0.052	0.050	0.054	0.080	0.12	
Asian USSR	*0.004*	*0.050*	*0.75*	*1.76*	*12.4*	*18.0*	*21.6*
Siberia	–	0.009	0.083	0.60	9.98	14.7	18.2
Tyumen' Oblast	–	0.004	0.050	0.40	9.25	13.8	
Tomsk-Novosibirsk Oblast	–	–	–	0.054	0.23	0.26	
Krasnoyarsk Kray	–	–	–	–	0.15	0.30	
Yakut ASSR	–	–	0.021	0.078	0.26	0.32[a]	
Central Asia	0.004	0.041	0.66	1.16	2.43	3.32	3.4
Uzbek SSR	0.004	0.005	0.61	0.67	0.80	0.95	
Turkmen SSR	–	0.036	0.036	0.38	1.52	2.16	

[a]Reported at 0.8 trillion m^3 in 1977.

Sources: 1950–73 from A. D. Brents et al. *Ekonomika gazodobyvayushchey promyshlennosti* [Economics of the Gas Industry]. (Moscow: Nedra, 1975), p. 25; 1975 from S. A. Orudzhev. *Gazovaya promyshlennost' na puti progressa* [The Gas Industry on the Road of Progress]. (Moscow: Nedra, 1976), p. 12.

billion m^3, of the Oil Ministry share consisted of natural gas not associated with oil production and the other half represented the share of associated gas that was not flared; the current utilization rate of associated gas is around 65-66% (see sections below on the utilization of associated gas and on gas condensate).[6 5]

Early Evolution of the Natural Gas Industry

Before World War II, there was no natural gas industry in the Soviet Union, properly speaking, and 87% of the gas extraction was associated with oil production, which was concentrated in the Baku district of the Azerbaijan SSR. As Table 18 shows, 2.8 billion m^3 of the total 1940 gas production of 3.2 billion was associated gas, and 2.5 billion m^3 of this was contributed by Baku, where it was used mainly as a local power-station fuel. Most of the nonassociated gas originated in the Carpathian gas field of Dashava in the West Ukraine, which had been in production since the 1920s and was acquired by the Soviet Union from Poland in 1939 (Fig. 5).

Table 18. Regional Natural Gas Production in the USSR[a]
(billion m^3)

Region	1940	1945	1950	1955	1960	1965	1970	1975	1976	1977	1978 est.	1980 plan
USSR	3.22	3.28	5.76	8.98	45.3	128	198	289	321	346	372	435
	2.81	*1.08*	*1.76*	*3.09*	*7.7*	*16.5*	*23.0*	*28.6*	*30.6*	*31.5*		
RSFSR	0.21	1.49	2.87	4.29	24.4	64.3	83.3	115	136	158	186	251
	0.15	*0.11*	*0.56*	*1.81*	*5.8*	*10.6*	*16.5*	*17.7*				
EUROPEAN USSR	3.20	3.18	5.56	8.52	44.2	109	139	154	164	166	167	157
	2.81				*7.2*	*15.0*	*20.5*	*23.5*				
European RSFSR	0.21	1.42	2.78	4.10	24.1	63.7	72.3	75	84	85	89	91
	0.16				*5.5*	*10.3*	*15.5*	*15.8*				
Komi ASSR	–	0.45	1.08	1.08	1.0	0.83	6.9	18.5	19.5			22
					0.06	*0.26*	*0.41*	*0.47*				
North Caucasus	0.195	0.12	0.32	0.60	13.7	40.0	47.0	23				
	0.142	*0.05*	*0.21*	*0.20*	*1.2*	*2.4*	*6.1*	*4.9*				
Krasnodar Kray	0.079	0.023			5.1	23.1	24.7	7.86				
	0.079				*0.8*	*0.8*	*0.65*	*0.67*				
Stavropol' Kray	–	–	–	–	8.2	15.3	16.4	11.4				6
						0.2	*0.69*	*0.84*				
Chechen-Ingush ASSR	0.087	0.041			0.35	1.36	4.25	3.5				
	0.062				*0.35*	*1.36*	*4.25*	*3.1*				
Dagestan ASSR	0.029	0.057			0.058	0.15	1.60	1				
Volga-Urals	0.015	0.86	1.38	2.43	9.4	22.9	18.4	34				
	0.015	*0.03*	*0.38*	*1.07*	*4.1*	*7.6*	*9.4*	*10.5*				
Tatar ASSR	–	–			1.43	2.79	3.88	4.4				
					1.4	*2.8*	*3.9*	*4.4*				
Bashkir ASSR	0.013	0.058			1.34	3.04	1.85	1.47				
					1.3	*1.7*	*1.4*	*1.5*				
Kuybyshev Oblast	–	0.26			1.0	1.97	2.26	2.06				
					0.9	*1.8*	*2.2*	*2.0*				

Table 18. (cont'd)

Region	1940	1945	1950	1955	1960	1965	1970	1975	1976	1977	1978 est.	1980 plan
Saratov Oblast	–	0.51			2.5	6.4	3.4	1.0				
					0.2	*0.2*	*0.2*	*0.3*				
Volgograd Oblast	–	–			2.6	7.1	4.0	3.0				
					0.3	*0.5*	*0.87*	*1.0*				
Astrakhan' Oblast	–	–			0.022	0.50	0.75	0.48				
Perm' Oblast						0.6	0.9	1.1				
						0.6	*0.9*	*1.1*				
Orenburg Oblast	–	–			0.5	0.56	1.3	20.1	31.8			
					0.5	*0.6*	*0.5*	*0.5*	*0.4*			
Azerbaijan SSR	2.50	0.98	1.23	1.49	5.84	6.18	5.5	9.9	11.0	11	11	11.9
	2.50	*0.98*	*1.00*	*0.99*	*1.30*	*3.15*	*2.54*	*4.18*				
Ukrainian SSR	0.50	0.78	1.53	2.93	14.3	39.4	60.9	68.7	68.7	67	64	53.8
	0.15		*0.087*	*0.094*	*0.48*	*1.58*	*2.16*	*2.88*				
Belorussian SSR	–	–	–	–	–	–	0.18	0.57				
							0.18	*0.57*				
ASIAN USSR	0.016	0.10	0.21	0.46	1.1	18.5	59.0	135	157	180	205	278
	0.016				*0.5*	*1.5*	*2.5*	*5.1*				
Siberia	–	0.07	0.09	0.19	0.34	0.6	11	40	52	73	97	160
Tyumen' Oblast	–	–	–	–	–	–	9.3	35.7	47.8	68	92	155
							0.1	*1.7*	*3.6*	*5.2*	*7*	*15*
Krasnoyarsk Kray	–	–	–	–	–	–	0.44	2.6	*3.2*	*3.5*		
Yakut ASSR	–	–	–	–	–	–	0.18	0.5				
Sakhalin Oblast	–	0.068	0.085	0.19	0.34	0.6	0.8	1.0				
		0.068	*0.085*	*0.12*	*0.28*	*0.32*	*0.39*	*0.2*				
Kazakhstan	0.004	0.005	0.007	0.025	0.039	0.029	2.1	5.2	5.2	5.5	5.5	5.9
Central Asia	0.012	0.03	0.12	0.24	0.72	17.9	46.0	89.7	99.3	101	102	112
					0.30	*1.2*	*1.3*	*2.0*				
Uzbek SSR	0.001	0.010	0.052	0.10	0.45	16.5	32.1	37.2	36.1	36	35	36.3
Turkmen SSR	0.009	0.015	0.065	0.14	0.23	1.16	13.1	51.8	62.6	64	66	75
					0.23	*1.05*	*1.34*	*1.92*				
Kirghiz SSR	–	–	–	–	0.041	0.16	0.37	0.29	0.26			
Tadzhik SSR	0.002		0.002	–	–	0.052	0.39	0.42	0.32			

[a] Associated gas component in italics.
Source: Data assembled from scattered Soviet sources.

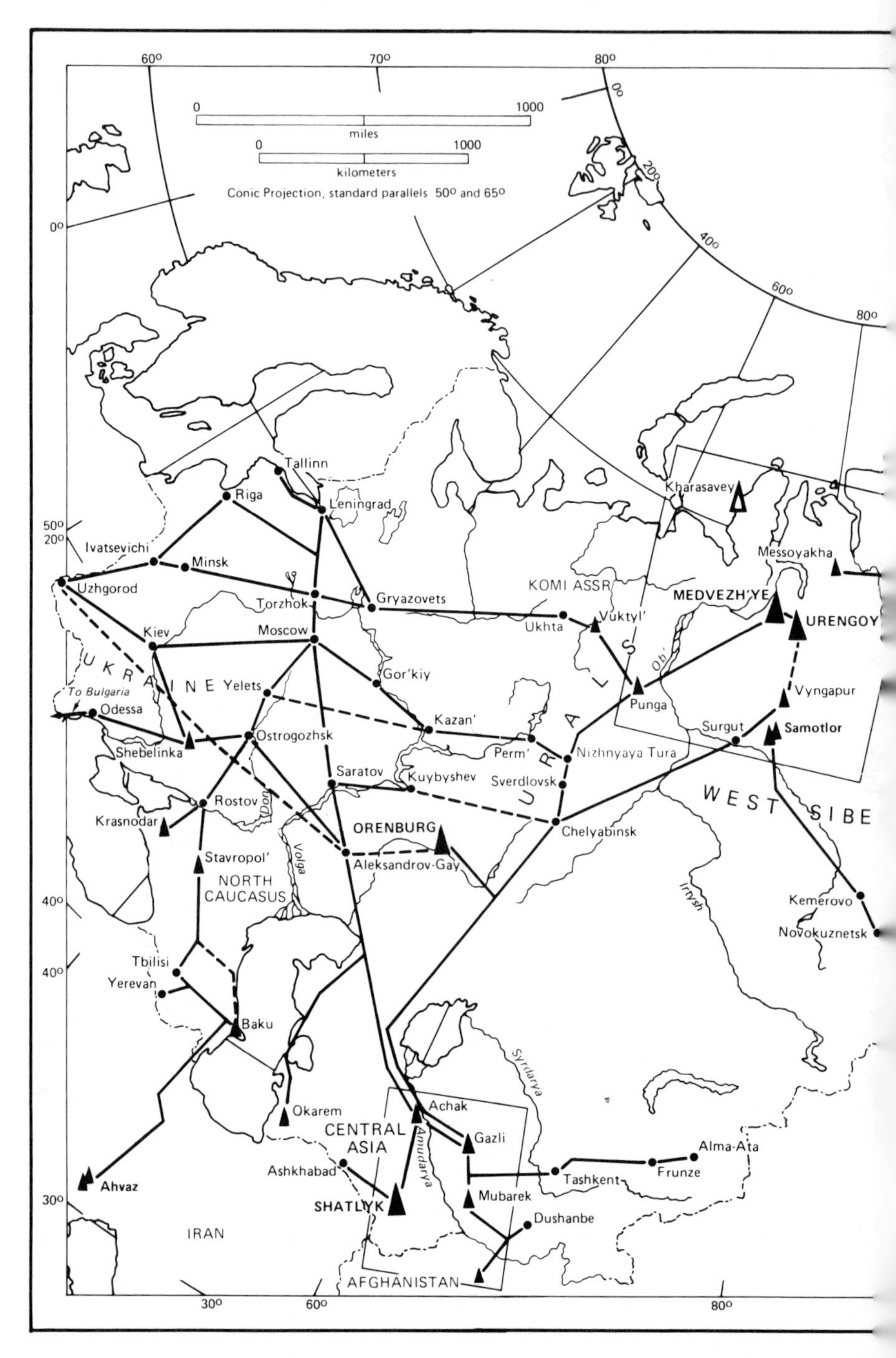

Fig. 5. Major gas fields and pipelines.

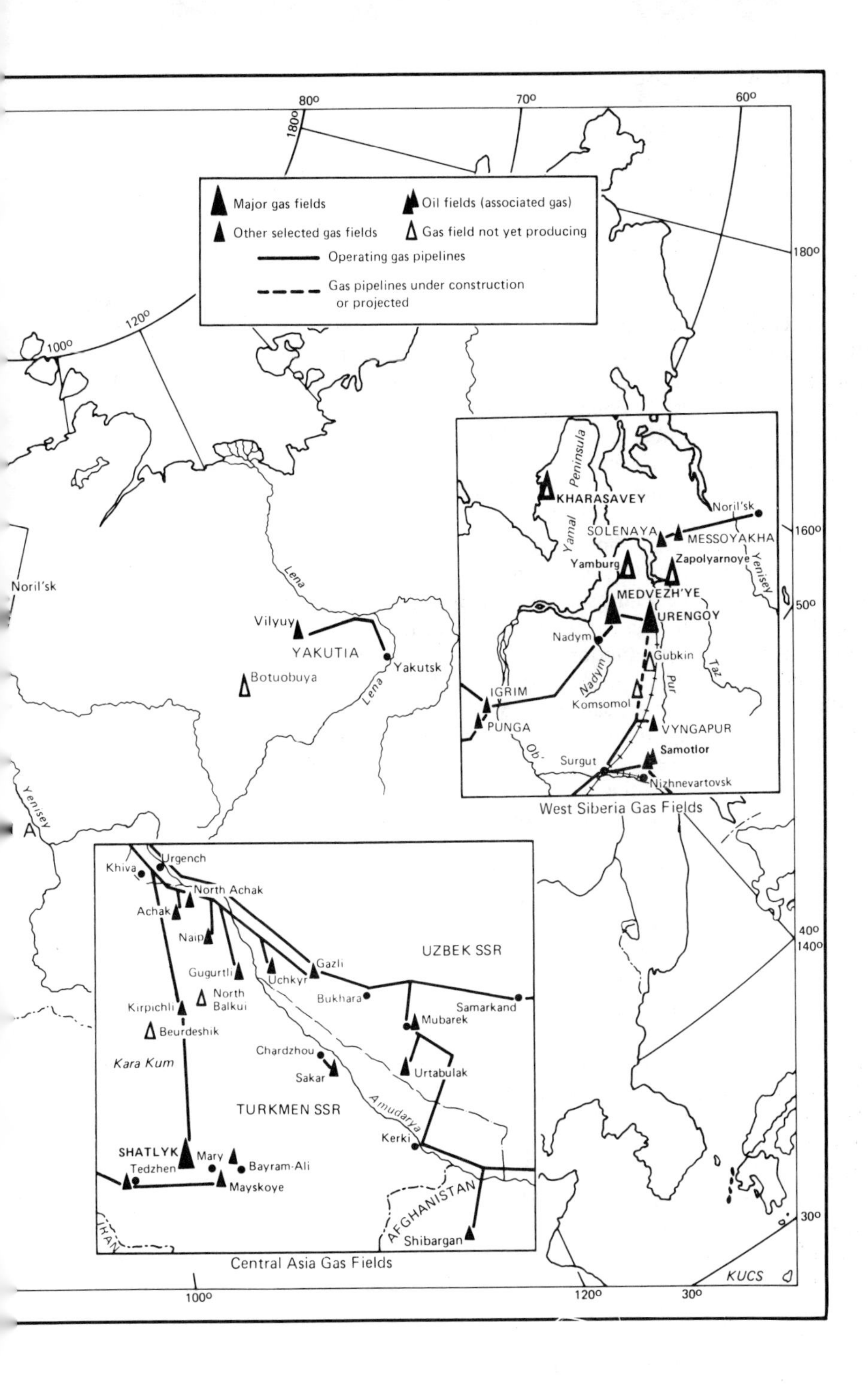
Major gas fields
Oil fields (associated gas)
Other selected gas fields
Gas field not yet producing
Operating gas pipelines
Gas pipelines under construction or projected
Noril'sk
Lena
Vilyuy
YAKUTIA
Yakutsk
Botuobuya
Yenisey
KHARASAVEY
Yamal Peninsula
SOLENAYA
MESSOYAKHA
Yamburg
Zapolyarnoye
Yenisey
MEDVEZH'YE
URENGOY
Nadym
Gubkin
Taz
Pur
IGRIM
Komsomol
PUNGA
VYNGAPUR
Samotlor
Surgut
Nizhnevartovsk
Ob'
West Siberia Gas Fields
Urgench
Khiva
North Achak
Achak
Naip
UZBEK SSR
Gazli
Gugurtli
Uchkyr
North Balkui
Bukhara
Samarkand
Kirpichli
Mubarek
Beurdeshik
Chardzhou
Kara Kum
Sakar
Urtabulak
TURKMEN SSR
Amudarya
Kerki
SHATLYK
Mary
Bayram-Ali
Tedzhen
Mayskoye
AFGHANISTAN
IRAN
Shibargan
Central Asia Gas Fields
KUCS

In the Azerbaijan SSR, associated gas dominated until the mid-1950s, when the new interest in natural gas production led to the development of the Karadag field, 20 miles southwest of Baku. The field, which was placed in production in 1956, had limited reserves (about 30 billion m^3) and, having peaked at 2.9 billion m^3 in 1958, gradually declined and became depleted in the late 1960s. Its operation prompted the construction in 1960 of a 28-inch transmission line supplying Karadag gas to the capitals of the two other Transcaucasian republics, Tbilisi in Georgia and Yerevan in Armenia. After the decline of Karadag, increased offshore oil exploration in the Baku district also uncovered the Bakhar gas field on the Makarov Bank, 25 miles southeast of Baku. The Bakhar field, which also yields gas condensate, was discovered in 1969 and was linked to the mainland by a number of pipelines of up to 56-inch diameter.[66] In 1975 it yielded 5 billion m^3, or one-half of all Azerbaijan gas.[67] The five-year plan 1976–80 suggests that the current production level is to be sustained into the early 1980s.

While natural gas production in Azerbaijan continued to evolve in conjunction with the petroleum industry, an entirely separate gas industry developed in the mid-1960s north of the Caucasus, in Stavropol and Krasnodar krays. The North Stavropol' field, one of the early giant fields in the Soviet Union, was discovered in 1951 and put into production in 1956 with explored reserves of 230 billion m^3. It maintained a high production level of more than 10 billion m^3 a year from 1962 to 1970, with a peak of 15.5 billion in 1966–67. As it declined in the 1970s, its diminishing reserves were supplemented to some extent by the Mirnoye field, with reserves of 60 billion m^3, situated about 80 miles east of Stavropol' city. By 1975, the old North Stavropol' field was down to 5.4 billion m^3, and Mirnoye was supplying about 5 billion of the total Stavropol' Kray output of 10.5 billion m^3 of nonassociated gas. The five-year plan 1976–80 envisaged a continuing decline to 5.2 billion by 1980.[68] Future prospects for a continued pipeline supply of natural gas from Stavropol' Kray thus appear dim, although an important deposit of wet natural gas in easternmost Stavropol' Kray has been set aside as the basis for a chemical industry. The wet gas, found in the oil producing district of Neftekumsk, will be providing natural gasoline for a polyethylene plant approaching completion in 1979 at the town of Budennovsk.[69]

There was no giant equivalent of the North Stavropol' field in the other main gas producing region of the North Caucasus, the Krasnodar Kray, where a number of middle-size fields also added up to a substantial portion of the Soviet Union's gas supply in the 1960s and early 1970s. Typical of the evolution of the Krasnodar fields was the Maykop field, with 85 billion m^3 of reserves, the largest single accumulation in this region. Maykop was placed in production in 1960, reached a peak of 8.2 billion m^3 in 1968, and was virtually exhausted by 1975. Among other Krasnodar fields, Berezanskaya peaked at 5.6 billion m^3 in 1967 and Leningradskaya at 3.6 billion in 1963; the maximum output of lesser fields ranged from 1.1 to 2.7 billion m^3.

The combined production of Stavropol' and Krasnodar krays, accounting for 28–29% of total Soviet gas production in the mid-1960s, fostered the construction of the Central Pipeline System, carrying natural gas from the North Caucasus north to Moscow. From the first 28-inch, 790-mile pipeline completed in 1956, the system was expanded to four strings of up to 40-inch diameter in the early 1960s, capable of transmitting more than 30 billion m^3 a year. The North Caucasus–Moscow system was extended in 1959 to Leningrad through a 28-inch line, with a 40-inch string added in 1968. In addition to giving rise to the Soviet Union's largest gas flow to Central Russia, the North Caucasus fields also generated a gas flow southward to Transcaucasia. A 28-inch line from Stavropol' reached Groznyy in 1959, and was extended across the Caucasus Mountains to Tbilisi in 1963, connecting with the pipeline laid from the Karadag field at Baku in 1960.

Apparently anticipating the decline of North Caucasus production and the limited gas potential of the Baku oil district, the Soviet Union signed an accord with Iran in 1966 providing for the importation of Iranian natural gas. The imported gas, which began to flow in 1970, reached 9 to 10 billion m^3 a year in the middle 1970s. To accommodate the increased flow into Transcaucasia, a second pipeline, of 40-inch diameter, was put into operation in 1971 between Baku and Tbilisi. The flow of Iranian gas is scheduled to increase further in the 1980s under the terms of a trilateral arrangement concluded in 1975 and involving Iran, the Soviet Union and three Western European countries (Austria, West Germany and France). Under the accord, 13 billion additional m^3 of Iranian gas are to enter the Soviet Union starting in 1981, and the Soviet Union will re-export 11 billion m^3 to the West European partners. The additional Iranian gas is to be accommodated by a third Transcaucasian line that, instead of following the difficult route between Tbilisi and Groznyy through the Caucasus Range, will run along the west coast of the Caspian Sea, skirting the eastern end of the mountains, and then feed into the Central Pipeline System at Stavropol'.[70]

Another early gas-producing region was the Ukrainian SSR, where, as noted above, the Dashava field yielded nonassociated gas in 1940 after having been acquired from Poland. Under Soviet control, this Carpathian district was expanded after the war with the operation of additional small fields (Opary, opened in 1940; Ugersko, in 1946; Bil'che-Volitsa, in 1949, and Rudki, in 1957). The combined Carpathian production, which rose to 1.5 billion m^3 in 1950 and 2.9 billion in 1955, was judged important enough to justify the construction of long-distance transmission mains; a 20-inch, 800-mile line to Moscow was completed in 1949, and an 860-mile line of diameters ranging to 32 inches was laid northward, reaching Minsk in 1960, Vilnius in 1961 and Riga in 1962.

The limited capacity of these fields and transmission lines was overshadowed in the mid-1950s by the development of the Shebelinka field, the largest in the European USSR, 35 miles southeast of Khar'kov. Put into operation in 1956 with explored reserves of 530 billion m^3, Shebelinka made the Ukraine the Soviet Union's principal gas-producing region in the 1960s and early 1970s, giving rise to a high-capacity Ukrainian gas transmission system, and generating the first significant exports of natural gas to Eastern and Western Europe. For a decade and a half, from 1963 through 1976, Shebelinka maintained output levels in excess of 20 billion m^3 a year, reaching a peak of 30–31 billion in the period 1967–72 before starting a steep decline. Ukrainian production as a whole was sustained through the mid-1970s by the development of two other, middle-size fields in the Khar'kov region—Yefremovka, 25 miles west of Shebelinka, and Krestishche, another 25 miles beyond. Yefremovka, with reserves of 70 billion m^3, was put in production in 1967 and peaked at 7 billion m^3 in 1970; Krestishche, with more than 100 billion m^3 in reserves, was placed in operation in 1970, and in 1976 surpassed the declining Shebelinka production at about 20 billion m^3.

The gas fields in the Khar'kov region generated transmission in two basic directions: westward to Kiev (first line in 1964) and beyond to the Czechoslovak border at Uzhgorod for export, and southwest through Dnepropetrovsk, Krivoy Rog, Odessa and Kishinev to the Rumanian border at Izmail for export through Rumania to Bulgaria. The westward system through Kiev now has a triple string of pipelines, ranging from 40 to 56 inches in diameter, with a transmission capacity of about 40 billion m^3 a year. The export transmission system through Czechoslovakia, known as the Bratstvo (Brotherhood) system, opened in 1967 with a 32-inch pipeline that carried gas from the West Ukrainian fields to Czechoslovakia and Austria; a separate line also transmitted gas to Poland. The Bratstvo system was expanded in 1974 as an enhanced supply from the Khar'kov region began to generate exports through Czechoslovakia to East Germany, West Germany and Italy; the designed capacity of the enlarged Bratstvo system is about 28 billion m^3 a year. The southwestern trunk-pipeline system, first opened in 1966 (to Odessa), was extended to Bulgaria in 1974.

After having long maintained a high level of production, thanks largely to the Shebelinka field, the Ukrainian gas industry was declining in the late 1970s. The depletion of available reserves was acknowledged in the five-year plan 1976–80, which envisaged a drop of production from 68.7 billion m^3 in 1975 to 53.8 billion in 1980. After having kept the peak level for another year, output began to decline in 1977, when secrecy on regional gas production statistics was imposed, and is estimated to have been about 67 billion m^3. The 1978 plan envisaged a drop to 62.8 billion m^3;[71] actual output was about 64 billion m^3.

The Volga-Urals region, which played such an important role in petroleum production, was not significant in natural gas development on a scale commensurate with that of the North Caucasus and the Ukraine until the discovery of the giant Orenburg gas field in 1966. Previously, the Volga-Urals as a whole had been in third place among the gas-producing regions, after the Ukraine and the North Caucasus, with associated gas accounting for one-third to one-half of total production. In contrast to the oil-producing areas (Tatar ASSR, Bashkir ASSR, Kuybyshev Oblast, Perm' Oblast) in the northern tier of the Volga-Urals, the nonassociated natural gas occurred mainly in the southern tier, including Saratov Oblast and Volgograd Oblast. Saratov was noteworthy in the early history of the gas industry for having transmitted gas to Moscow from the Yelshanka-Kurdyum field (initial reserves of 15 billion m^3) over the Soviet Union's first long-distance transmission main, a 12-inch line, completed in 1946.

When gas development began in earnest in the mid-1950s, one middle-size field each sustained much of the output of Saratov and Volgograd oblasts. In Saratov Oblast, it was Stepnoye on the left bank of the Volga, whose reserves of 30 billion m^3 exceeded those of any single right-bank field. Production at Stepnoye, starting in 1959, led to the construction of a 735-mile pipeline northward to the industrial cities of Gor'kiy and Cherepovets, completed in 1961. But the field was short-lived, peaking in 1962–66 at about 3.5 billion m^3, and became exhausted by the mid-1970s. At its peak it contributed about 60 percent of Saratov Oblast gas output. In Volgograd Oblast, the Korobki field at the settlement of Kotovo, 110 miles north of Volgograd, was placed in production in 1962 with 68 billion m^3 of reserves. It peaked at 5 million m^3 in 1965, and was down to less than 2 billion in the mid-1970s.

The Orenburg Gas Project

The discovery of the Orenburg field, with explored reserves of 1,792 billion m^3, was significant not only for the addition to the gas-producing potential, but for its favorable location near economically developed parts of the Soviet Union. While the old gas-producing regions had all been situated in a relatively hospitable environment, the new reserves discovered in the 1960s were either in the Arctic tundra or in the Central Asian desert, with the exception of Orenburg. Though favorably located, the Orenburg field was technically difficult to develop because the gas contained not only condensate but also corrosive sulfur, both of which had to be separated before the residual gas could be fed into long-distance transmission lines. The Orenburg project called for development in three stages, each centered on a gas-processing plant with a capacity of 15 billion m^3, for an ultimate total of 45 billion. The first two stages were to feed gas into the Soviet domestic gas distribution system, and the third was to be coordinated with the completion of a 1,700-mile 56-inch trunk pipeline from Orenburg to the Czechoslovak border at Uzhgorod. There Orenburg gas was to feed into the Bratstvo system for distribution to some of the East European countries that had participated in the construction by providing both labor and materials. The participating countries were to be repaid for their investment in the form of gas deliveries.

In the early 1970s, in anticipation of the jump in production from the Orenburg field, the Soviet Union connected the future producing area with domestic markets. A 330-mile, 40-inch line was completed in 1971 to the gas-burning Zainsk thermal power station in the Tatar ASSR to the north; it was followed in 1972 by a 150-mile, 15-inch condensate line to the Bashkir petrochemical center of Salavat, and in 1973 by a 250-mile, 40-inch gas pipeline to Kuybyshev.[72] Pending the start of operations in the Orenburg gas-processing complex, some of these lines were filled with gas provisionally from the small Sovkhoznoye field, 60 miles NNE of Orenburg near the Bashkir border. Even before the processing complex went into operation, Orenburg Oblast's smaller fields were producing 3.6 billion m^3 of natural gas in 1973 (including gas associated with oil fields).

The first 15-billion m^3 stage of the complex went into operation in February

1974, raising 1974 gas production to 11.8 billion m^3. The second stage, inaugurated in September 1975, brought gas output in Orenburg Oblast to 20.1 billion m^3 in 1975. As the designed capacity of the two stages was gradually broken in, production rose to 31.8 billion in 1976 and 34.6 billion in 1977.[73] To accommodate the increasing output, yet another domestic pipeline was completed in 1976, a 745-mile, 40-inch transmission main, running parallel to the Comecon line and connecting Orenburg with the Central Asia–Central Russia pipeline system at Aleksandrov-Gay and with the North Caucasus–Moscow system at Novopskov.

In addition to about 30 billion m^3 of residual gas, the first two stages of the Orenburg complex also yielded about 2.5 million tons of condensate (one-fourth of the nation's gas condensate production) and 700,000 tons of recovered sulfur (one-sixth of the Soviet sulfur output).

The Orenburg-Uzhgorod pipeline, which was completed in late 1978 and is being fed by the third stage of the gas-processing complex (placed in operation at the same time), is designed to carry 28 billion m^3 of gas at full capacity after all compressor stations are installed, probably in the early 1980s. Of this amount, 24 billion m^3 is expected to be delivered at the border, with 11.2 billion divided annually among Czechoslovakia, East Germany, Hungary and Poland, leaving the balance for export to Western Europe; two other participating Comecon countries–Rumania and Bulgaria–will receive their payback in gas through the export line completed in 1974 from the Ukraine through Rumania to Bulgaria. The amount of gas transported through the Orenburg-Uzhgorod line is expected to build up gradually as compressor capacity is installed, with 8 to 9 billion m^3 planned to move through the pipeline in the first year of operation in 1979.[74]

In addition to the condensate and the sulfur, which have been recovered since the beginnings of the project, Orenburg gas is also rich in helium and ethane, which were envisaged as by-products of the third stage of the processing complex. A helium extraction plant, with an initial throughput capacity of 3 billion m^3 of natural gas, was placed in operation in February 1978.[75] Orenburg gas contains 0.044 to 0.146% inert gases,[76] mainly helium, suggesting a potential helium production of up to 4.4 million m^3 a year. A United States natural gas delegation that visited the Soviet Union in 1969 learned that Soviet helium production capacity then was about 2 million m^3 a year.[77] This would suggest that helium output in the Soviet Union may have tripled as a result of the construction of the Orenburg extraction unit. The five-year plan 1976-80 did in fact call for an increase of Soviet helium production in 1980 by 3.6 times the 1975 level.[78]

The high ethane content of Orenburg gas–about 50 grams per cubic meter–spurred early plans for an ethane extraction unit with a potential yield of 2.25 million tons (from 45 billion m^3 of natural gas). The ethane (via the ethylene route) was to have formed the basis of a vinyl resins industry in the Orenburg area in combination with chlorine from a chlorine–caustic soda plant using the nearby rock salt of Sol'-Iletsk.[79] However, plans for an ethane-based chemical industry in the Orenburg area were canceled in 1978, and any ethane likely to be extracted is now to be transmitted by pipeline to existing petrochemical centers to the north, particularly Kazan'. There are also plans for additional transmission lines for Orenburg gas condensate to natural gasoline plants in the northern oil fields, where the raw material base is becoming depleted; mention has been made, in particular, of condensate pipelines from Orenburg to natural gasoline plants at Tuymazy and Shkapovo, in Bashkir ASSR, and to Minnibayevo near Al'met'yevsk, in Tatar ASSR.[80]

Central Asian Gas Region

The importance of the Central Asian region as a potential producer of natural gas emerged in the middle 1950s, but began to be translated into actual production for the Central Russian market only a decade later. The role of Central Asia can best be put into perspective in the evolution of the Soviet natural gas industry by considering it as an intermediate producing region after the early limited resources of the European USSR (North Caucasus, Ukraine) had been developed and before the vast potential of West Siberia could be fully exploited. Central Asian production thus became a crucial factor in the Soviet industry in the late 1960s, contributing more than 20% of national gas output through the 1970s, with a maximum share of 30–31% during the middle of the decade. Two of the Central Asian republics were involved in this development: First, the Uzbek SSR, where the gas industry developed on a significant scale in mid-1960s and peaked a decade later at an annual output level of 36–37 billion m^3; second, the larger gas resources of the Turkmen SSR, which started to come into play in the early 1970s and stabilized at about 70 billion m^3, or twice the Uzbek peak, toward the end of the decade.

Until the 1950s, Uzbekistan had been a minor producer of natural gas for local consumption, from small fields in the eastern Fergana Valley, a cotton-growing center. The new interest in gas production spurred the development of further small fields in that area, but the main developments were to be concentrated in the western deserts—the Kyzyl Kum, on the right bank of the Amudarya, and Turkmenia's Kara Kum, on the left bank. The Uzbek developments were centered in the area of the ancient oasis city of Bukhara. The discovery of a relatively small field, at Dzharkak, 30 miles southeast of Bukhara, led to the construction of the first Central Asian long-distance gas pipeline, which reached Tashkent in 1960 and was extended to Chimkent in 1961. The output of the Dzharkak field, which never exceeded 400 million m^3 a year, proved inadequate to fill the 28-inch line with a transmission capacity of 4.5 billion m^3, and the actual source of gas for the pipeline turned out to be the giant Gazli field, discovered in 1956 north of Bukhara and put into production in 1962.

But most of the output of Gazli (see below) was earmarked for transmission to the Urals and to Central Russia, and for an additional source of gas for the Central Asian market, Soviet planners turned to another small cluster of gas fields in the Mubarek district, 50 miles southeast of Bukhara. A second Central Asian regional pipeline, originating at Mubarek, reached Tashkent in 1968 and was extended to Frunze in 1970 and Alma-Ata in 1971. Here again the original source of gas proved inadequate, largely because much of the natural gas in the Mubarek district was high in sulfur, which tends to corrode steel pipe and must be removed before pipeline transmission. Pending the completion of a sulfur-recovery plant at Mubarek in 1972, nonsulfurous gas was fed into the new pipeline by the South Mubarek field, which was put in production in 1966 and peaked in 1971 at 2.3 billion m^3. About 1 billion m^3 was supplied by the North Mubarek field, which was developed in 1968 but required sulfur recovery for full-scale production. The rest of the gas flow into the Mubarek–Alma-Ata line, with a transmission capacity of 5.5 billion m^3, consisted of imports from Afghanistan. The flow of Afghan gas, from the Shibarghan field developed by the Soviet Union, began in 1967 and has been ranging between 2.5 and 3 billion m^3 a year. The gas originally crossed into the Soviet Union at Kelif through an underwater pipeline laid across the Amudarya, the border river. However, the line was subject to breaks from shifting stream channels and was replaced in 1974 by a more secure suspended pipeline above the river.[81]

The Mubarek sulfur-recovery plant had been projected as a smaller counterpart of the Orenburg complex consisting of three stages with a gas throughput capacity of 5 billion m^3 and a sulfur yield of 170,000 to 200,000 tons each.[82] While the Orenburg field is estimated to contain 60% of the Soviet Union's high-sulfur gas reserves (totaling 2.9 trillion m^3), southwest Uzbekistan in the Mubarek area accounts for 25% and Turkmenian fields, just across the Uzbek border, 11%. The Mubarek sulfur-recovery complex was judged essential for the development of these Central Asian reserves. The first stage began operations in early 1972 using local North Mubarek gas. The raw-material base was expanded the following year when the high-sulfur gas field of Urtabulak, 50 miles west of Mubarek, was linked to the processing plant by a pipeline that had been especially treated to reduce corrosion. Natural gas production in the Mubarek district thus rose from 3.23 billion m^3 in 1970 to 7.68 billion in 1975 as a result of the high-sulfur operations.[83] The sulfur-recovery plant encountered problems in achieving its designed capacity, and only 89,000 tons of sulfur was reported to have been recovered in 1975 out of an ultimate capacity of 200,000 tons.[84] Since a large portion of Uzbekistan's undeveloped reserves are of the sulfur-bearing type, much appeared to depend on insuring smooth operation of the sulfur-recovery technology at the Mubarek plant, the completion of the two other projected stages and the planned construction of yet another sulfur-recovery complex at Alat, southwest of Bukhara.[85]

While the Uzbek gas producers were struggling with the problem of developing the sour gas resources in the southwest, most of the republic's production came from the giant sweet gas field of Gazli, whose initial explored reserves of 456 billion m^3 were of the same order of magnitude as those of Shebelinka in the Ukraine. Shortly after its discovery in 1956, the Gazli field was earmarked as a source of gas for the Urals industrial region, which was suffering a growing fuel deficit. Two 40-inch lines, extending more than 1,200 miles to the Urals around the west shore of the Aral Sea, were completed in 1963 and 1965, providing an annual transmission capacity of 20 billion m^3, which Gazli has been filling since 1966. The field, which reached a peak production level of 26 billion m^3 in

1968–1971, had used up about 60% of its initial reserves by the end of 1977, and a gradual decline of production became evident in the late 1970s. In addition to supplying the entire flow of Central Asian gas to the Urals, Gazli has also been contributing to the transmission system carrying Central Asian gas, mainly from the newer Turkmenian fields, to Central Russia. In 1970, of a total Gazli production of 26.1 billion m^3, 19 billion moved through the Urals pipeline, 3 billion was fed into the transmission system to Central Russia, and 4 billion was transported through the regional Central Asian distribution system serving Tashkent and other major cities. In recent years, as Gazli began its decline, Uzbek gas production has been supplemented by two middle-size sweet gas fields of more than 40 billion m^3 in reserves each. One is Uchkyr, 30 miles west of Gazli, which was put into production in 1968; the other is Shakhpakhty, started in 1971 about 130 miles west of Kungrad, in the Ustyurt Plateau, in what is perhaps the most remote location of any of the Uzbek gas fields. The two fields, producing at a rate of 2 to 3 billion m^3 a year each, are feeding into the Central Asia–Central Russia transmission system.

This major gas transport system was inaugurated in 1967 with the opening of the first 40-inch line from Gazli to Moscow, a distance of 1673 miles, and has been expanding in four stages. It is based almost entirely on production from the Turkmen SSR, with the Uzbek SSR contributing a relative small share. Until the middle 1960s, Turkmenia's gas production consisted of small amounts of gas (up to one billion m^3 in 1965) associated with oil fields in the western part of the republic, on the Caspian Sea shore. Exploration in the Kara Kum desert of eastern Turkmenia had resulted in the discovery of gas fields as early as 1959 in the Darvaza area of the central Kara Kum. Despite relatively small reserves (totaling no more than 87 billion m^3),[86] low reservoir pressures and remote location, the Darvaza fields were once considered a potential source area for long-distance gas transmission to Central Russia.[87] However, before these plans could be carried out, larger and more accessible fields were discovered, and the Darvaza area, much bruited about in the early 1960s, is no longer a factor in Soviet gas prospects.

The pre-eminence of Turkmenia in gas production began with the development of the Achak field, in the northeast Kara Kum, 40 miles southeast of Khiva, the ancient oasis town. Achak was promptly designated as a supply source for the gas transmission system to Central Russia, and was placed into production in 1966, with gas reserves of 152 billion m^3 and condensate reserves of 3.4 million tons.[88] The valuable condensate began to be recovered in 1968 after completion of a special pipeline to the Pitnyak rail station, 40 miles away. After Achak had been put into production in November 1966, it first helped supply the Gazli–Urals pipelines for one year and, starting in late 1967, became a major source of gas for the first stage of the Central Asia–Central Russia transmission system. Achak peaked at more than 12 billion m^3 a year in the early 1970s, and then began to decline. However, the supply of gas in the Achak region was supplemented by the development of nearby fields; in 1970, the Gugurtli field, south of Dargan-Ata, with reserves of 85 billion m^3 and an annual output of 3 billion m^3; and in 1972, the North Achak field, with 20 billion m^3

in reserves, and the large Naip field, with about 170 billion m^3 in reserves and a designed production capacity of 15 billion a year, reached in 1974. The growing Turkmenian production led to the completion of the second stage of the Central Asia–Central Russia transmission system, with a 48-inch line in 1970 bringing the combined transport capacity of the first two stages to 25 billion m^3. The actual transmission out of Central Asia in 1970 was about 35 billion m^3, or 18% of the total Soviet gas production, with 19 billion flowing to the Urals and 16 billion to Central Russia.

The next phase in the development of the Turkmenian gas resources was focused on the Mary oasis of southeast Turkmenia, where the relatively small Bayram-Ali field (52 billion m^3 of reserves) had been discovered in 1962 and had been earmarked in early plans as a potential source of gas for Central Asia. The first gas site to be actually developed in the Mary district was Mayskoye, 20 miles southeast of Mary. Its small reserves (18 billion m^3), discovered in 1964, were considered adequate as a regional supply source, and in 1970 the Mayskoye field began feeding gas to Ashkhabad, the Turkmen capital, through a 20-inch, 300-mile line with a capacity of about one billion m^3 a year. These developments were overshadowed by the discovery in 1968 of the giant Shatlyk field (originally called Shekhitli), 30 miles southwest of Mary, with explored reserves of 876 billion m^3. These were judged sufficient to sustain an annual output of 35 billion m^3 over a period of 16 years, and Shatlyk was designated as the point of origin of a third stage of the Central Asia–Central Russia pipeline system. Production in the Shatlyk field began in 1973 when the first of two 56-inch, 290-mile pipelines was laid from Shatlyk northward across the Kara Kum to Khiva to join the Central Asia–Central Russia transmission system; the second string, completed in 1975, brought the capacity of the Shatlyk–Khiva pipeline segment to 40 billion m^3 a year. Shatlyk reached its designed production level of 35 billion m^3 in 1977. In that year, the Bayram-Ali field, the first to be discovered in the Mary district, was finally developed, with an annual output of 3 billion m^3, joining the flow to Central Russia. Other gas developments in 1977 included the start of production in the Sakar field, near Chardzhou, for local use, and the development of a field at Tedzhen, which feeds into the Mayskoye–Ashkhabad pipeline. Meanwhile, the Shatlyk–Khiva pipeline had provided an outlet for gas fields deeper in the Kara Kum. The largest, Kirpichli, started production in 1978, with reserves of 150 billion m^3 and a designed annual output of 8 billion m^3, and work was under way in 1979 on the development of two nearby fields–North Balkui and Beurdeshik–which would raise the annual production capacity in the Kirpichli cluster to 12 billion m^3.[91]

By 1978, the gas fields of eastern Turkmenia, which were operated by the Gas Ministry's Turkmen Administration, were producing at the rate of 60 billion m^3 a year, of which all but perhaps 1.5 billion m^3, consumed within Turkmenia, was transmitted to Central Russia.

The development of gas fields in the oil-producing district of western Turkmenia followed a separate course under the aegis of the Oil Ministry. Although a small gas field, Kyzylkum, had been discovered in 1952 southeast of Nebit-Dag, the old Turkmen oil field, the gas potential of the region did not

Table 19. Length and Diameters of Gas Trunk Pipelines

Diameter (in.)	Optimal annual capacity (bill. m^3)	Length of pipelines (thou. of km at year-end)								
		1940	1950	1955	1960	1965	1970	1975	1976	1977
< 28		0.33	2.3	4.9	11.8	20.2	29.9	36.2		
28	4.0				6.2	10.6	12.9	16.1		
32	6.2				2.3	4.0	5.0	7.0		
40	8.7				0.7	7.5	15.9	20.6		
48	13.3						3.8	15.7		} 23 (48 and 56)
56	20.2							3.6		
Total length of pipelines		0.33	2.3	4.9	21.0	42.3	67.5	99.2	103.5	111.7

Sources: Ekonomika stroitel'stva magistral'nykh truboprovodov [Economics of Construction of Trunk Pipelines] (Moscow: Nedra, 1977), p. 10. S. A. Orudzhev. *Gazovaya promyshlennost' po puti progressa,* op. cit., pp. 45, 46. A. K. Kortunov, *Gazovaya promyshlennost' SSSR* (Moscow: Nedra, 1967), p. 96. V. M. Gal'perin et al. *Razvitiye i perspektivy transporta gaza po magistral'nym truboprovodam* [Development and Prospects of Gas Transport by Trunk Pipelines] (Moscow: Nedra, 1968), p. 88.

become evident until the late 1950s and early 1960s when substantial gas reserves were found to be associated with the newly discovered oil fields of Kotur-Tepe and Barsa-Gel'mes and with the more southerly fields of Kamyshldzha and Okarem, where gas proved to be more significant than oil. The combined gas reserves of these fields were put at more than 150 billion m^3, enough to justify the construction of a separate western branch of the Central Asia–Central Russia gas transmission system running north along the east shore of the Caspian Sea to a junction with the triple-string eastern branch carrying gas from eastern Turkmenia and Uzbekistan's Gazli field. Although the plan for a western pipeline branch was put forward as early as 1967,[92] construction efforts remained focused on the more accessible and larger eastern fields, particularly Shatlyk, and work on the western branch did not begin in earnest until late 1972, when it reached the Mangyshlak oil district in northeast Kazakhstan, with an annual gas producing potential of 5 billion m^3. Advancing south along the Caspian shore, the western branch reached the Kotur-Tepe and Barsa-Gel'mes oil fields in late 1974, and the Kamyshldzha and Okarem fields, 150 miles far south, in 1976. The 48-inch pipeline was slow to attain its designed transmission capacity of 15 billion m^3 because of delays in the installation of compressor stations along the way and in the completion of gas-processing plants at the producing fields. In late 1977, the western fields were said to be producing at the annual rate of 9 billion m^3,[93] with perhaps 8 billion moving through the pipeline.

By the middle 1970s, the Central Asia–Central Russia gas transmission system had thus been completed, with a potential capacity of 68 billion m^3. Actual gas movements in 1975 were 55 billion m^3, with 4 billion coming from the Mangyshlak fields of northwest Kazakhstan, 46 billion from eastern Turkmenia, 3 billion from western Turkmenia and 3 billion from Uzbekistan. With the Shatlyk field producing at full capacity and an improvement in the gas flow from western Turkmenia, the system appeared to be approaching its designed capacity in 1978. The system consists of four parallel lines from the junction point of Beyneu, in northwest Kazakhstan, where the three eastern feeders and the western feeder converge, for a distance of 465 miles to Aleksandrov-Gay, in Saratov Oblast, where the system divides into two lines going northwest to Moscow and two others proceeding 470 miles westward across the Volga River to the North Caucasus–Moscow transmission system at Ostrogozhsk. The gas flow from Central Asia can thus replenish the southern supply of gas that has become depleted by the exhaustion of the North Caucasus fields.

The construction of the Central Asia–Central Russia system, with an aggregate pipeline length of 13,750 km, represented an important factor in the spurt of Soviet gas pipeline construction after 1965 as more distant sources of gas began to be developed. The length of the Soviet gas pipeline network rose from 42,300 km in 1965 to 67,500 km in 1970, or by 60%, and then by 47% to 99,200 km in 1975 (Table 19). Moreover, the need for large transmission capacities prompted a shift toward pipelines of increasingly large diameters. In 1965, the largest diameter was 40 inches, and it represented about 18% of the total pipeline length; by 1975, 48-inch and 56-inch diameters had been introduced, and the share of large pipeline diameters (40 inches and larger) had increased to 40% of the Soviet transmission network.

Komi Gas Development

While the desert regions of Central Asia served as a buffer in providing a growing supply of natural gas to the Soviet economy after 1965, the future of the industry lies in the country's northern taiga forest and tundra, including the Komi ASSR and particularly West Siberia. The general impact of the geographical shift of the Soviet gas industry into remote producing areas, whether southern desert or northern forest and tundra, has been to reduce the return on capital and to raise the costs of production, both because of unfavorable climatic conditions and because of the need to provide an infrastructure in areas far from settlement and transportation. Overall in the gas industry, the return on capital declined from 294 m^3 of gas per ruble of fixed assets in 1965 to 173 m^3 in 1970 and an estimated 127 m^3 in 1975, with one Soviet author predicting 83 m^3 for 1980.[94] A comparison of cost factors for the largest gas fields appears to show that northern gas development involves a lower return on capital and higher costs of production than the development of desert fields, not to speak of fields in the economically developed and settled European part of the country. At the peak production periods, the return on capital and the production costs at two major Central Asian fields (Gazli and Achak) were of the same order of magnitude as two major fields in the European USSR (North Stavropol' and Shebelinka). By comparison, the Vuktyl' field in the Komi ASSR of northern European Russia showed a substantially lower return on capital and high costs of production (Table 20). The higher costs in Komi ASSR have been attributed to the need for building up a regional infrastructure, including roads, power

Table 20. Economic Parameters of Selected Major Gas Fields

Field and region	Years[a]	Output/capital ratio (m^3 per ruble)	Production cost (rubles per thou. m^3)
North Stavropol'–North Caucasus	1964–68	750 to 850	0.14
Shebelinka–Ukraine	1961–65	450 to 500	0.24 to 0.26
Gazli–Uzbekistan	1966–72	575 to 700	0.15 to 0.16
Achak–Turkmenia	1970–72	320 to 450	0.30 to 0.31
Vuktyl'–Komi ASSR	1972	135	1.35

[a]Years are at or near peak output, when the output/capital ratio tends to be at a maximum and the cost of extraction at a minimum.

Source: A. D. Brents, V. Ya. Gandkin, G. S. Urinson. *Ekonomika gazodobyvayushchey promyshlennosti* [Economics of the Gas-Extraction Industry] (Moscow: Nedra, 1975), pp. 43, 54, 61, 66, 69.

transmission lines and other facilities, representing 25% of the total value of fixed assets.[95]

Before the development of the Vuktyl' field, which moved the Komi ASSR to the forefront of Soviet gas producers, this region of northern European Russia had been a minor, but significant element in the gas industry. In World War II, a carbon-black plant was evacuated from Maykop in the Caucasus to Sosnogorsk, near Ukhta in Komi ASSR, to make use of the small supplies of natural gas in the area.[96] In the 1950s, when Komi ASSR was producing one billion m^3 of natural gas a year, the Sosnogorsk plant was manufacturing around 40,000 tons of carbon black annually, or one-half of Soviet production.[97] The Sosnogorsk plant also turned out a substantial part of the Soviet Union's helium.[98] Available gas reserves in the early period never exceeded 20 billion m^3, and no significant new discoveries were made during the 1950s and early 1960s. There was a brief flurry of excitement in 1956, when what appeared to be a large gas field, with a substantial condensate component, was discovered at Dzhebol' on the upper Pechora River. Construction began promptly on a new gas town named Komsomol'sk, and plans were announced for the long-distance transmission of gas to Perm', 500 miles to the south.[99] Further exploration showed reserves to be negligible, and the project was abandoned. The surviving lumbering settlement of Komsomol'sk remains a witness to this Soviet example of boom-and-bust.

The discovery of the giant Vuktyl' gas field in 1964 opened a new phase in the gas industry of the Komi ASSR. The field, 120 miles east of Ukhta and on the right bank of the Pechora River gained additional significance because it contained large amounts of gas condensate, estimated at 185 million tons, in addition to its explored gas reserves of 388 billion m^3. The field was put into production in 1968, reaching 6.2 billion m^3 in 1970 and 17.8 billion in 1975. The production of condensate, which began moving to the Sosnogorsk gas-processing plant in 1969, reached 2 million tons (48 percent of all Soviet condensate) in 1970 and, after having peaked at 4.5 million tons in 1973, declined to 4 million (44% of the national total) in 1975.[100] The development of the Vuktyl' field gave rise to the construction of yet another major gas transmission system, the Northern Lights, which was also designed to transport part of the vast expected flow from West Siberia southwestward clear across the European USSR through Torzhok (at the intersection with the Moscow-Leningrad pipelines) and Minsk to the export point of Uzhgorod on the Czechoslovak border. The first 48-inch line from Vuktyl' reached the Torzhok junction point in 1969, and Minsk in 1974, linking up with the old pipeline that had reached Minsk from the western Ukraine in 1960; the segment to the western Ukraine was doubled in 1977. A second Vuktyl'–Torzhok string, also of 48-inch diameter, was completed in 1973, and the first 56-inch line from West Siberia began feeding into the Northern Lights system in 1976.[101] By the winter of 1978–79 the Northern Lights transmission system was carrying close to its capacity of 55 billion m^3, including about 20 billion from the Komi ASSR (mostly Vuktyl' gas) and more than 30 billion from West Siberia. A portion of the flow is to be channeled to Leningrad through a 40-inch, 385-mile line that was nearing completion in 1979 between Gryazovets, on the Northern Lights system, and the Soviet Union's second largest city.

The future of the gas industry in Komi ASSR beyond the development of the Vuktyl' field appeared uncertain. The intensive exploration effort for oil and gas in Nenets Autonomous Okrug of Arkhangel'sk Oblast, to the north, has identified the Layavozh gas field, 50 miles east of Nar-yan-Mar, as well as the Yareyyu field nearby, and exploration by winter-road convoys from Nar'yan-Mar had been reported under way in the winter of 1974-75.[102] No plans for development have been announced. A more likely prospect is the development of the giant Kharasavey gas field on the west coast of West Siberia's Yamal Peninsula. This remote field, discovered in 1974, is situated administratively within Tyumen' Oblast, but is being developed by the Komi Gas Administration because of its proximity to the Komi ASSR and the Northern Lights transmission system. The first development team was landed at Kharasavey in April 1976 from an icebreaker-escorted cargo ship that also unloaded supplies on the fast ice off the Yamal coast. These supply expeditions, of steadily increasing magnitude, have been repeated annually and the construction of a pipeline from Kharasavey to Ukhta on the Northern Lights transmission system has been tentatively scheduled to begin in 1980.[103] Whatever the course of development, long-term projections of Komi gas production call for an increase from a planned 1980 level of 22 billion m^3 in 1980 to 40 billion by 1990, suggesting that the Komi ASSR is expected to be producing only a fraction of the output projected for West Siberia.[104] Of the projected Komi total, 17 billion m^3 is expected to originate from the newly discovered fields in Nenets Autonomous Okrug.

The Supergiant Gas Fields of West Siberia

The huge potential of the West Siberian gas fields became evident in the middle to late 1960s when a series of deposits were discovered of a size unlike any previously identified in the Soviet Union. The explored gas reserves of Tyumen' Oblast, in which these supergiants were situated, jumped from 400 billion m^3 at the end of 1965 to more than 9,000 billion in 1970 and about 16,000 billion in 1975 (Table 17). The four largest fields alone represented a total of 10,000 billion m^3, or nearly 40% of the Soviet Union's reserves. They were: Urengoy, discovered in 1966, with 3,900 billion m^3; Yamburg, 1969, with 2,500 billion; Zapolyarnoye, 1965, with 2,000 billion, and Medvezh'ye, 1966, with 1,550 billion. The Medvezh'ye field, relatively more accessible than the others, was the first to be developed, with production starting in 1972 and reaching its designed annual level of 65 billion m^3 in 1977. Urengoy, the largest, started production in the summer of 1978, and is expected to reach its designed capacity of 100 billion m^3 a year in the early 1980s. No target dates have been announced for the development of the two other supergiants, which are evidently being held in reserve, but the exploitation of a few "smaller" fields–commensurate with the largest in the European USSR–is under way or planned. They are: Vyngapur, discovered in 1968, with 291 billion m^3, the southermost of the West Siberian gas fields, where production began in late 1978; the

Gubkin–Komsomol field cluster, discovered in 1965-66, with reserves of 457 billion m^3 in the Komsomol field and 353 billion in Gubkin, which are slated for development next,[105] and the Yubileynoye field, discovered in 1969 between Urengoy and Medvezh'ye, where the start of development was also scheduled under the five-year plan 1976-80.[106]

The development dates have been continually delayed by the problem of transporting millions of tons of supply and equipment to the remote Arctic regions, providing a basic infrastructure in once uninhabited swampy forest, and laying thousands of miles of wide-diameter pipeline to transmit the gas to distant consuming regions. The initial phase of West Siberian gas development in the early 1970s was slow, and the region made only a modest contribution to the national gas supply while the Central Asian desert areas functioned as an intermediate buffer. By 1975, West Siberia, meaning essentially Tyumen' Oblast, yielded 12% of the Soviet gas compared with 33% for Central Asia and Kazakhstan. But the pace of development appeared to pick up in the second half of the 1970s, as Medvezh'ye reached its designed capacity and Urengoy went into production. Gas production plans, which were being underfulfilled in the late 1970s because the difficulties of development, called for 155 billion m^3 by 1980, or 36% of planned national output, including 139-140 billion from the northern dry gas fields of West Siberia and 15-16 billion m^3 of associated oil gas from the southern oil fields.

The West Siberian gas potential was not yet apparent when gas workers first penetrated the region in the 1950s and made the first strike at Berezovo in 1953. Although this was a very modest find, with reserves of 4 billion m^3, it appeared important enough at that early stage to justify plans for the construction of a pipeline from Berezovo, an old Siberian settlement on the lower Ob' River, 350 miles south to the northern Urals industrial town of Serov. Before these plans could be implemented, exploration teams discovered additional fields that were larger in reserves and closer to the Urals–the Igrim group, discovered in 1959-61, with 15 billion m^3, and the Punga field, found in 1961, with 85 billion m^3. Priority was given to Punga, and the field was put in production in 1966 with the completion of a 40-inch, 456-mile pipeline to Serov and Nizhniy Tagil in the Urals, where it joined the gas flow that had reached the Urals from Central Asia in 1963. The head of the pipeline was extended in 1969 from Punga to the nearby Igrim fields, and in 1970 the Punga-Igrim fields peaked at 9.2 billion m^3, including 7.7 billion from Punga, which had already started its decline, and 1.5 from the Igrim group, which rose to 2.1 billion the following year before beginning its descent. By the late 1970s, these small fields ceased to be a significant factor in the West Siberian gas industry (Table 21).

As the drilling of production wells began in 1970 in the Medvezh'ye field, the Punga–Igrim area served as the jumping-off point for the first 48-inch pipeline to be laid to that supergiant, 350 miles to the northeast. Along the way the transmission line tapped another small field, Pokhroma (reserves of 28 billion m^3), which was put in production in 1971 and produced more than 2 billion m^3 for a few years, feeding the gas into the trunk line over a 28-inch, 35-mile branch. The 48-inch transmission main was completed to the Medvezh'ye field in

Table 21. Regional Distribution of Gas Production in West Siberia
(billion m^3)

Region	1970	1971	1972	1973	1974	1975	1976	1977	1978 est.	1980 plan
USSR	198	212	221	236	261	289	321	346	372	435
Khanty-Mansi Auton. Okrug	9.27	9.40	9.64	7.39	6.1	5.8	6.3	6	7	16
Nonassociated	9.17	9.26	9.37	6.57	4.6	3.6	2.7	1	–	–
Punga	7.72	6.80	5.50	3.07						
Igrim	1.45	2.10	1.76	0.66						
Pokhroma	–	0.36	2.12	2.85						
Ob' oil gas	0.10	0.14	0.27	0.82[b]	1.5[b]	2.2[b]	3.6	5.2	7	16
Yamal-Nenets Auton. Okrug[a]	0.25	0.95	3.27	10.7	20.6	32.2	44.7	65	89	139
Medvezh'ye	–	–	1.97	9.2	18.6	29.9	41.5	62	73	
Urengoy	–	–	–	–	–	–	–	–	12	58
Vyngapur	–	–	–	–	–	–	–	–	–	15
Gubkin-Komsomol	–	–	–	–	–	–	–	–	–	–
TYUMEN' OBLAST	9.3	9.4	11.6	16.5	24.7	35.7	47.8	67.9	92	
Total gas fields	9.17	9.26	11.3	15.8	23.2	33.5	44.2	62.7	85	
Total oil gas	0.10	0.14	0.3	0.8	1.5	2.2	3.6	5.2	7	
Noril'sk area[a]	0.44	1.69	2.16	2.30	2.48	2.6	3.2	3.5	4	
Yamal-Nenets AO	0.25	0.95	1.30	1.55	1.99	2.3				
Taymyr AO	0.19	0.74	0.86	0.75	0.49	0.3				

Notes:

[a]The Noril'sk area is supplied by the Messoyakha field (developed in 1970) and the Solenaya field (developed in 1973) astride the administrative border between Tyumen' Oblast's Yamal-Nenets Autonomous Okrug and Krasnoyarsk Kray's Taymyr Autonomous Okrug. Although extracted by a single gas administration, the output appears divided between the two civil administrative areas in the annual statistical handbook of the RSFSR. For example, in 1970, total production in the Noril'sk district was 0.44 billion m^3; the RSFSR yearbook lists 0.19 for Taymyr AO, and a breakdown of Yamal-Nenets AO production by fields leaves a residual of 0.25 billion m^3.

[b]Includes the following amounts from the gas cap of the Fedorovsk oil field: 0.17 in 1973; 0.60 in 1974 and 0.48 billion m^3 in 1975.

Sources: Narodnoye Khozyaystvo RSFSR v 1975 g. (Moscow: Statistika, 1976), pp. 147 and 148. *Geologiya nefti i gaza Zapadnoy Sibiri* (Moscow: Nedra, 1975), pp. 8–9. A. D. Brents et al. *Ekonomika gazodobyvayushchey promyshlennosti* (Moscow: Nedra, 1975), p. 39. *Sibirskaya neft'* (Moscow: Nedra, 1977), pp. 89–90.

1972 as production began. It reached 30 billion m^3 in 1975 and the designed capacity of about 65 billion m^3 in 1977 after the last of eight 8 billion m^3 field facilities had been completed. A total of 200 production wells had been drilled from 1970 through 1977 at an investment cost of 710 million rubles.[107]

While the development of the Punga-Igrim gas fields had been aided to some extent by the construction of the lumber railroad from Ivdel', in the Urals, to Sergino, on the Ob' River, the Medvezh'ye project posed far more complex transport problems. No year-round overland route existed, and the equipment and supplies needed for gas development had to be transported either by water down the Ob' River to Ob' Gulf and then up the shallow Nadym River during the brief summer shipping season of barely three months, or by truck over temporary cross-country winter roads. The need for a year-round overland access route to Nadym, the base town for the Medvezh'ye field, gave rise to a controversy over whether the abandoned Salekhard–Igarka railroad should be rehabilitated. This line was under construction by forced labor at the end of the Stalin era, from 1949 to 1953, and was abandoned after Stalin's death when the massive use of forced labor in construction projects was abolished. By that time it had reached the area of the gas field that was to be later discovered, 250 miles from Salekhard on the lower Ob' River. Despite pressure from the Gas Ministry in 1970–71 for rehabilitation of the route, the Railroad Ministry resisted the project on the ground that it was uneconomical, and the dilapidated track and bridges remained in disrepair. However the Gas Ministry, at its own expense, restored a segment for its own development purposes from the river landing at Nadym to Pangody, a forward settlement in the Medvezh'ye gas field, 65 miles ENE of Nadym. Traffic on the line opened in 1976.[108]

Aside from the problem of gaining access to the Medvezh'ye field, there was also controversy over the optimal routes to be followed by the projected gas pipelines out of West Siberia. Early plans called for the transmission of most Siberian gas across the northernmost section of the Urals at the latitude of Salekhard (about 67°N) and then southwest to Central Russia by the Northern Lights route through the Komi ASSR. Out of a West Siberian gas production plan of 280 billion m^3 then envisaged for 1980–it has since been cut into half because of development delays–50 billion m^3 was to be transmitted to the Urals and as much as 230 billion m^3 to Central Russia via the Northern Lights route. The movement of so much gas was to be accomplished by extra large diameter pipelines of up to 80 and 100 inches, which are no longer being seriously considered.[109] Opponents of the Northern Lights route argued that it passed through northern, uninhabited regions that would result in higher construction costs, and they favored transmission of most West Siberian gas through the industrialized Middle Urals where the gas flow could be tapped by major consumers.[110] The second view prevailed in the early 1970s and, instead of laying three pipeline strings across the northern Urals and one into the Middle Urals, as originally planned, the priorities were reversed. After the first 48-inch line had reached Medvezh'ye in 1972, a second 48-inch line through Punga and Ivdel' to Nizhnyaya Tura in the Middle Urals was laid in 1974 and was extended westward through Perm' and Gor'kiy to Moscow later that year.[111] The transmission system from Medvezh'ye into the Middle Urals handled more than 40 billion m^3 of gas before the first crossing of the northern Urals was attempted. Moreover, it developed that the gas-pipeline strategists had altogether abandoned the original northernmost crossing point, a mountain gap along the railroad from Labytnangi (on the Ob' River opposite Salekhard) to the Vorkuta area and then southwest along the Northern Lights route. In an effort to avoid a 150-mile stretch of permafrost in the north, the planners selected a more southerly route across the Urals at the latitude of Punga (Lat 63°N) where the greater height of the mountain range (about 3,000 feet) at any rate caused an increase in pipeline construction costs compared with the northern gap.[112] The first northern Urals crossing from Punga to the Northern Lights system via Vuktyl' was completed in 1976,[113] and a second string appeared to be projected. However, it had become plain that the main gas flow from the Siberian fields, notably Urengoy, would not move across the northern Urals and along the Northern Lights route as envisaged in 1970, but along an entirely new southerly route through the Middle Urals industrial city of Chelyabinsk.

The development of the Urengoy field, which began in 1974, was even more difficult than that of Medvezh'ye. Even before development started, the field had been much in the news in 1972-73 in connection with the so-called North Star venture of a United States consortium involving the liquefaction of Urengoy gas for tanker shipment to the American East Coast. After 1974, when Congress limited United States Export-Import Bank credits to the Soviet Union–the required Urengoy credits have been estimated at $8 million–discussion of the project was dropped. Negotiations were briefly picked up in 1976 by a West European group, but no accord was reached. Urengoy development thus proceeded entirely as a Soviet domestic project. The huge field, stretching 100 miles north-south, lies 130 miles east of Nadym along the west (left) bank of the Pur River. When the first winter-road truck convoy delivered a drilling rig to the Urengoy field in early 1974, it took four days to make the 76-mile cross-country run over the frozen tundra from Pangody, the forward settlement in the Medvezh'ye field. Supplies and equipment continued to be delivered to Urengoy development workers both over the seasonal winter road and, during the 75-day navigation season, by barges hauled up the Pur River to a landing that was still 60 miles from the field itself. Although work was reported under way on extending the Nadym-Medvezh'ye gas field railroad an additional 80 miles east to Urengoy, it had not reached its destination by the time the field was put in production in the spring of 1978. The new gas town rising in the field, 50 miles west of the Nenets reindeer herders' old trading post of Urengoy, was initially referred to as Yagel'nyy, for *yagel'*, a Russian term for the local reindeer moss. Since 1977, the name Novyy Urengoy (New Urengoy) has been used.

Extension of a short feeder pipeline from Nadym-Medvezh'ye began in the winter of 1975-76, and the Urengoy gas started to flow through the line in April 1978 when the first 10 billion m^3 field gathering facility was completed. In that first year of production, about 12 billion m^3 was extracted, with the 1980 plan set at 58 billion m^3. The first phase of development aims at the completion of 10 field facilities by 1982-83, with a combined output of 100 billion m^3.[114]

While the first Urengoy gas flowed through Medvezh'ye into the transmission system already in place, work was under way in 1978 on the first of a series of pipelines approaching Urengoy from the south. This 56-inch line running from Chelyabinsk through Surgut reached the intermediate field of Vyngapur, 165 miles northeast of Surgut, in late 1978, enabling that field to produce its first gas. The pipeline was to be extended 200 miles north to Urengoy in 1979, tapping the Gubkin-Komsomol fields along the way. In addition to the approaching pipeline, the first of a projected three-string transmission system, the development of Urengoy was also expected to be accelerated by the construction of the Surgut–Urengoy railroad, scheduled to reach the Urengoy field in 1980 (see page 59). This 400-mile line, serving oil fields along the way, had been regarded by Soviet planners as a more effective way of reaching the northern gas fields than the rehabilitation of the abandoned railroad from Salekhard.

The long-term strategy for the transmission of Urengoy gas to the European USSR and for export appeared to envisage three basic pipeline routes: (1) the Northern Lights system diagonally across the European USSR to the Czechoslovak border for export; (2) the existing transmission system for Medvezh'ye gas southwest to the Urals and on to Perm' and Kazan' with a projected extension to Yelets, joining the old North Caucasus–Moscow system, and on to Kiev; (3) the new southerly route through Surgut and Chelyabinsk with an extension to Kuybyshev.[115] This system may ultimately also be extended to the Czechoslovak border to provide additional export capacity. If three 56-inch lines are completed along the Surgut–Chelyabinsk route, as appears to be planned,[116] the combined transmission capacity of this system would exceed 100 billion m^3, or the entire production capacity envisaged for Urengoy in the early 1980s. Over the short run, the transmission of West Siberian gas to the European USSR, which began in 1966 from the Punga-Igrim fields, rose to 9 billion m^3 in 1970, 30 billion in 1975 and is expected to reach 135–140 billion m^3 in 1980, depending on fulfillment of the gas-field development plan.

While West Siberian production far overshadows other Siberian gas fields, two developments of local significance bear mention—the gas supply for the important north Siberian nonferrous center of Noril'sk and the development of a Yakutian gas region.

Until the later 1960s, the Noril'sk mining and metallurgical complex relied largely on a local coal deposit for fuel. The deposit, at Kayyerkan, 17 miles west of Noril'sk, consisted mainly of low-calorific, high-ash coal, which was mined through seven small shaft mines and, after 1960, one strip mine, with a combined capacity of 3 million tons. The discovery of the first northwest Siberian gas field, at Tazovskiy, in 1962 prompted plans for the laying of a 400-mile pipeline to Noril'sk. But these plans were shelved in 1967 when another gas field, Messoyakha, was discovered on the boundary between Tyumen' Oblast's Yamal-Nenets Autonomous Okrug and Krasnoyarsk Kray's Taymyr Autonomous Okrug, only 160 miles from Noril'sk. The discovery of the Messoyakha field, with reserves of 44 billion m^3, was followed in 1969 by the Pelyatka field and in 1970 by the Solenaya field, each with about 100 billion m^3. The cluster of fields assured Noril'sk a reliable supply of gas for a considerable period. The first 28-inch pipeline was completed in December 1969, with a capacity of 2.5 billion m^3 a year, followed by an identical second line in 1973 and a third in 1978 for a potential transmission capacity of 5.5 billion m^3 a year. In addition, a gas condensate component of the Solenaya field, put in production in 1972, is being piped to a natural gasoline plant at Dudinka, completed in 1978. By supplying fuel to motor vehicles in the Noril'sk district, the plant is expected to reduce the need for long-haul shipments of petroleum products to the area.[117] Actual natural gas transmission to Noril'sk rose to 2.6 billion m^3 in 1975, moving into the 3-4 billion m^3 range in the late 1970s (Table 21). The high-cost, low-quality coal operation, which employed 6,000 workers until 1970, has been virtually shut down since the arrival of natural gas, releasing the coal miners for other jobs in the labor-short Noril'sk complex. By 1972, coal output had already dropped to 900,000 tons.[118]

In Yakutia, gas production for local consumption began in 1968 when the Tas-Tumus field at the mouth of the Vilyuy River started operation. The field, discovered in 1956 with 26 billion m^3, transmitted gas to the city of Yakutsk, for use in a thermal power station, and beyond to a small cement plant at Mokhosogollokh, between Pokrovsk and Bestyakh, on the Lena River south of Yakutsk. However, the gas flow from the Tas-Tumus field peaked at 184 million m^3 in 1970, and output began to decline. In an effort to maintain a gas supply, a 120-mile extension was laid westward in 1973 to the larger Mastakh field, on the Vilyuy River at Kysyl-Syr, 30 miles east of the town of Vilyuysk. As a result of the tapping of the Mastakh field, with reserves of 59 billion m^3, the gas flow topped 500 million m^3 in 1975. A second, parallel line was being laid in 1978.[119]

Despite this modest production for local use, Yakutia has been viewed as a gas-bearing region with a vast potential, inferred by Soviet geologists to be as great as 13 trillion m^3 of reserves. This prospect led in 1974 to an agreement in principle among the Soviet Union, Japan and the United States, calling for the export of liquefied natural gas from Yakutian deposits provided that at least one trillion m^3 could be proved up. This was considered to be the minimum reserve required to justify a costly development project and the annual delivery of 10 billion m^3 of natural gas (equivalent to 7.5 million tons of LNG) to Japan and the United States each over a period of 25 years. Japanese and United States nonruble investment in the project has been estimated at several billion dollars. In 1976, Japan and a United States consortium (El Paso Company and Occidental Petroleum Corporation) extended loans of $25 million each for the exploration phase of the project. At a project meeting in Tokyo in May 1978, the partners agreed tentatively to aim for a start of production of Yakutian natural gas by 1985. Japan, which looks to LNG imports as a growing component of its total energy supply in the 1980s, continues to be interested in the Yakutian project, but appears reluctant to proceed without United States participation, which remains in doubt because of Congressional limitations on Export-Import Bank loans to the Soviet Union.

At the May 1978 meeting of the project partners, a Soviet Deputy Minister of Foreign Trade, Nikolay G. Osipov, reported that about 800 billion m^3 of

Yakutian gas had been proved by the end of 1977 and that the one trillion m^3 goal was expected to be met by the end of 1979. The reserves were said to be concentrated in two widely separated areas–460 billion m^3 in the Middle-Vilyuy (Sredne-Vilyuy) field, originally identified in 1963 west of the Mastakh field, and 340 billion m^3 in the Middle Botuobuya (Sredne-Botuobuya) field, discovered in the early 1970s on the Botuobuya River, a right tributary of the Vilyuy, 350 miles southwest of the town of Vilyuysk. The Middle Vilyuy reserves occur in the center of the great Vilyuy-Lena sedimentary basin and are associated with Lower Triassic beds at a depth of about 8,000 feet; the Middle Botuobuya field lies on the southwest edge of the basin, adjoining the Tunguska platform, and its reserves are in the Lower Cambrian at 4,000 feet.[121]

Associated Gas and Condensate Processing

While most Soviet natural gas is dry or lean, i.e., contains few or no hydrocarbons commercially recoverable as liquid product, wet oil-well gas associated with oil fields can yield heavier fractions in gas processing plants. Gas plant products include ethane, the raw material for ethylene manufacture; propane and butane, the bottled liquefied petroleum gases, and natural gasoline, a mixture of heavier hydrocarbons. Associated gas is utilized by being collected at the wellhead; otherwise it would be vented or flared. The collected gas may be used either in unprocessed form, say, as a fuel in nearby power stations, which is a wasteful use because of the high value of the natural gas liquids as petrochemical feedstocks; or it may be processed to separate the liquids. Such processing is usually required to insure efficient long-distance transmission of the residual gas.

On drilling ever more deeply in the search for hydrocarbon reserves, the Soviet Union has been developing gas-condensate reservoirs, in which some hydrocarbons, originally in the gaseous phase under the heavy pressure of the deep reservoirs, tend to condense on reduction of pressure when extracted and form a liquid termed "condensate." Like the natural gas liquids contained in associated gas, condensate can also be separated from the gas for use in the petrochemical industry.

Utilized associated gas, whether processed or unprocessed, is included in Soviet gas production statistics (see Table 18). Before the development of the Soviet natural gas industry on a large scale in the middle 1950s, associated gas represented a major component of total production–87% in 1940, 33% in 1945 and 30% in 1950. As the production of gas fields expanded, the share of associated gas in total production declined further, to 10% in 1975 and 9% in 1977. As would be expected, associated gas production has been concentrated in oil-producing regions. Before World War II, the dominant Baku fields accounted for 89% of all oil-well gas utilization; in the late 1950s and through the 1960s, the Volga-Urals region accounted for 40 to 50% of the production, and in the late 1970s, West Siberia, after a sluggish start, began to assume an increasing share.

Traditionally the Soviet Union has been slow to utilize gas withdrawn from oil reservoirs in connection with oil production, and slow to construct processing plants for the gas that was utilized. As a result, a large share of the extracted gas has been flared or, if utilized, has been wastefully burned in power stations without recovery of the valuable natural gas liquids. The need for greater utilization and, particularly, greater processing of oil-well gases started to become a major issue in the 1950s, with the development of the Volga-Urals oil-producing region. An intensive program of gas-plant construction raised the amount of processed oil-well gas to 2.4 billion m^3 in 1960 and 5.7 billion in 1965, but this represented only 18–24% of the total resources of recoverable associated gas, the rest being either utilized without processing or flared. An intensive effort to maximize utilization of oil-well gases in the Volga-Urals raised the utilization rate from 58% in 1960 to 70% in 1965, and kept flaring losses reasonably under control. Most of the new gas-processing capacity was concentrated in the Tatar ASSR (at Minnibayevo near Al'met'yevsk), Bashkir ASSR (Tuymazy and Shkapovo) and Kuybyshev Oblast (Otradnyy and Neftegorsk). The gas utilization rates in some of these mature oil-producing areas have been quite high, with 95% of the oil-well gas used in 1976 in the Tatar ASSR; the associated gas output in that area represented about 15% of all associated gas utilized in the USSR in the middle 1970s, and most of the gas was processed in the Minnibayevo plant, then the nation's largest with a capacity of 3.7 billion m^3.[123]

The situation deteriorated dramatically in the late 1960s and early 1970s as the rapid growth of West Siberian oil production began and the recovery of associated gas, as has been customary in the early stages of oil development, failed to keep pace with petroleum production. As increasingly large amounts of oil-well gas were being flared in the West Siberian fields, the utilization rate declined dramatically from 70% in 1965 to 58% in 1973. A study of nighttime images of the earth from space, in which the gas flares of oil fields are a characteristic feature, showed the flare of the giant Samotlor oil field to be one of the largest as recently as January 1975.[124] As oil production in West Siberia rose from 44 million tons in 1971 to 148 million in 1975, with 75% of the increment coming from Samotlor, the flaring of oil-well gas increased from 2.3 billion m^3 to 8.2 billion m^3 in the absence of gas-processing plants and long-distance gas transmission lines (Table 22). West Siberian gas flaring in 1975 exceeded that of the entire Soviet oil industry in 1965, before the development of the Siberian fields, and represented about 43% of all the oil-well gas flared in 1975 (Table 22).

During the 9th five-year plan (1971–75), the construction of additional gas-processing capacity proceeded mainly outside West Siberia, at Groznyy in the North Caucasus, Perm' in the Volga-Urals and in the Mangyshlak district of Kazakhstan, while the West Siberian projects lagged; out of a construction plan of 13.4 billion m^3 of additional capacity, only 9 billion was completed. The completions improved the processing of utilized oil-well gas outside West Siberia, raising the throughput of gas from 11 billion m^3 in 1970 to 21 billion in 1975 while reducing unprocessed utilization from 12 billion m^3 in 1970 to 7.6 billion in 1975.

In view of the tremendous waste of oil-well gas, the Soviet authorities

Table 22. Disposition of Associated Natural Gas Supplies
(billion cubic meters)

Year	Total extraction	Utilized Share of Associated Natural Gas				Flared Gas Share	
		Processed	Unprocessed	Total	Rate (%)	West Siberia	Total
1960	13.3	2.4	5.3	7.7	58.9		5.6
1965	23.6	5.7	10.8	16.5	69.9		7.1
1970	37.6	11.0	12.0	23.0	61.1		14.6
1971	40.4			25.0	61.8	2.3	15.4
1972	43.8	11.6	14.2	25.7	58.8	4.5	18.1
1973	46.0			26.5	57.6	6.2	19.5
1974	47.0			27.6	58.7	6.9	19.4
1975	47.6	21.0	7.6	28.6	60.0	8.2	19.0
1976	47.6			30.6	64.3		17.0
1977	47.3			31.5	66.6		15.8
1980 plan				43.5			

Sources: Neftyanoye Khozyaystvo, 1974, No. 3; 1975, No. 3; 1977, No. 2; *Neftyanik*, 1976, No. 3; 1978, No. 3; *Sotsialisticheskaya Industriya*, April 6, 1975; *Ekonomicheskaya Gazeta*, 1977, No. 8; *Gosudarstvennyy pyatiletniy plan razvitiya narodnogo khozyaystva SSSR na 1971–1975 gody* [State Five-Year Plan of Economic Development of the USSR for 1971–1975] (Moscow: Politizdat, 1972), p. 107; *Gazovaya Promyshlennost'*, 1978, No. 3; *Ekonomika neftyanoy promyshlennosti*, 1976, No. 7; A. K. Kortunov, *Gazovaya Promyshlennost'* SSSR [The Gas Industry of the USSR] (Moscow: Nedra, 1967), pp. 192-201; Yu. I. Bokserman, *Puti razvitiya novoy tekhniki v gazovoy promyshlennosti* [Ways of Developing New Technology in the Gas Industry] (Moscow: Nedra, 1964), p. 64. West Siberian utilization data from: *Sibirskaya neft'* [Siberian Oil], edited by V. I. Muravlenko and V. I. Kremneva (Moscow: Nedra, 1977), pp. 89-90.

appeared determined to improve the situation in West Siberia in the 10th five-year plan (1976–80). Out of a planned gas-plant construction of 15.7 billion m^3, 13.5 billion was to be completed in West Siberia, including 10 billion in the Samotlor district, raising processing capacity in West Siberia to 16 billion m^3 by the end of 1980. This would represent 43% of the oil-well gas processing capacity planned for the entire Soviet oil industry.

The first 2 billion m^3 gas plant in the Samotlor district had gone on stream at the end of 1975, followed by the second plant in December 1976. The two plants, which were slow to be broken in, were designed to transmit residual gas to the Surgut thermal power station, 110 miles to the west. The Surgut station, the electric power source for the entire West Siberian oil-producing region, had begun generating power in late 1972 and, pending the delayed completion of the two Samotlor gas plants at Nizhnevartovsk, had burned both unprocessed oil-well gas from the West Surgut and Ust'-Balyk fields nearby and cap gas from the Fedorovsk field, also in the Surgut district. The station will be using more than 4 billion m^3 of gas when it reaches its designed capacity of 2,400 MW in 1980.

The third Samotlor gas plant, which went into operation in November 1977, and two others yet to be completed will be feeding their residual gas into a long-distance transmission line running 700 miles southeast along the Ob' River to the Kuznetsk Basin. The 40-inch line, with a carrying capacity of 10 billion m^3 a year, started functioning in conjunction with the third Samotlor gas plant, but has been slow to transmit a substantial volume of gas both because of delays in installing compressor stations along the line and because consuming industries have not been ready to use the gas. The pipeline project was designed to reduce urban pollution by converting coal-fired heat and power stations in the Kuzbas to gas and to serve as a basis for ammonia capacity expansion at the Kemerovo nitrogen-chemical complex, thus saving coke for metallurgical uses. The Nizhnevartovsk–Kuznetsk Basin pipeline represents the final realization of a gas transmission project that goes back to 1962 when in an early stage of West Siberian gas exploration, the Okhteur'ye gas deposit was discovered on the left bank of the Vakh River in northern Tomsk Oblast. The discovery of what seemed then a significant find promptly gave rise to plans to supply gas to the Kuznetsk Basin. Two years later the proposed source of this supply line shifted to the newly discovered Myl'dzhino field in the Vasyugan swamp, which was both closer to the Kuzbas and, with 81 billion m^3, was also endowed with larger reserves. But the discovery of the oil resources in the Nizhnevartovsk area, with the huge associated gas potential, again superseded these plans, and Samotlor was finally chosen as the head of the transmission system.[125]

Compared with the utilization of associated natural gas, the production of condensate has been a relatively recent development, assuming significance since 1970. As a mixture of liquid hydrocarbons separated from natural gas, condensate is included in Soviet petroleum statistics. In the national aggregate, condensate plays a negligible role, with 9 million tons of condensate in 1975 out of a total liquid hydrocarbons output of 491 million tons, or 1.8%. However, at the level of individual producing areas, condensate can be a significant component of the total. For example, in the Komi ASSR of northern European Russia, one of the major producers, condensate represented 40% of liquid hydrocarbons production in 1973 (6.65 million tons of crude oil and 4.55 million tons of condensate making up a total of 11.2 million tons of liquid hydrocarbons). Condensate is particularly valuable as a chemical feedstock because, in addition to natural gasoline, it often contains aromatic fractions such as benzene and toluene, which are used in making a wide range of synthetics.

Condensate output began in the Soviet Union on a commercial basis in 1955 as part of the operation of the Karadag natural gas field in the Baku district of Azerbaijan. In 1960, this and other gas-condensate fields yielded 501,000 tons of condensate, or 75% of Soviet production (Table 23). In the 1960s, the Baku district was superseded by Krasnodar Kray of the Northern Caucasus, where most major gas fields also yielded condensate. In the middle of the decade, when the Krasnodar gas fields began their peak production of 23-24 billion m^3, they also produced 660,000 tons of condensate, or 57% of the national output.

But the principal upsurge of condensate production, beginning in 1970, has been associated with two giant gas fields—the Vuktyl' field in Komi ASSR and

Table 23. Production of Gas Condensate
(million metric tons)

Region	1960	1965	1970	1973	1975	1977	1980 plan
USSR	0.67	1.18	4.2	7.65	9.0		
RSFSR	0.12	0.74	2.9	5.1	5.7		
Komi ASSR	–	–	1.99	4.55	4.0		2.6
Orenburg Oblast	–	–	0.021	0.039	1.7		2.5
Krasnodar Kray	0.075	0.66	0.56	0.21			
Stavropol' Kray	0.048	0.076	0.043	0.098			
Tyumen' Oblast	–	–	0.014	0.002		0.6	
Ukrainian SSR	0.044	0.24	0.79	1.48			
Azerbaijan SSR	0.50	0.20	0.32	0.57			
Uzbek SSR	–	–	0.17	0.22			
Turkmen SSR	–	–	0.057	0.14			
Kazakh SSR	–	–	0.026	0.10			

Sources: N. V. Mel'nikov (ed.), *Toplivno-energeticheskiye resursy* [Fuel and Energy Resources] (Moscow: Nedra, 1968), pp. 460–461; Brents et al., op. cit., p. 37; Data for 1977 (*Literaturnaya Gazeta*, Jan. 18, 1978).

the Orenburg field. Komi condensate production began in 1969 and, after reaching its peak of 4.55 million tons in 1973, has been gradually declining, with a level of 2.6 million envisaged for 1980 in the current five-year plan.[126] In the Orenburg field, the first two stages of the three-stage complex, which went into operation in February 1974 and September 1975, produced 1.7 million tons of condensate in 1975, reaching their designed capacity of more than 2 million tons in 1977. The third stage opened in November 1978.

NOTES

[1]L. M. Umanskiy and M. M. Umanskiy. *Ekonomika neftyanoy i gazovoy promyshlennosti* [Economics of the Oil and Gas Industry] (Moscow, 1974), p. 22.

[2]*Neftyanik.* 1977, No. 11.

[3]*Chechen-Ingushskaya ASSR, 1917–1977* (Groznyy, 1977), p. 12.

[4]*Ekonomicheskaya Gazeta*, 1970, No. 38; *Neftyanoye Khozyaystvo*, 1971, No. 3.

[5]S. M. Lisichkin. *Ocherki razvitiya neftedobyvayushchey promyshlennosti SSSR* [Essays on the Development of the Oil Industry of the USSR] (Moscow, 1958), p. 286.

[6]*Bakinskiy Rabochiy*, March 16, 1971.

[7]*Izvestiya*, Feb. 18, 1975; *Pravda*, May 4, 1975.

[8]*Bakinskiy Rabochiy*, June 20, 1975.

[9]*Bakinskiy Rabochiy*, Jan. 30, Nov. 25 and 26, 1976.

[10]*Bakinskiy Rabochiy*, Oct. 8, 1975; April 20, 1977; July 11, 1978.

[11]*Neftyanik*, 1976, No. 1, p. 2.

[12]*Narodnoye Khozyaystvo SSSR v 1967 g.* (Moscow, 1968), p. 316.

[13]*Narodnoye Khozyaystvo Kuybyshevskoy Oblasti za 1971–75 gg.* (Kuybyshev, 1976), p. 52.

[14]*Ekonomika Sovetskoy Ukrainy*, 1978, No. 1; *Pravda Ukrainy*, Oct. 25, 1978.

[15]*Sovetskaya Belorussiya*, Feb. 2, 1976.

[16]*Sovetskaya Belorussiya*, Jan. 27, 1977.

[17]*Neftyanoye Khozyaystvo*, 1976, No. 3.

[18]*Turkmenskaya Iskra*, March 21, 1968; *Pravda*, May 12, 1968.

[19]*Ekonomicheskaya Gazeta*, 1977, No. 8; *Soviet Geography*, March 1979.

[20]*Bakinskiy Rabochiy*, Feb. 25, 1977; *Izvestiya*, Dec. 30, 1977.

[21]*Bakinskiy Rabochiy*, Aug. 11, 1970.

[22]*Pravda*, July 6, 1970.

[23]*Sotsialisticheskaya Industriya*, Jan. 24, 1974.

[24]*Neftyanoye Khozyaystvo*, 1976, No. 1.

[25]*Kazakhstankaya Pravda*, May 21, 1971.

[26]*Neftyanik*, 1976, No. 3.

[27]*Kazakhstanskaya Pravda*, Sept. 7, 1975 and Jan. 4, 1976; *Izvestiya*, July 9, 1976.

[28]*Sotsialisticheskaya Industriya*, July 23, 1975 and Feb. 18, 1976.

[29]*Soviet Geography*, Nov. 1974, p. 594; map, p. 595.

[30]*Izvestiya*, March 11, 1978.

[31]*Sotsialisticheskaya Industriya*, Sept. 21, 1974 and Sept. 10, 1975.

[32]*Neftyanoye Khozyaystvo*, 1971, No. 5; *Komi ASSR v devyatoy pyatiletke* (Syktyvkar, 1976), p. 48.

[33]1976 output was 22% over 1975 (*Neftyanoye Khozyaystvo*, 1977, No. 4); 1977 output was almost 28% over 1976 (*Sovetskaya Rossiya*, March 1, 1978); the 1978 rate was 28% over 1977 (*Sovetskaya Rossiya*, Aug. 4, 1978).

[34]*Sotsialisticheskaya Industriya*, May 17, 1978.

[35]*Polar Geography*, 1977, No. 2, p. 175, and 1978, No. 2, p. 133.

[36]*Sovetskaya Rossiya*, July 13, 1976.

[37]*Ekonomicheskaya Gazeta*, 1968, No. 20.

[38]*Sovetskaya Rossiya*, June 12, 1970.

[39]I. M. Zaytseva, "Economic evaluation of the results of oil and gas exploration and improvements in planning," in *Nekotoryye voprosy predplanovogo obosnovaniya osvoyeniya neftegazovykh resursov Zapadnoy Sibiri* [Some Issues in the Preplanning Study of the Development of Oil and Gas Resources of West Siberia] (Novosibirsk: Institute of Economics, 1973), p. 67.

[40]*Sovetskaya Rossiya*, Feb. 28, 1970.

[41]*Bakinskiy Rabochiy*, May 28, 1976; *Neftyanoye Khozyaystvo*, 1977, No. 3.

[42]*Gudok*, June 10, 1978; *Izvestiya*, July 2, 1976; *Stroitel'naya Gazeta*, July 2, 1976.

[43]*Gudok*, Sept. 5, 1976.

[44]*Stroitel'stvo Truboprovodov*, 1976, No. 10.

[45]P. V. Vasnev, "Physical-economic conditions of industrial development in the Ob' North," in *Geograficheskiye usloviya osvoyeniya Obskogo Severa* [Geographical Conditions of Development in the Ob' North] (Irkutsk: Institute of Geography, 1975), pp. 35–36.

[46]*Neftyanoye Khozyaystvo*, 1978, No. 4.

[47]*Izvestiya*, April 19, 1978; *Pravda*, July 14, 1978.

[48]A. M. Seregin; B. A. Sokolov; Yu. K. Burlin. *Osnovy regional'noy neftegazonosnosti SSSR* [Fundamentals of Regional Oil and Gas Occurrence in the USSR] (Moscow University, 1977), p. 62.

[49]*Izvestiya*, April 19 and May 16, 1978.

[50]*Sovetskaya Rossiya*, Dec. 28, 1967.

[51] *Stroitel'naya Gazeta*, Oct. 9, 1974.

[52] *Izvestiya Akademii Nauk SSSR, seriya geograficheskaya*, 1968, No. 4, pp. 57–63.

[53] *EKO (Ekonomika i organizatsiya promyshlennogo proizvodstva)*, 1977, No. 6, p. 40.

[54] *Pravda Vostoka*, June 4, 1974.

[55] *Sovetskaya Litva*, Sept. 3, 1977; June 11, 1978; Dec. 14, 1978.

[56] Theodore Shabad and Victor L. Mote, *Gateway to Siberian Resources (the BAM)* (New York: Halsted, 1977), pp. 88, 118–119, 132.

[57] I. I. Kantor et al. "Choice of locomotive capacity for the BAM," *Zheleznodorozhnyy Transport*, 1978, No. 3, pp. 72–75.

[58] *Soviet Geography*, October 1978, pp. 583–584.

[59] *Turkmenskaya Iskra*, Aug. 3, 1978.

[60] The flow of crude oil from the Asian USSR to the European USSR was 15 million tons in 1970 and 113 million in 1975, with a projected flow of 242 million in 1980 (*Energetika SSSR v 1976–1980 godakh* (Moscow: Energiya, 1977), p. 148; Ya. Mazover, "Location of fuel-extracting industries," *Planovoye Khozyaystvo*, 1977, No. 11, pp. 139, 145). The westward flow of Siberian crude oil is estimated by deducting the flow of crude oil from Kazakhstan and Central Asia, estimated at 15 million tons in 1970, 25 million in 1975 and 20–25 million in 1980.

[61] *Izvestiya*, Nov. 7, 1974.

[62] *Transportnoye Stroitel'stvo*, 1974, No. 10.

[63] *Sotsialisticheskaya Industriya*, July 3, 1977 and Aug. 4, 1977.

[64] *Pravda*, Nov. 16, 1977; *Sovetskaya Rossiya*, Nov. 30, 1977.

[64a] *Vneshnyaya Torgovlya*, 1978, No. 11, No. 6.

[65] *Neftyanik*, 1978, No. 3.

[66] *Pravda*, Aug. 24, 1974; *Izvestiya*, Nov. 27, 1974; *Sovetskaya Rossiya*, Sept. 19, 1975.

[67] *Sotsialisticheskaya Industriya*, Oct. 2, 1975.

[68] *Gazovaya Promyshlennost'*, 1976, No. 10.

[69] *Izvestiya*, April 24, 1975; *Sovetskaya Rossiya*, Aug. 2, 1977.

[70] *Pravda Ukrainy*, May 6, 1978.

[71] *Soviet Geography*, 1978, No. 4, p. 278.

[72] *Sel'skaya Zhizn'*, Sept. 30, 1971; *Sovetskaya Rossiya*, Jan. 27, 1972; *Stroitel'stvo Truboprovodov*, 1973, No. 11.

[73] *Orenburgskaya oblast' v devyatoy pyatiletke* [Orenburg Oblast in the 9th Five-Year Plan], statistical handbook (Chelyabinsk, 1976), pp. 42–43; *Ekonomicheskaya Gazeta*, 1977, No. 6; *Sovetskaya Rossiya*, Jan. 6, 1978.

[74] *Ekonomicheskaya Gazeta*, 1978, No. 9.

[75] *Gazovaya Promyshlennost'*, 1977, No. 6; *Izvestiya*, Jan. 4, 1978 and Feb. 15, 1978.

[76] *Gazovaya Promyshlennost'*, 1969, No. 3, p. 1.

[77] U.S. Bureau of Mines. *Mineral Facts and Problems. 1970 edition* (Bureau of Mines Bulletin 650), p. 80.

[78] *Gazovaya Promyshlennost'*, 1976, No. 11.

[79] *Sovetskaya Rossiya*, June 25, 1974 and Aug. 15, 1974.

[80] *Pravda*, June 3, 1975.

[81] *Turkmenskaya Iskra*, July 16, 1974.

[82] *Sotsialisticheskaya Industriya*, Jan. 12, 1971; *Pravda Vostoka*, July 14, 1976.

[83] *Narodnoye khozyaystvo Uzbekskoy SSR v 1975 g.* [The Economy of the Uzbek SSR in 1975], statistical yearbook (Tashkent, 1976), p. 84.

[84] *Pravda Vostoka*, Feb. 18, 1976; *Gazovaya Promyshlennost'*, 1977, No. 1.

[85] *Pravda Vostoka*, Nov. 27, 1973 and Feb. 1, 1974.

[86] *Zakonomernosti razmeshcheniya i poiski zalezhey nefti i gaza v Sredney Azii i Kazakhstana* [Patterns of Distribution and the Prospecting for Oil and Gas Deposits in Central Asia and Kazakhstan] (Moscow: Nauka, 1973), p. 138.

[87] *Turkmenskaya Iskra*, Dec. 22, 1964, Sept. 10, 1965 and Jan. 7, 1968.

[88] A. D. Brents et al. *Ekonomika gazodobyvayushchey promyshlennosti* [Economics of the Gas-Extracting Industry] (Moscow: Nedra, 1975), p. 69.

[89] Brents, op. cit., p. 73.

[90] *Gudok*, Aug. 9, 1977.

[91] *Soviet Geography*, November 1978, p. 672.

[92] *Turkmenskaya Iskra*, June 9, 1967 and Dec. 5, 1967.

[93] *Turkmenskaya Iskra*, Dec. 2, 1977.

[94] Brents, op. cit., p. 108; V. A. Smirnov, "The Gas Industry," *EKO*, 1975, No. 5, p. 55.

[95] R. D. Margulov; Ye. K. Selikhova; I. Ya. Furman. *Razvitiye gazovoy promyshlennosti i analiz tekhniko-ekonomicheskikh pokazateley* [Development of the Gas Industry and An Analysis of Technical-Economic Parameters] (Moscow, 1976), pp. 31–32.

[96] *Neftyanoye Khozyaystvo*, 1967, No. 11.

[97] *Sovetskaya Rossiya*, May 7, 1957; *Pravda*, Aug. 21, 1957.

[98] G. I. Granik. *Ekonomicheskiye problemy razvitiya i razmeshcheniya proizvoditel'nykh sil Yevropeyskogo Severa SSSR* [Economic Problems of Development and Location of the Productive Forces of the European North of the USSR] (Moscow: Nauka, 1971), p. 131.

[99] *Sovetskaya Rossiya* Jan. 22, 1957; Feb. 1, 1958; Oct. 26, 1958.

[100] *Komi ASSR v devyatoy pyatiletke*, op. cit., p. 48.

[101] *Sotsialisticheskaya Industriya*, May 20, 1976.

[102] *Pravda*, March 28, 1970, and March 18, 1975; *Sotsialisticheskaya Industriya*, May 5, 1974; *Izvestiya*, Dec. 8, 1974.

[103] *Polar Geography*, 1977, No. 1, p. 89; No. 2, pp. 172–173; No. 3, pp. 243–244; No. 4, p. 326; 1978, No. 1, p. 54; No. 2, pp. 131–132; S. A. Orudzhev, *Gazovaya promyshlennost' po puti progressa* (Moscow: Nedra, 1976), p. 61; for a map of Kharazavey, see: Central Intelligence Agency, *Polar Regions Atlas*, May 1978, p. 24.

[104] Akademiya Nauk SSSR. Komi filial. *Timano-Pechorskiy territorial'noproizvodstvennyy kompleks* [Timan-Pechora Territorial-Production Complex] (Syktyvkar, 1976), p. 37, Table 12; p. 46, Table 16.

[105] *Izvestiya*, Oct. 2, 1975; *Gazovaya Promyshlennost'*, 1976, No. 8.

[106] *Gazovaya Promyshlennost'*, 1976, No. 7.

[107] *Sotsialisticheskaya Industriya*, Dec. 15, 1977, and Jan. 6, 1978; *Stroitel'stvo Truboprovodov*, 1978, No. 3.

[108] *Stroitel'naya Gazeta*, July 17, 1970, and July 16, 1971; *Sotsialisticheskaya Industriya*, Feb. 18, 1976.

[109] *Pravda*, April 24, 1969; *Gazovaya Promyshlennost'*, 1969, No. 11.

[110] V. S. Bulatov, "On possible pipeline routes for Tyumen' gas," *Soviet Geography*, 1972, No. 3, pp. 153–162.

[111] *Soviet Geography*, February 1975, p. 121.

[112] *Stroitel'stvo Truboprovodov*, 1976, No. 5, p. 7.

[113] *Soviet Geography*, 1976, No. 9, p. 492.

[114] *Pravda*, March 7, 1977; *Ekonomicheskaya Gazeta*, 1978, No. 24.

[115] *Orudzhev*, op. cit., pp. 59, 61; *Stroitel'naya Gazeta*, Jan. 13, 1978.

[116] *Sotsialisticheskaya Industriya*, Jan. 23, 1976.

[117] For reports on the Dudinka natural gasoline plant, see: *Pravda*, Sept. 9, 1975; *Gudok*, May 14, 1976; *Sovetskaya Rossiya*, May 11, 1978. On the third Messoyakha-Noril'sk gas line, see *Sovetskaya Rossiya*, Sept. 17, 1978.

[118] *Stroitel'stvo Truboprovodov*, 1975, No. 5; *Problemy rasseleniya v rayonakh Severa* (Leningrad: Stroyizdat, 1977), p. 95.

[119] *Izvestiya*, March 27, 1976; *Sovetskaya Rossiya*, Nov. 16, 1977; *Izvestiya*, March 22, 1978.

[120]Foreign Broadcast Information Service. *Japan*, May 23, 1978, C3; May 30, 1978, C3; *East-West Trade News*, May 31, 1978, p. 2.

[121]Seregin et al., op. cit., pp. 128–130; *Neftegazonosnyye provintsii SSSR* [Oil and Gas Bearing Provinces of the USSR] (Moscow: Nedra, 1977), pp. 100–131.

[122]For brief review of the gas-processing industry, see Leslie Dienes, *Locational Factors and Locational Developments in the Soviet Chemical Industry* (University of Chicago, Department of Geography, 1969, Research Paper No. 119), pp. 79–86; *Soviet Geography*, October 1974, pp. 521–523.

[123]*Neftyanik*, 1976, No. 1, and 1977, No. 8.

[124]Thomas A. Croft, "Nighttime images of the earth from space," *Scientific American*, July 1978, p. 94; also Theodore Shabad's letter to the editor, *Scientific American*, 1978, No. 10, p. 8.

[125]*Soviet Geography*, February 1978, p. 149. For initial Okhteur'ye pipeline proposal, see map in *Izvestiya*, March 2, 1965; for Myl'dzhino pipeline project, see: *Pravda*, March 26, 1967; *Sovetskaya Rossiya*, Sept. 13, 1969.

[126]M. P. Chukichev. *Komi ASSR v desyatoy pyatiletke* [Komi ASSR in the 10th Five-Year Plan] (Syktyvkar, 1976), p. 66.

Chapter 4

ENERGY SUPPLIES: SOLID FUELS

Until the upsurge of the hydrocarbons in the 1950s, solid fuels, particularly coal, played a key role in the Soviet energy supply, providing more than 70% of the total fuel production in equivalent standard fuel units. Coal accounted for as much as 66% of all fuel produced in the early 1950s; peat, which ranged around 5 to 6% in the 1940s, was significant as an auxiliary fuel in some of the industrial western regions of the USSR, especially Central Russia, and oil shale, though much less important on a national scale, played a more localized role in Estonia and in adjoining Leningrad Oblast of the RSFSR.

Although the production of solid fuels continued to increase in absolute terms, the rapid growth of the oil and natural gas industries drastically reversed the relative significance of coal, peat and oil shale in the national fossil fuel supply by the middle 1970s. Within a span of a quarter century, it was the hydrocarbons that accounted for two-thirds of the fuel production while coal dropped to less than 30% and the national contribution of peat and oil shale became negligible.

The realization in the Soviet Union that reserves of solid fuels, especially coal, were more significant than those of oil and gas and that a greater share of thermal power generation should be converted from oil to coal led to the proclamation of the new fuels policy in the middle 1970s, allotting again a more important role to coal. The intention appeared to be to arrest the 25-year decline of the relative share of coal in the national fuels supply, to stabilize its level at around 30% of total fuel output and, over the long run, perhaps raise the contribution of coal again by a few percentage points. The relative resurgence of coal has proved to be a slower and more difficult process than anticipated by the

Soviet planners. The reasons apparently include problems of converting some of the thermal power stations from the use of fuel oil to coal, a reorientation of the psychology of producing agencies in keeping with the proclaimed new policy, and the long lead times required in developing the distant eastern strip mines and in building the mineside power-generating complexes that are to be the principal consumers. In fact, after having increased at a slow, but steady rate while oil and gas were receiving priority, the incremental growth of coal production actually slowed down under the avowedly coal-oriented policy after 1975, and virtually stagnated in 1977–78 in what was for the coal-minded Soviet Union an unprecedented development. The five-year plan of 805 million tons in 1980 was likely to be missed by a big margin.

COAL AND LIGNITE

The history of development and the distribution of Soviet coal production and reserves can be discussed both according to coal type and according to mining method. Soviet coal statistics distinguish the following basic coal types:

Table 24. Soviet Coal Production by Coal Type and Mining Method (million metric tons)

Year	USSR total	Coal type						Mining method	
		Hard coal categories							
			Bituminous						
		Total	Total	Coking	Non-coking	Anthracite	Lignite	Surface mining	Underground mining
1940	166	140	104	35.3	69.0	35.7	26.0	6.3	160
1945	149	99	82	29.8	52.7	16.9	49.9	17.8	131
1950	261	185	145	51.7	93.4	40.2	75.9	27.1	234
1955	390	277	219	77.4	141	57.8	113	64.5	325
1960	510	375	301	110.0	191	74.1	135	102	408
1965	578	428	351	139	212	76.5	150	141	437
1970	624	476	400	165	235	75.8	148	167	457
1975	701	538	461	181	280	77.0	164	226	475
1976	712	548	470	186	284	77.7	164	232	480
1977	722	555	477	186	295	78.0	167	244	478
1978 est.	724							252	472
1980 plan	805								

Sources: Nar. khoz. SSSR, various issues.

(1) the higher-ranking hard coals, which in turn may be divided into anthracite and bituminous coals, with the bituminous coals being either coking coal or noncoking coal; (2) the lower-ranking lignite, which contains less fixed carbon and therefore has a lower heating value than the higher-ranking hard coals. In the late 1970s, the hard coals represented 77% of coal production, with anthracite accounting for 11%, bituminous coking coal 26% and noncoking coal, used mainly as a power-station fuel, 40%; the remaining 23% of production consisted of the lower-ranking lignite (Table 24). Although the Soviet Union has been exceeding the United States since the middle 1950s in the total physical tonnage of coal produced, Soviet output is lower than that of the United States in terms of heating value because it contains substantial amounts of lignite, which represents only a small part (about 4% in 1977) of American production.

In terms of mining methods, there has been a trend of expansion of surface mines that has paralleled the increasing share of lignite production because most lignite is being strip-mined. At the start of the period of oil and gas dominance in the early 1950s, surface mines yielded 10 to 15% of all Soviet coal. In the late 1970s, they were contributing more than 30%. The share of surface mining was expected to increase in the 1980s with the proposed expansion of two major eastern strip-mining districts–Ekibastuz, a producer of subbituminous coal in northeast Kazakhstan, and especially the Kansk-Achinsk lignite basin of southern Siberia, where most Soviet hopes for a major upsurge of coal production over the long term rest.

Coal Reserves

According to a reassessment of Soviet coal resources carried out in 1968,[1] the total geological reserves have been estimated at the huge figure of 6,790 billion metric tons, including 2,090 billion tons of lignite (to a depth of 600 meters) and 4,700 billion tons of hard coals (to a depth of 1800 meters. The different depth criteria were adopted because it is considered economical to mine hard coals, with their higher heat value, at greater depths than the lower-ranking lignite. Of the total geological reserves, 5,700 billion tons are considered recoverable with present technology, but these recoverable reserves fall into several classes of decreasing geological reliability. The highest productive categories ($A+B+C_1$), which constitute the basis of long-term Soviet planning, are 255 billion tons; the prospective category C_2 is 170 billion tons, and the bulk of the recoverable reserves–more than 5,000 billion tons–fall into the predicted (D) categories. The present discussion will be restricted to the productive $A+B+C_1$ categories, which are the so-called measured (proved and probable) reserves and represent the most realistic basis for considering the resource base of the Soviet coal industry. Most of the predicted reserves fall into two huge inferred coal basins in Siberia that are usually shown on Soviet maps of coal reserves but have not been explored and are not likely to be developed in the foreseeable future. They are the Tunguska basin, with more than 2,000 billion tons of hard coal, and the Lena basin, with more than 1,500 billion tons of hard coal and lignite.

The more realistic, explored reserve component of 255 billion tons that will

Table 25. Regional Distribution of Soviet Coal Reserves ($A+B+C_1$ categories; billion metric tons)

Region, coal type and mining method	Reserves ($A+B+C_1$)		1975 Production	
	Bill. tons	(%)	Mill. tons	(%)
USSR total	255	100	701	100
Coal types:				
Hard coal (HC)	147	58	538	77
Coking coal	65.5	26	181	26
Noncoking coals	81	32	357	51
Lignite (Li)	108	42	164	23
Mining method:				
Surface mining	110	43	226	32
Underground mining	145	57	475	68
EUROPEAN USSR	61	24	357	51
Donets Basin (HC)	40.4	16	222	32
Moscow Basin (Li)	4.8	1.9	34.1	4.9
Pechora Basin (HC)	7.9	3.1	24.2	3.5
Urals coal fields (HC, Li)	3.6	1.4	45.3	6.5
ASIAN USSR	194	76	344	49
Siberia	163	64	242	35
Kuznetsk Basin (HC)	60	23	137	20
Kansk-Achinsk (Li)	73	29	28	4
South Yakutia (HC)	2.6	1	0.3	0.04
Kazakhstan	26	10	92	13
Karaganda (HC)	7.6	3	46	6.5
Ekibastuz (HC)	3.7	1.5	46	6.5
Central Asia (HC, Li)	5	2	10	1.4

Sources: V. A. Shelest. *Regional'nyye energoekonomicheskiye problemy SSSR* [Regional Energy-Economic Problems of the USSR] (Moscow: Nedra, 1975), pp. 112–131; *Optimizatsiya razvitiya i razmeshcheniye ugledobyvayushchey promyshlennosti* [Optimization of the Development and Location of the Coal-Mining Industry] (Novosibirsk: Nauka, 1975), pp. 10–16.

be considered here in detail falls into 108 billion tons of lignite and 147 billion tons of hard coal, including 65.5 billion tons of coking coal and 81 billion tons of noncoking coal (mainly bituminous and some anthracite). A regional breakdown of coal reserves and recent production shows that the European USSR (including the Urals) still produces about one-half of Soviet coal on the basis of 24 percent of the nation's reserves (Table 25). But the regional distribution of Soviet coal production has been shifting increasingly eastward, particularly with the development of the Kuznetsk Basin in southern Siberia and the Ekibastuz Basin in northeast Kazakhstan; the eastward trend is to be intensified in the 1980s with the projected development of the Kansk-Achinsk lignite basin, which contains 29% of the measured coal reserves of the Soviet Union and as much as two-thirds of the lignite reserves and of the reserves suitable for surface mining.

For purposes of more detailed discussion of Soviet coal supplies, we will first consider the major hard-coal basins, which provide virtually all the anthracite and high-ranking bituminous coal (both coking and noncoking); then the important Ekibastuz and Kansk-Achinsk basins, which yield lower-ranking coals, but are also of national significance as fuel sources, and finally a number of smaller coal producers of local importance.

Major Hard-Coal Basins

These basins produce the Soviet Union's most valuable coals, with the highest heat value, making them suitable for long-haul transportation to serve large marketing areas. The oldest and still the largest producer is the Donets Basin of the southern European USSR, followed by the expanding Kuznetsk Basin of southern Siberia. Two others–Karaganda and Pechora–are substantially smaller both in reserves and in production, and a fifth, the South Yakutian basin, is in the early stages of development and, in contrast to the others, will serve mainly an export market (Fig. 6).

The *Donets Basin*, which has been producing since the middle of the 19th century, has the most favorable location of all the hard-coal basins with respect to the market of the industrially developed and populated European USSR and is endowed with fairly large reserves of high-quality coal. But an unfavorable coal geology, distinguished by thin, discontinuous seams, and the need for building ever deeper mines as the upper coal beds become depleted have tended to raise production costs and made Donbas coal less competitive with Kuznetsk Basin coal. Because of stagnating production, around 210 to 220 million tons, and the higher costs of Donets coal, its marketing zone has shrunk, extending basically through the cis-Volga region of the European USSR, with Kuznetsk coal shipments making inroads even into this area.

Administratively, the Donets Basin lies astride the boundary between the eastern Ukraine and Rostov Oblast of the RSFSR, with the Ukrainian part yielding about 85 percent of the total Donbas production, and the Rostov Oblast part 15% (Table 26). The Ukraine's contribution includes about 45 million tons of anthracite, more than 80 million tons of coking coal and about 55 million tons of bituminous steam coals, while the Rostov portion consists mainly of anthracite (about 30 million tons) and 3–4 million tons of coking coal.

Because of the depletion of the higher coal seams and the need for increasingly deep mines in the main Ukrainian producing oblasts–Donetsk Oblast with 105 million tons of production and Voroshilovgrad with 75 million–an effort has been under way since the late 1950s to develop shallower outlying portions of the Donets Basin. The main development has been concentrated in the so-called Western Donbas, in the Pavlograd area of Dnepropetrovsk Oblast. The first mines opened in 1963–64 giving rise to the new coal towns of Pershotravensk and Ternovka, with populations of about 25,000 in the late

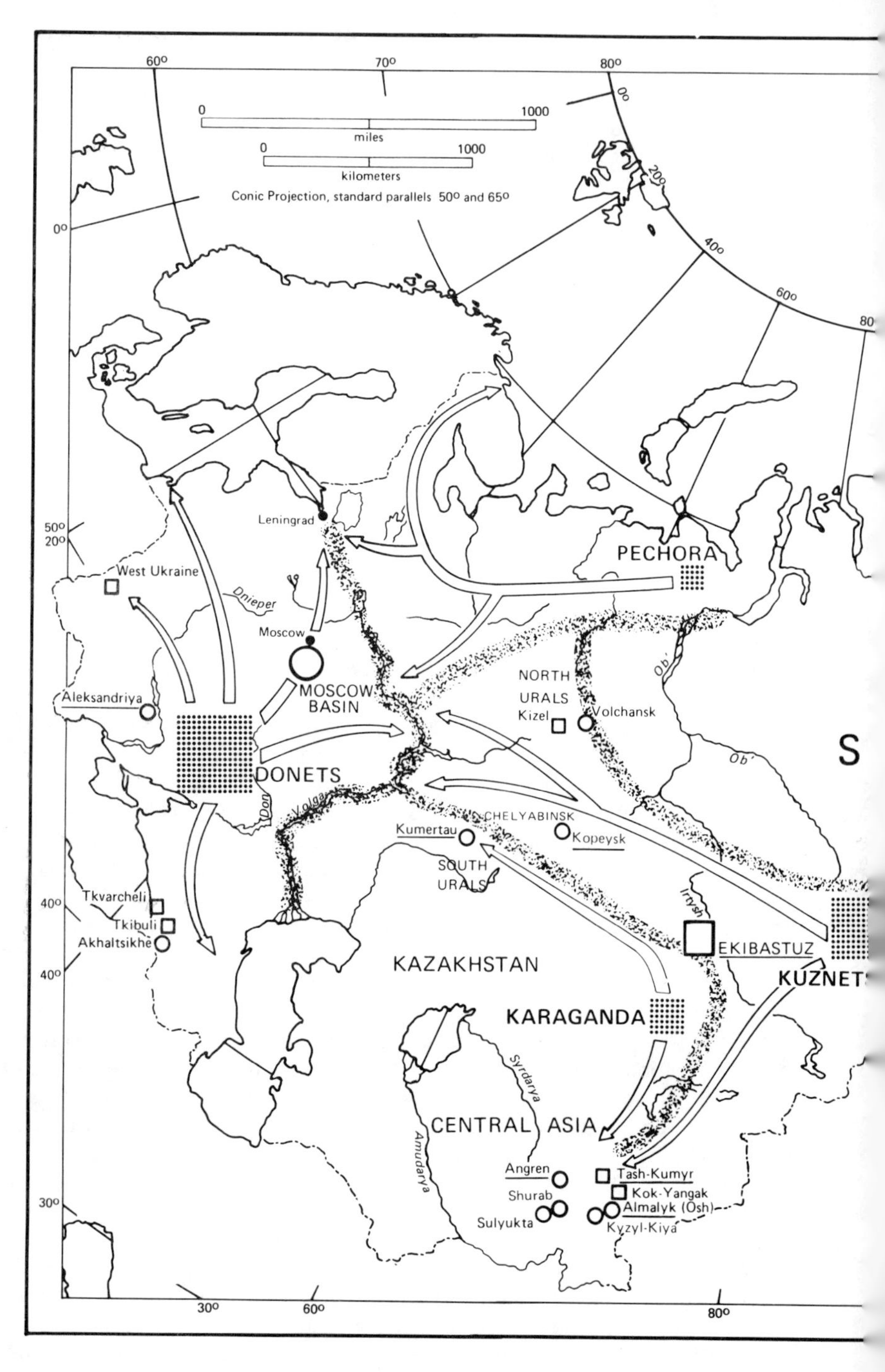

Fig. 6. Map of coal-producing centers and marketing zones.

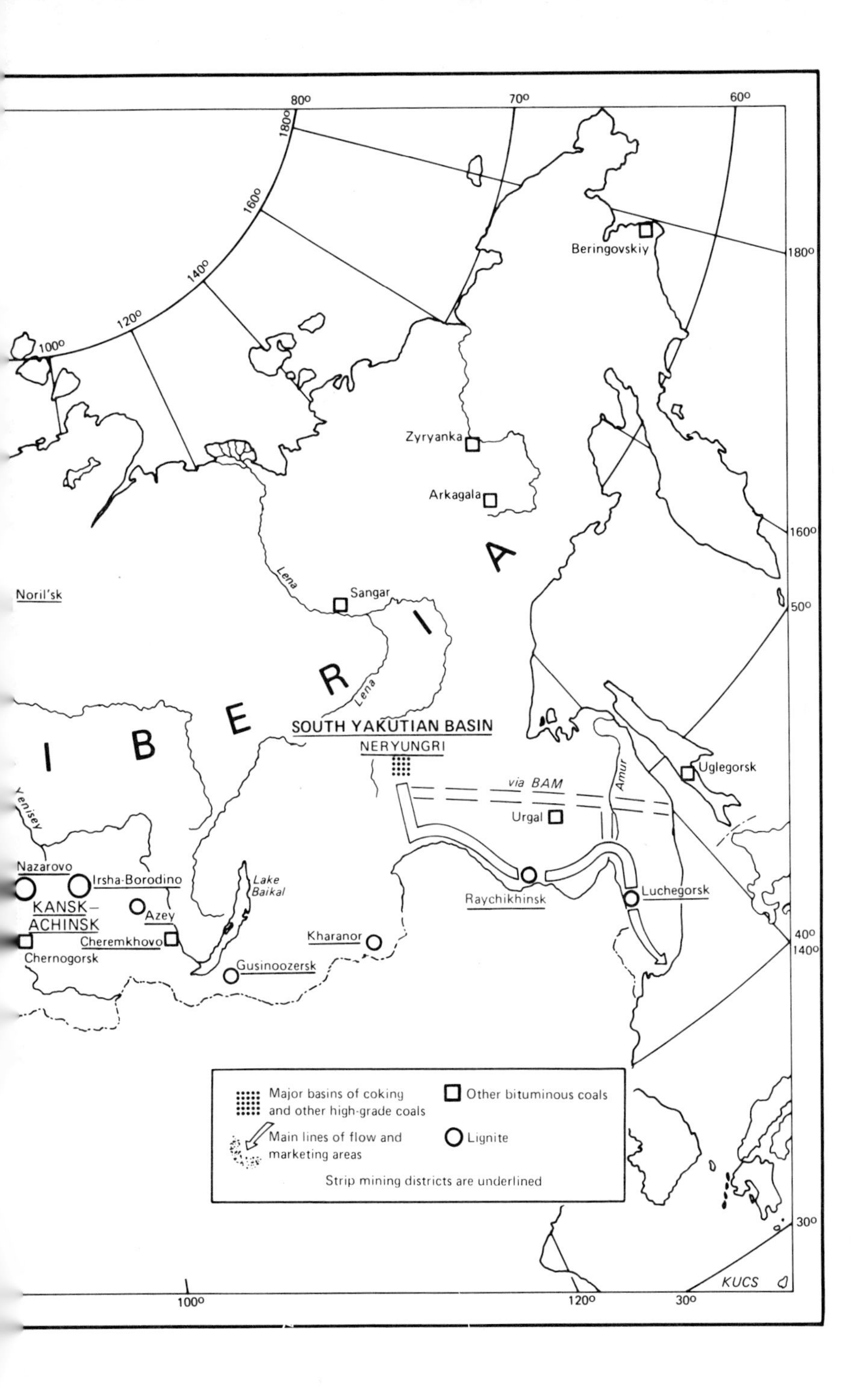
Beringovskiy
Zyryanka
Arkagala
Noril'sk
Sangar
Lena
S I B E R I A
SOUTH YAKUTIAN BASIN
NERYUNGRI
via BAM
Amur
Uglegorsk
Urgal
Yenisey
Nazarovo
Irsha-Borodino
KANSK-ACHINSK
Azey
Lake Baikal
Raychikhinsk
Luchegorsk
Cheremkhovo
Chernogorsk
Kharanor
Gusinoozersk
Major basins of coking and other high-grade coals
Other bituminous coals
Main lines of flow and marketing areas
Lignite
Strip mining districts are underlined
KUCS

Table 26. Regional Distribution of Soviet Coal Production (1940–80)
(in million metric tons)

Region	1940	1945	1950	1955	1960	1965	1970	1975	1976	1977	1978	1980 plan
USSR total	166	149	261	390	510	578	624	701	712	722	724	805
RSFSR total	72.8	105	160	229	295	326	345	381	387	394	397	436
EUROPEAN USSR	118	88.8	170	255	327	351	355	354	356			
Pechora Basin	0.27	3.3	8.7	14.2	17.6	18.1	21.5	24.2	25.8	26.7	28	31.0
Moscow Basin	10.1	20.3	30.9	39.5	42.8	40.8	36.2	34.1	30.9	29.5	28	
Donets Basin	94.3	38.4	94.6	141	188	206	216	222	224	222	212	235
RSFSR (Rostov) part	11.0	8	18.2	24.9	32.0	32.5	32.2	32.7	34	34	33	
Ukrainian part	83.3	30	76.5	116	156	173	184	189	190	188	179	
Ukrainian SSR total	83.8	30.3	78.0	126	172	194	207	216	218	217	211	229
L'vov-Volhynia	–	–	–	–	4.0	9.8	12.3	14.5	15			
Aleksandriya	0.5	0.1	1.6	9.7	12.0	11.1	10.8	12.8	11.8			
Georgian SSR	0.63	0.66	1.73	2.71	2.85	2.6	2.3	2.05	1.93			2.17
Urals (including Bashkir ASSR)	11.9	25.7	32.5	47.1	58.5	61.6	53.5	45.3	45			
Bashkir ASSR	–	–	–	1.8	3.6	6.7	6.9	9.4	10			
Perm' Oblast	4.6	8	10.2	11.1	12.0	9.9	8.7	6.8	6.4			
Sverdlovsk Oblast	1.6	6	9.8	16.4	20.3	21.2	17.2	10.0	10			
Chelyabinsk Oblast	5.7	11	12.4	17.7	22.6	23.7	20.8	19.1	19			
ASIAN USSR	47.6	60.5	91	135	183	227	269	347	356			
Siberia	38.9	47.1	69.2	102	143	172	199	243	252			
West Siberia	22.5	30.0	38.5	58.6	84.1	96.9	113	137	142	145	148	162
Kuznetsk Basin	22.4	29.9	38.4	58.3	83.6	96.4	113	137	142	145	148	162
East Siberia	9.1	9.1	17.3	26.0	36.0	46.8	54.9	71	76			
Krasnoyarsk Kray	1.7	2.5	5.3	8.8	14.6	21.7	27	36	37	39		
Kansk-Achinsk	0.4	0.6	1.7	4.3	8.5	13.9	18.2	27.9	29.1	31.6	31.9	42.3

Table 26. (cont'd)

Region	1940	1945	1950	1955	1960	1965	1970	1975	1976	1977	1978	1980 plan
East Siberia (cont'd)												
Irkutsk Oblast	5.1	4.5	8.5	13.4	16.6	19.4	21.9	24				
Cheremkhovo	5.0	4.2	8.0	12.9	15	17	16	14				
Azey	–	–	–	–	1	2	6	10				
Buryat ASSR	0.04	0.19	0.58	0.64	1.1	1.1	1.4	1.5				
Gusinoozersk	0.04	0.15	0.52	0.55	1.0	1.0	1.3	1.4				
Chita Oblast	2.2	1.8	2.8	3.1	3.4	4.1	4	7.5				
Kharanor	–	–	–	–	0.4	1	2	5.5				
Tuva ASSR	–	–	–	0.04	0.11	0.28	0.52	0.6				
Far East	7.4	8.0	13.4	17.9	22.8	28.3	31.1	35	34			
Yakut ASSR	0.14	0.32	0.28	0.58	0.92	1.5	1.6	2	2			
Amur Oblast	2.4	2.7	4.4	6.5	8.7	12.3	12.3	13.5	14			
Khabarovsk Kray	–	–	0.15	0.49	0.91	1.2	1.3	1.5	1.5			
Maritime Kray	3.7	3.6	5.0	5.5	6.4	7.2	9.0	10.1	11			
Sakhalin Oblast	0.5	0.6	2.3	3.7	4.6	4.6	4.7	5	5			
Magadan Oblast	0.1	0.6	0.9	1.0	1.1	1.3	1.5	2.2	2.7			
Kazakhstan	7.0	12.0	17.4	28.0	32.4	45.8	61.6	92.2	93.7	98.4	105	126
Karaganda	6.3	11.3	16.4	24.7	25.8	30.9	38.4	46.3	47.4	48.2	49	52
Ekibastuz	–	–	–	2.3	6.0	14.3	22.7	45.9	46.2	50.2	56	74
Central Asia	1.7	1.4	3.8	5.9	7.8	9.1	8.3	10.3	10.5			
Uzbek SSR	0.003	0.10	1.48	2.6	3.4	4.5	3.7	5.3	5.4			
Angren	–	–	1.4	2.5	3.3	4.4	3.6	5.2	5.3			
Kirghiz SSR	1.48	1.05	1.85	2.7	3.5	3.7	3.7	4.1	4.3			
Tadzhik SSR	0.20	0.24	0.45	0.58	0.85	0.90	0.89	0.87	0.80			

Source: Compiled from wide range of Soviet sources.

1970s. By that time eight mines had been put into operation, with a combined production of 9 million tons (4% of the Donbas total). Four more mines were under construction and were planned to raise Western Donbas output to 16 million tons.[2] In the late 1960s, publicity also surrounded a project to develop a so-called South Donbas, a southern spur of the basin in Donetsk Oblast, with reputed reserves of 1.8 billion tons. The first mine, the South Donbas No. 1, opened in 1973 at the new coal town of Ugledar, 30 miles southwest of Donetsk city, with a designed capacity of 2 million tons a year, but no further development has taken place.[3] Plans have even been announced for development of a northern wing of the Donbas, in the Starobel'sk area of Voroshilovgrad Oblast.

While these relative small developments are being pressed on the fringes of the Donets Basin, the strategy in the main producing areas is to build larger and deeper mines with annual capacities of the order of 3 to 4 million tons each. Such mines take five years or more to build, and their construction has not been keeping up with the depletion of shallower coal seams. Of the 40.4 billion tons of measured reserves in the Donets Basin, about 52% lie at depths to 2,000 feet, and 48% between 2,000 and 6,000 feet. In 1972, nearly half the mines were already operating at depths of 2,000 feet or more, and an increasing number were penetrating beyond 3,000 feet. The Krasnaya Zvezda mine, which went into operation in early 1974 near the town of Thorez with an annual capacity of 2 million tons, was described as the deepest Soviet anthracite mine, penetrating to 4,300 feet.[4]

In the face of the deteriorating mining conditions, a high-level Soviet directive in early 1976 asked greater priority for further development of the Donets Basin, setting production goals of 231-233 million tons for 1980 and 245-250 million tons for 1985.[5] The optimism behind these goals became clear in late 1976, when the official five-year plan for 1980 was scaled down to 229 million tons.[6] But even this goal appeared to be far beyond reach as Donets Basin output began to slip after having reached a high point of 223.7 million tons in 1976. The problem seemed to lie particularly in the Ukrainian portion of the basin as indicated by a decline of total Ukrainian production, and especially its coking-coal component (see Tables 26 and 28).

The *Kuznetsk Basin*, the second largest hard-coal producer of the Soviet Union, presents a sharp contrast to the Donets Basin in mining geology and ease of development. Kuznetsk coal seams are thick and continuous, and 90% of the basin's measured reserves, which are 50% greater than those of the Donbas, lie at depths to 2,000 feet, with 64% of all Kuzbas reserves at up to 1,000 feet. A substantial portion of Kuzbas coal, in further contrast to the Donbas, is accessible by surface mining. The favorable mining conditions tend to lower the costs of Kuznetsk coal to such an extent that it is competitive with Donets coal in large areas of the European USSR despite the long rail haul, of 1,500 to 2,000 miles, that Kuzbas coal must cover to reach markets west of the Urals.

While the Donets Basin appears to be having difficulties even maintaining its production level, the Kuzbas has been expanding at a rapid pace, adding output increments of about 20-25 million tons every five years (Table 26). Coking coal, which is extracted mainly in underground mines, accounts for 40% of total Kuzbas production; bituminous steam coals, with 60%, are mined both underground and in strip mines in roughly equal proportions. Beyond 1980, when the five-year production goal of 162 million tons, including 64.8 million tons of coking coal, may just about be reached, long-term projections look toward a Kuznetsk Basin output of 275-315 million tons by 1990, including 75-90 million tons of coking coal and 200-225 million tons of steam coals.

The development of the Kuznetsk Basin, which began in the late 19th century around Anzhero-Sudzhensk on the Trans-Siberian Railroad, proceeded southward as the volume of mining increased. In the 1930s, new mine construction shifted to Leninsk-Kuznetskiy, Kiselevsk and Prokop'yevsk, and after World War II to the southern portion of the basin, including the Tom'-Usa river district where the mining city of Mezhdurechensk arose. While the population of most Kuzbas coal-mining cities leveled off or even declined during the 1960s and 1970s, Mezhdurechensk grew from 55,000 in 1959 to 91,000 in 1977. It was near Mezhdurechensk that the huge Raspadskaya No. 1 underground mine, with a designed capacity of 7.5 million tons, went into operation in late 1973.

The strip-mined portion of Kuznetsk production has been increasing as large excavating and hauling equipment has become available, in part from Western manufacturers. In 1977, 16 surface mines were in operation, producing 41.6 million tons of coal, or 29% of Kuzbas output. In addition, 8 million tons was produced by strip-mine sections associated with underground mines.[9] The largest surface mines are scattered throughout the basin, and include the Kedrovskiy–Chernigovskiy–Novokolbinskiy cluster (ultimate capacity 15 million tons) near Berezovskiy, north of Kemerovo; the Kiselevskiy strip mine (ultimate capacity 6 million tons) near Kiselevsk, and the Siberginskiy mine (ultimate capacity 10 million tons) near Mezhdurechensk.

Long-term projections into the 1980s call for the next major expansion phase to be focused on the undeveloped Yerunakovo district, 25 miles north of Novokuznetsk, on the left bank of the Tom' River. This coal field has reserves of 5.2 billion tons of coal suitable for surface mining, or roughly one-half of all the strip-minable reserves of the Kuznetsk Basin.[10] Although the development of the district has been envisaged since the 1960s, actual work has been postponed pending expansion of production in existing mining areas. The latest designs for Yerunakovo provide for a 30 million ton strip mine, the Taldinskiy mine, and a 6 million ton deep mine, named Il'yinskiy.

In view of the good prospects for further development of the Kuznetsk Basin and the threat of stagnation in the production of the Donets Basin, the outlook is for steadily increasing westward flows of Kuzbas coking coals and steam coals to the Urals and beyond into the European USSR. Coal makes up by far the largest rail freight item, accounting for 20% of all rail loadings in the Soviet Union, and the rail lines leading westward from the Kuznetsk Basin to the Urals are among the most heavily used coal-hauling routes in the country. Kuzbas coking coals, after meeting the requirements of the iron-steel and coke-chemical industries of the Kuznetsk Basin itself, supply two-thirds of the needs of the Urals iron and steel industry, and also provide part of the coking coal consumed in European Russia (at the Cherepovets and Lipetsk iron and steel plants). In 1975, the Kuznetsk Basin is estimated to have shipped 30-35 million tons of coking coal (50-60% of total production) westward, with about two-thirds going to iron and steel plants in the Urals and one-third farther west into European Russia. Steam coal movements in 1975 may have amounted to 20 million tons, with 8 million going to the Urals and 12 million beyond. It has been suggested that more Kuzbas steam coals be made available for shipment to the western regions by using the lower-ranking lignite from the nearby Kansk-Achinsk Basin as a fuel for the central electric stations and other power-intensive industries within the Kuznetsk Basin itself.[11] The high value of Kuzbas coal is also demonstrated by the fact that this basin, deep in the Eurasian landmass, provides 15% of the Soviet coal exports, or about 4 million out of 26 million tons, going mainly eastward to the Pacific coast for shipment to Japan, but also to some countries in Western Europe (Belgium, Finland, West Germany and Sweden).[12]

The *Karaganda Basin* in central Kazakhstan has a more limited potential than the two larger hard-coal basins, with considerably smaller reserves (7.6 billion tons; see Table 25) and shaft-mined coals with a higher ash content (up to 30%) that must be cleaned before shipment. The basin resembles the recent development of the Donets Basin in its declining incremental growth, particularly in coking coals, which represent 40% of the total output. The modern history of the Karaganda Basin began in the early 1930s, when it was first reached by a railroad, and it was earmarked as a supplier of coking coal for the growing iron and steel industry of the Urals. Coal production in Karaganda expanded fairly steadily until the middle 1970s, when the annual growth rate slowed. Since the development of a local iron and steel industry in the 1960s, a growing share of the coking coal has been used in the Karaganda area itself, and out of the 19 million tons being produced annually in the late 1970s, only about 10 million tons is moving to the Urals. Karaganda's steam-coal production, ranging around 29 million tons, is consumed mainly within Kazakhstan and Central Asia, with perhaps 6-7 million tons moving to the Urals and partly beyond into the European USSR.

The *Pechora Basin*, aside from limited reserves, is handicapped by an unfavorable location within the Arctic Circle in northeasternmost European Russia and by the presence of permafrost, which makes mine construction and exploitation technically more difficult and costly. The development of the Pechora Basin, like that of Karaganda effected in its early stages with forced labor, began in 1941 when it was reached by a railroad, and the basin made an important contribution to the war effort by supplying coal to the unoccupied northern half of European Russia. Production, particularly of coking coal, began to increase in the middle 1950s, when Pechora was selected as the source of coking coal for the Cherepovets iron and steel plant. Although Pechora coking coal also needs cleaning to reduce the ash content, production has been gradually rising despite the high mining costs induced by the hostile environment. In an effort to modernize the mining operations and achieve some economies of scale, some of the smaller old mines have been shut down, and larger mines have been built. The Vorgashor mine, opened in late 1975 in the Vorkuta district, has been described as the largest shaft mine in the European USSR, with a designed capacity of 4.5 million tons.[13] Pechora coals are mined in two widely separated districts, producing coal of different qualities. Two-thirds of the output stems from the Vorkuta district and consists almost entirely of coking coal; one-third is mined at Inta, 150 miles southwest of Vorkuta, and consists wholly of steam coal. Pechora coking coal remains the basic fuel source for the Cherepovets iron and steel plant, although for technical reasons it must be mixed with Kuznetsk coal to yield a good coke. The relatively high yield of coking coal in the Pechora Basin (close to two-thirds of total production) has also generated a flow to the Lipetsk iron and steel plant in Central Russia.[14]

A fifth, prospective hard-coal supplier is the *South Yakutian Basin*, which, in its initial stages at least, will have an export orientation. In contrast to the other hard-coal basins, with the partial exception of the Kuznetsk Basin, South Yakutian coal is accessible through surface mining. The resulting low cost makes it an important asset in the Soviet coal industry, but its remote Far Eastern location is unlikely to make it a significant factor in the Soviet domestic supply. Surface mining began in the South Yakutian Basin on a small scale in 1963 to provide fuel for a small local power station at Chul'man, and was yielding about 250,000 tons a year by the middle 1970s. Large-scale development is proceeding under a Soviet-Japanese agreement of June 1974, providing for Japanese credits, to be paid back in the form of coking-coal deliveries. Japan has been importing coking coal from the Kuznetsk Basin since the early 1960s, and its interest in South Yakutian coal was motivated largely by a desire to develop an equivalent Siberian source of coking coal closer to the Pacific coast. Moreover, the construction of the Baikal-Amur Mainline, undertaken at Soviet expense, was expected to provide access to the South Yakutian Basin and insure a reliable transport route. The coal will move by rail to the Pacific port of Nakhodka, where a deepwater coal-leading terminal, with a first-stage capacity of 6.2 million tons a year, went into operation in December 1978. It will be handling Kuznetsk coal exports to Japan until the start of South Yakutian hauls in 1983.[14a]

The South Yakutian Basin, where the new coal town of Neryungri was founded in 1975, was reached two years later by a spur from the Baikal-Amur Mainline known as the "Little BAM." The first steam coal moved out of the basin in 1978 for local use in the BAM zone. The first-stage capacity of the South Yakutian project, to be reached in 1983, is 13 million tons of coal, including 9 million tons of coking coal and 4 million tons of steam coal. Some of the steam coal, which overlies the coking-coal seams, is to be used in the Neryungri area at a thermal power station under construction at Serebryannyy Bor, with a first-stage capacity of 640 MW, to be expanded ultimately to around 2,000 MW. The power plant will also use refuse coal discarded by a large cleaning plant that is designed to upgrade coking coal for export.[15]

Ekibastuz Basin

While the Soviet Union's hard-coal producers play a crucial role in the nation's coal industry in providing both unreplaceable coking coal to the iron and steel industry and long-haul, high-grade steam coal for power generation, there are two additional sources of lower-ranking surface-mined coals that play a significant role in the national energy supply. One is the ongoing development of the Ekibastuz Basin of northeast Kazakhstan; the other is the more problematical future development of the Kansk-Achinsk Basin in southern Siberia.

The Ekibastuz Basin, situated on the so-called South Siberian Railroad between the Kazakhstan cities of Tselinograd and Pavlodar, had its origins in the early 1950s in a forced labor camp depicted by Alexander Solzhenitsyn in the third volume of *The Gulag Archipelago*. Though classified as a hard coal in the Soviet Union, the Ekibastuz product would rank in the United States as a subbituminous coal, with a low heating value (4.2 million kilocalories per ton, or 60% of standard fuel) and a high ash content of up to 40%. However, because of its very low extraction cost, resulting from the surface mining of thick seams by large-capacity dragline excavators, Ekibastuz coal was judged suitable as a medium-haul steam coal that could be transported up to 600–700 miles as far west as the Urals, where energy resources lagged behind economic development. Two of the largest Urals thermal power stations—Troitsk, built in the 1960s, with a designed capacity of 2,500 MW, and Reftinskiy, built in the 1970s, with a designed capacity of 3,300–3,800 MW—have been burning Ekibastuz coal, whose shipments to the Urals in 1975 amounted to 23 million tons.

In its first stage of development, Ekibastuz has been operating two huge strip mines: the old Nos. 1, 2 and 3 mines, first to be developed in the northwest portion, were joined by 1970 into the so-called Central mine with an annual capacity of 20 million tons; the No. 5/6 mine, subsequently renamed Bogatyr (Hero), which opened its first 5 million ton section in 1970, reached a capacity of 35 million tons in 1978, and is to be expanded to 50 million by 1980.[16] With a combined production of about 55 million tons in 1978, Ekibastuz thus accounted for 22% of all Soviet surface-mined coal. The basin supplied steam coal to more than a dozen thermal power stations (totaling 12,000 MW) in northern Kazakhstan, adjoining parts of West Siberia as well as the Urals. In addition to the two large Urals stations at Troitsk and Reftinskiy, Ekibastuz coal users included the nearby 2,400 MW Yermak station, south of Pavlodar, and heat and power plants in the cities of Omsk, Petropavlovsk and Kurgan. In 1975, out of an Ekibastuz production of 46 million tons, 18 million was consumed by power stations within northern Kazakhstan, 23 million in the Urals and 5 million in West Siberia.

In light of the new coal orientation of Soviet energy policy beginning in the mid-1970s as well as the low production costs of Ekibastuz coal, plans were formulated for further expansion of the basin, looking to an output growth from 74 million tons in 1980 to 115 million in 1985 and 170 million in 1990.[17] As part of the 1985 program, two more strip mines were to be excavated—an Eastern mine, with a capacity of 20 million tons, and a Northern mine with 30 million tons. The second half of the 1980s was to see further expansion of the Bogatyr', Eastern and Northern pits as well as the development of a subsidiary basin, Maykyuben, 40 miles south of Ekibastuz, with a strip-mining capacity of 20 million tons a year.

Continued expansion of rail hauls of Ekibastuz coal to distant power stations were ruled out in light of these ambitious production plans. Each million tons of Ekibastuz coal a year was equivalent to more than 50 carloads a day, and the 3,000 carloads being produced daily in the late 1970s were putting a strain on rail terminal facilities. The alternative was the construction of a series of giant mine-mouth power stations within the Ekibastuz Basin itself, and the transmission of the electricity over extra high voltage lines to the distant consumers. A prolonged research and development effort to produce the complex technology for the long-distance transmission of large amounts of electricity appeared to be nearing a successful conclusion in the middle 1970s after more than a decade of work, and long-formulated plans for a huge power-generating complex were ordered to be carried out in the late 1970s and through the 1980s.[18] The plans call for the construction of four 4,000 KW onsite stations, each equipped with eight 500 KW generating units. These stations, to be situated within a radius of 10 to 20 miles from the city of Ekibastuz, would each consume 16 million tons of coal and draw cooling water from the Irtysh-Karaganda Canal, which passes through the area. The generating complex is to transmit 40% of the projected annual electricity generation of 42 billion KWH to the unified European power system over a 1,500-mile, 1,500 KV DC intertie extending from Ekibastuz to the Tambov area in Central Russia. Plans have also been announced for a 1,150 KV AC line running from Ekibastuz about 550 miles to the Urals grid at Kustanay.[19] Although construction on the DC line had been ordered to begin in 1978,[20] it is doubtful whether either of these transmission projects will be completed much before the middle of 1980s.

Meanwhile, work is proceeding on the first of the four proposed onsite power stations on the north shore of Lake Zhangel'dy, a small salt lake that has been drained and refilled with freshwater from the Irtysh-Karaganda Canal to serve as a cooling pond. The first 500 KW generating unit had been scheduled to go into operation in 1978, but construction ran over into 1979. Pending the construction of the high-capacity intertie, the Ekibastuz station will be transmitting electricity over the existing 500 KV grid that already links North Kazakhstan with the Urals and Siberia.[21]

Kansk-Achinsk Basin

While the immediate prospects of large-scale construction of mine-mouth generating stations and extra high voltage long-distance power transmission are linked mainly with the development of the Ekibastuz Basin, the longer-term expectations rest in the Kansk-Achinsk Basin, situated farther east in southern Siberia. This basin, which is named for the Siberian towns of Kansk and Achinsk and stretches 400 miles along the Trans-Siberian Railroad, benefits from a favorable location in one of the more developed regions of Siberia and from an easy mining geology, with a thick, horizontal seam of lignite close to the surface. However, Kansk-Achinsk lignite, aside from a low heating value of 3.6 million kilocalories per ton (one-half of standard fuel), which makes it uneconomic to transport over long distances, also suffers from other shortcomings that make it difficult to use. It tends to self-ignite when exposed to air, and has some troublesome combustion characteristics, such as the formation of a calcium slag on boiler heating surfaces. Research and development aimed at upgrading the lignite to an improved fuel such as a char or semicoke have not been conclusive.

Notwithstanding these unresolved problems, the Soviet Government has announced plans to go ahead with the development of a vast lignite-mining and onsite power-generation project in the Kansk-Achinsk Basin, of even greater magnitude than the one in Ekibastuz. These plans have been spurred both by the long-term decision to give renewed priority to coal-generated electricity and by the fact that the Kansk-Achinsk Basin contains measured reserves of 73 billion tons, or two-thirds of all the lignite reserves or strippable reserves of the Soviet Union. The Kansk-Achinsk project announced by Gosplan, the Soviet economic planning agency, envisages the development of as many as nine huge strip mines with a combined capacity of up to 300 million tons.[22] The lignite would be fed directly into a series of mine-mouth thermal stations–as many as six to eight have been mentioned–each with an installed generating capacity of 6,400 KW, consisting of eight 800 MW units. Each station, according to present calculations, would consume 25 million tons of lignite a year and generate about 40 billion KWH.[23] Although such a concentration of power-generating capacity would presumably give rise to a cluster of power-intensive industries in the region of the Kansk-Achinsk Basin, the success of the project, as in the case of Ekibastuz, would be dependent on the long-distance transmission of electricity westward to the European USSR, where the real need lies. As a first step in such a transmission system, plans have been announced for the construction of an experimental 1,150 KV AC line running from the Kansk-Achinsk Basin for 170 miles southwest to Novokuznetsk in the Kuznetsk Basin. Such a line would carry 6,000 MW of electricity, or barely the capacity of just one of the proposed generating stations.

These ambitious development plans come at a time when the Kansk-Achinsk Basin, unlike Ekibastuz, is not yet a significant factor in the Soviet energy supply. The start of mining operations in the Kansk-Achinsk Basin goes back to 1940, when small shaft mines in the Irsha-Borodino district, southwest of Kansk, at the eastern end of the basin, began to yield a few hundred thousand tons a year (Table 26). Production picked up with the start of surface mining, which began in 1950 at Irsha-Borodino and two years later at Nazarovo, south of Achinsk, in the basin's western wing. By 1977, Irsha-Borodino was producing 18 million tons and Nazarovo 14 million tons, for a combined Kansk-Achinsk output of 32 million tons of lignite. This represented 4% of Soviet coal production, far out of proportion to the basin's explored reserve potential of 29% (Table 25). About one-half of Nazarovo lignite was being burned by a 1,400 MW mine-mouth thermal station forming part of the Siberian electricity grid; the rest of the Kansk Achinsk lignite was being hauled over relatively short distances to other power plants in the southern portion of Krasnoyarsk Kray, the major civil division in which the basin is situated.

As in Ekibastuz, the implementation of the ambitious development program appeared to be slow getting started in the Kansk-Achinsk Basin in the late 1970s.

The first stage of the program focuses on the Berezovskoye No. 1 strip mine, the associated Berezovskoye No. 1 mine-mouth station and the new town of Sharypovo, all clustered in the western part of the basin, 50 miles southwest of Nazarovo, on the border of Kemerovo Oblast. After a small pilot mine had been excavated in 1975 on the Berezovskoye site to extract the first few hundred thousand tons of lignite for testing of combustion quality, the go-ahead for full-scale construction was apparently given and a ceremony in September 1978 marked the official start of stripping of the overburden of the projected 55 million ton mine. The combustion tests, carried out at the Krasnoyarsk city heat and power station, were intended to provide data for the special design of boilers needed to cope with Kansk-Achinsk lignite in the projected mammoth power stations.[24]

Early in 1978, the Soviet Government was said to have ordered the Ministry of Electric Power to release its blueprints for the start of construction of the Berezovskoye No. 1 power station, which is to be located at the projected town of Sharypovo, on the site of a Siberian village by that name. On the start of operation of the mining and power-generation project, expected sometime in the mid-1980s, the mined lignite will be moving to the electric station over a 10-mile conveyor. The first buildings have been reported under construction at Sharypovo, which is planned for an ultimate population of 200,000.[25] The final element of the project, the 1,150 KV AC line to the Kuznetsk Basin had been scheduled for completion by 1980 under the 10th five-year plan, but the delivery of equipment was reported overdue in 1978, so that this aspect, too, was expected to run over into the early 1980s. Despite the hesitating start, there were indications that the Government was determined to go forward with Kansk-Achinsk development. It was announced in the summer of 1978 that construction was beginning on the northern outskirts of Krasnoyarsk city on a manufacturing plant that would turn out the huge walking draglines and other heavy excavation equipment required in the Kansk-Achinsk strip-mining operations. To suggest the magnitude of this large manufacturing project, the announcement stated that it would be commensurate with the Kama truck factory or the Atommash reactor-fabricating plant in the European USSR and was expected to give rise to a satellite city with a population of 100,000 north of Krasnoyarsk.[26]

Local Coal Supply Sources

The major hard-coal basins and the two power-oriented strip-mine complexes accounted for 80% of total coal production in the late 1970s. The rest of the industry was made up by a large number of scattered supply sources ranging in output from 30 million tons (Moscow Basin) to a few hundred thousands of tons and varying in quality from good hard coal to low-ranking lignite. Although these sources are generally designed to meet local needs, mainly in power generation, some of them cover substantial market areas. In some cases, supposedly local coals travel excessively long distances–mostly westward in the case of eastern coals–reflecting the fuel shortage in the Soviet Union's European regions.

By far the largest of these coal sources is the Moscow Basin, whose low-ranking lignite has long been a major fuel in nearby electric generating stations of Central Russia. Moscow Basin lignite is one of the poorest fuels being produced commercially in the Soviet Union, with a heating value of 2.5 million kilocalories per ton (35% of standard fuel), high sulfur and an ash content of 35%. Most of it, moreover, is shaft-mined, making it one of the most expensive coals in the Soviet Union. Yet it has played an important local role for many decades because of the proximity of a large power-generating capacity and the apparent desire on the part of the Soviet authorities to make use of local fuels if available, even if they are of poorer quality and higher cost than some long-haul coals. The Moscow Basin's share in Soviet coal production has consistently been greater than its relatively limited reserves (2% of the Soviet total), and in the 1940s and 1950s the basin accounted for more than 10% of the nation's output, reaching a peak production of 47.3 million tons in 1958. Since then, Moscow Basin output has been slowly declining as old mines have become depleted and new mines construction lagged. In an effort to maintain production, the coal industry has sought to develop the few strippable portions of the Moscow Basin, where close to one-fourth of the 30 million ton output in the late 1970s was surface-mined. In the mid-1970s, the basin still supplied six large central electric stations in the region, including the power plants of Kashira (2,018 MW), the Cherepet' plant at Suvorov (1,500 MW) and the Ryazan' station at Novomichurinsk (1,200 MW), but the declining lignite output required increasingly large shipments of coal from the outside.

In the Ukraine, a significant local coal source is the L'vov-Volhynian basin of the western Ukraine, where the shaft-mined bituminous coal has a fairly high heating value, but is also quite expensive. This basin, the westernmost in the Soviet Union, performs an important function in supplying steam coals to nearby power stations, notably the 2,400 MW Burshtyn station, that are part of the interconnected Mir (Peace) power grid feeding electricity to Eastern Europe. Farther east in the Ukraine, the Dnieper lignite basin, centered on Aleksandriya, yields a low ranking product that is upgraded by briquetting before shipment.

The local coal supplies of the Urals industrial region have always been crucial because regional energy resources have not been commensurate with the level of industrial development. As in the case of the Moscow Basin, the contribution of the Urals to the Soviet coal supply (more than 10% through the 1940–65 period) has been far greater than the region's reserve share (1.4%). After having reached an aggregate production peak of more than 60 million tons in the mid-1960s, the output of most Urals fields has been declining, and the region's production was down to less than 45 million tons in the late 1970s. The decline has been particularly pronounced (as much as 50%) in the Kizel basin of Perm' Oblast and the Karpinsk-Volchansk basin of northern Sverdlovsk Oblast. Kizel, one of the oldest coal mining districts in the Soviet Union, produces good bituminous coal, including some of coking grade, while the Sverdlovsk field yielded a low-ranking lignite in surface mines. The decline has been more gradual in the Chelyabinsk basin, where lignite with a somewhat higher heating value is produced both in deep mines and in surface operations. The only Urals basin that appears to have better prospects, at least over the short term, is the so-called South Urals lignite basin, which straddles the border between Bashkir ASSR and Orenburg Oblast. Its production, a very low-ranking lignite that is briquetted for use, was concentrated in the Kumertau strip mine of Bashkir ASSR and is planned to extend to the newly developed Tyul'gan surface mine in Orenburg Oblast, to the south.

In Siberia, several local coal sources extend along the Trans-Siberian Railroad east of the Kuznetsk Basin and of the Kansk-Achinsk Basin, which in itself ranks as a local producer pending its projected large-scale expansion. Historically, the most important source in East Siberia has been the Cheremkhovo Basin of Irkutsk Oblast, also one of the oldest coal producing districts that was developed originally in the late 19th century as a fuel source for the Trans-Siberian Railroad. As the old shaft mines have been worked out, production of the basin's bituminous steam coals has shifted increasingly to surface mines, which are also the basic mining mode in the Azey district, a western extension of the Cheremkhovo Basin at Tulun.[27]

New strip-mining operations are also in evidence elsewhere along the Trans-Siberian Railroad, usually in conjunction with nearby thermal power stations. They include the coal and electric power projects of Gusinoozersk in Buryat ASSR, Kharanor in Chita Oblast (under development) and Luchegorsk in Maritime Kray. The older surface mines of Raychikhinsk in Amur Oblast, yielding a tolerable lignite, have long functioned as a regional fuel source, as have the old bituminous mines of Artem and Partizansk (the former Suchan) in the Vladivostok area.

Table 27. Regional Distribution of Soviet Coking-Coal Reserves ($A+B+C_1$ categories; in billion metric tons)

Region	Reserves ($A+B+C_1$)		1975 Production	
	Bill. tons	(%)	Mill. tons	(%)
USSR total	65.5	100	181	100
EUROPEAN USSR	21.8	33	107	59
Donets Basin	18.2	28	88.5	49
Vorkuta	3.0	4.6	14.4	8
Kizel	0.39	0.6	2.2	1.2
Georgian SSR	0.26	0.4	1.7	0.9
ASIAN USSR	43.7	67	74.3	41
Kuznetsk Basin	32.5	50	56.2	31
South Yakutia	2.6	4	–	–
Karaganda	4.65	7	18.1	10

Sources: N. D. Lelyukhina. *Ekonomicheskaya effektivnost' razmeshcheniya chernoy metallurgii* [The Cost-Effectiveness of Location of the Iron and Steel Industry] (Moscow: Nedra, 1973), pp. 190–191; 1975 production from Table 28.

Table 28. Regional Distribution of Soviet Coking-Coal Production (1940–80) (million metric tons)

Region	1940	1945	1950	1955	1960	1965	1970	1975	1976	1977	1980 plan
USSR total	35.3	29.8	51.7	77.7	110	139	165	181	186	186	
RSFSR total	6.7	16.2	18.9	27.0	37.8	48.9	65.4	76.6	81.0	82.3	
EUROPEAN USSR	28.1	13.8	31.3	49.3	73.2	90.3	101	107	108	107	
Donets Basin	27.4	11.4	28.4	44.4	64.9	80.4	84.3	88.5	88.1	86.9	
Rostov	0.12	0.90	1.07	2.27	2.9	3.4	3.6	3.8	3.8	3.8	
Ukraine	27.3	10.5	27.3	42.1	62.0	77.0	80.7	84.7	84.3	83.1	
Vorkuta	–	–	0.17	0.93	3.8	5.2	12.7	14.4	16.0	16.7	20.7
Kizel	0.70	2.34	2.70	2.08	2.3	2.6	2.1	2.2	1.8	1.8	
Georgian SSR	–	–	–	1.93	2.17	2.1	1.8	1.7	1.7	1.8	
ASIAN USSR	7.2	16.0	20.4	28.4	37.1	48.7	63.9	74.3	78.4	79.1	
Siberia	5.90	12.9	14.9	21.7	28.8	37.7	46.9	56.2	59.4	60	64.8
Kuznetsk	5.90	12.9	14.9	21.4	28.5	37.5	46.9	56.2	59.4	60	64.8
Noril'sk	–	–	–	0.30	0.25	0.2	–	–	–	–	
Karaganda	1.34	3.10	5.5	6.7	8.3	11.0	17.0	18.1	19	19.1	

Sources: Compiled by author from wide range of Soviet sources.

Coking Coal and Coke

Coking coal production has traditionally received high priority in the Soviet Union because of its specialized use as a fuel for the iron and steel industry and, until the unsurge of petrochemicals, also in the coke-chemical industry. Although coking coal represents a substantial share of the Soviet coal reserves, the needs of Soviet industry have been so great that production has been consistently strained and the output share of coking coal has been in step with the reserve share, at about 26% (Table 25). While most coking-coal reserves are located in the Asian USSR (two-thirds of the total; Table 27), the proportion of actual coking-coal production has been just the opposite, with two-thirds of the coking coal mined in the European USSR. It is only in the late 1960s and in the 1970s, that the eastern share of production has begun to rise significantly, from 35% in 1965 to 43% in 1977 (Table 28). Coke production, like the iron and steel industry, has remained concentrated in the European USSR so that in the late 1970s only about 15-16 million tons of Soviet coke, or about 18% of the total, was being produced in the Asian USSR (Table 29). This discrepancy between the regional distribution of coking-coal production and the manufacture of coke has given rise to large westward flows of coking coal from the Asian to the European regions of the Soviet Union.[28]

Most of the westward flow of coking coal (one-fourth of the Soviet output, or 46 million tons) is destined for the Urals, which produces 23% of Soviet coke and must bring in virtually all its coking-coal requirements. These amount to 35 million tons, including two-thirds (23-24 million tons) from Kuznetsk, one-fourth (9-10 million tons) from Karaganda, with barely 2 million tons contributed by the diminishing output of the Urals' own Kizel Basin.

Table 29. Regional Distribution of Soviet Coke Production (1940-80)
(million metric tons)

Region	1940	1950	1960	1970	1973	1975	1976	1980 plan
USSR total	21.1	27.7	56.2	75.4	81.4	83.6	84.8	94
RSFSR total	5.4	12.7	25.3	34.4	38.5	40		
European Russia	0.23	0.22	4.3	8.0	9.7	10		
Urals	2.2	8.5	15.5	17.9	18.7	19		
Siberia	3.0	4.0	5.5	8.5	10.1	11		
Ukrainian SSR	15.7	15.0	30.1	37.5	38.9	40.4		44.1
Georgian SSR	–	–	0.7	0.7	0.7	0.7		
Kazakh SSR	–	–	0.1	2.8	3.3	3.5		

Sources: Promyshlennost' SSSR [Industry of the USSR], statistical handbook (Moscow: Gosstatizdat, 1957), p. 117; I. P. Krapchin. *Effektivnost' ispol'zovaniya ugley* [The Cost-Effectiveness of Coal Utilization] (Moscow: Nedra, 1976), p. 100.

The growing coking coal needs of the European USSR west of the Urals have generated an increasing flow of Kuzbas coking coal even beyond the Urals as the principal European mining district, the Donets Basin, appears to have ceased to grow, and the other coking-coal supplier, the Arctic district of Vorkuta, has not been able to make up for the lag of the Donbas. Vorkuta coking coal was originally intended to meet the requirements of the Cherepovets iron and steel complex in northern Russia, and still accounts for more than 80% of the steel plant's consumption of around 10 million tons of coking coal, with the balance coming from the Kuznetsk Basin.

But Vorkuta coking coal is now also supplying one-fourth or more of the coking-coal needs of Central Russia, which are concentrated in the iron and steel complex of Lipetsk. The needs of this region, estimated at 11 million tons, are also being met by Donets coal (30%) and increasingly by Kuznetsk coal (40%).

The Donets Basin's principal market has, of course, been the coke industry of the Ukraine, which in addition to the coke plants of iron and steel complexes also includes a number of specialized coke-chemical plants, such as the 6 million ton coke plant of Avdeyevka in the Donets Basin itself. But the basin has been having increasing difficulty supplying the needs of the coke industry of the Ukraine, now ranging around 77–78 million tons of coal, in addition to deliveries to Central Russia and to export commitments of about 3 million tons a year. As a result, Kuznetsk coking coal has begun making inroads even into the Ukraine since the middle 1970s.

OIL SHALE

The Soviet Union is one of the two countries—the other is China—that rely on oil shale for any significant portion of their domestic energy supply. The shale is a sedimentary rock containing a bituminous material known as kerogen; it can be used directly as a solid fuel for electric power generation, with a heating value of the order of low-ranking lignite (2 to 2.5 million kilocalories per ton) and a high ash content (up to 50%), or it can be heated to yield a shale oil suitable for further refining into oil products and petrochemicals. Soviet reserves of oil shale in the proved categories ($A+B+C_1$) are 6.6 billion tons, with three-fourths concentrated in the Baltic shale basin that extends from Estonia (3.9 billion tons) into adjoining Leningrad Oblast (1.1 billion tons). The only other large shale reserves lie in Kuybyshev Oblast on the Volga River (0.72 billion tons).[29]

Oil-shale production, reflecting the reserve distribution, is concentrated almost entirely in Estonia and Leningrad Oblast, with the rest being mined in the Kashpirovka strip mine in the southern outskirts of Syzran' (Kuybyshev Oblast). In 1977, out of a total Soviet production of 36 million tons, Estonia accounted for 29.7 million, Leningrad Oblast for 5.4 million, and the Syzran' operation for about 900,000 tons (Table 30).

Until the late 1950s, Estonian shale was used as a solid fuel in small local heat and power stations and was processed at Kohtla-Järve, the center of the mining district, to produce shale gas, fuel oil, lubricants, gasoline, road oil and electrode

Table 30. Oil-Shale Production in the Soviet Union (million metric tons)

Year	USSR total	Estonian SSR			RSFSR		
		Total	Deep mines	Strip mines	Total	Lenin-grad Obl.	Volga region
1940	1.68[a]	1.89[a]	1.89		0.73	0.40	0.31
1945	1.39	0.86	0.86		0.51		0.51
1950	4.72	3.54	3.54		1.16	0.34	0.81
1955	10.8	7.00	7.00		3.78	2.07	1.71
1960	14.2	9.25	8.4	0.8	4.90	3.48	1.42
1965	21.2	15.8	11.9	3.9	5.4	4.15	1.28
1970	24.3	18.9	12.6	6.3	5.4	4.25	1.17
1975	34.5	28.5	13.7	14.8	6.0	5.15	0.90
1976	35.1	29.0	14	15	6.1	5.21	0.9
1977	36.0	29.7	14	16	6.3	5.4	0.9
1978		30.3					
1980 plan		28.5					

[a]Estonia produced 0.94 in the first half of 1940 while still independent, and 0.95 after it became part of the USSR.

Sources: Toplivno-energeticheskiye resursy [Fuel and Energy Resources] (Moscow: Nauka, 1968), p. 618. *Eesti NSV rahvamajandus 1976. aastal* [Economy of the Estonian SSR in 1976] (Tallinn: Eesti Raamat, 1977), p. 69. *Narodnoye khozyaystvo Leningrad i Leningradskoy oblasti za 60 let* [Economy of Leningrad city and Leningrad Oblast Over the Last 60 Years] (Leningrad: Lenizdat, 1977), p. 69. *Narodnoye khozyaystvo Kuybyshevskoy oblasti za 1971-1975 gg.* [Economy of Kuybyshev Oblast in 1971-75] (Kuybyshev, 1976), p. 52.

coke. The use pattern changed dramatically after 1960 with the construction of two large shale-burning thermal power stations. The first, known as the Baltic station, went into operation in 1959 west of Narva and reached its designed capacity of 1,600 MW in 1966; the station was later rated at 1,435 MW. The second power plant, known as the Estonian central station and situated southwest of Narva, started up its first generators in late 1969 and reached its designed capacity of 1,600 MW in 1972. The two stations, using 80% of the vastly expanded shale output, raised Estonia's electric power production from 2 billion KWH in 1960 to 19 billion in the late 1970s. They provide an ample electricity supply for the republic's own economy and leave a surplus of 11 billion KWH for transmission to neighboring areas; these were the Latvian SSR, which received about 4 billion, and Leningrad Oblast 7 billion.

The need for a rapid expansion of shale mining for power generation stimulated the development of strip mines after 1960; by the late 1970s, four surface mines (Viivikonna, Sirgala, Narva and October) supplied slightly more than one-half of Estonian shale. However, surface-mined shale turned out to be generally of lower quality than the deep-mined product, and the construction of underground mines continued, with the large No. 9 (Estonia) mine opening in 1972; it has a designed capacity of 9 million tons of crude shale, which is beneficiated to 5 million tons of concentrate.

In view of the inefficiency of burning the low-calorific shale directly as a solid fuel, and the accumulation of piles of rock waste and ashes next to the power stations, work began in the middle 1970s on pilot-plant development of a more efficient retorting process that would yield shale oil for use under power-plant boilers as well as gas either as an additional fuel or for making chemical derivatives. If the 2 million ton pilot project being built at a cost of 29 million rubles proves successful, the shale-based generating capacity of Estonian power stations may be doubled from its present capacity of 3,000 MW.

The smaller shale production in the eastern wing of the Baltic basin, in Leningrad Oblast, around the town of Slantsy (named for a Russian word meaning "shales"), is derived from three shaft mines and is also used either for chemicals or as solid fuel in the two Estonian power plants as well as in a small heat and power station at Slantsy itself. Similar use is made of the negligible shale production at Syzran'.

PEAT

In addition to the unusual reliance on oil shale as a local power station fuel, the Soviet Union is also the world's largest peat producer, both for fuel and for agriculture. While horticultural use of peat is not uncommon in many of the advanced countries, including the United States, the only other country that has been making significant use of peat for fuel is Ireland. The total Soviet peat harvest has been variously estimated at 200 million tons or more, but only about one-fourth is harvested by the centralized peat industry, with most of the peat being gathered by collective and state farms for agricultural purposes, including soil improvement and stable litter. Until the late 1950s, all of the industrial harvest was allocated to fuel purposes; since then a growing portion has been allocated to agriculture; in the first half of the 1970s, roughly two-thirds of the industrial harvest was fuel peat and one-third was intended for agricultural uses (in addition to the far greater, but unrecorded amount harvested by farms for their own use). Soviet peat statistics cover largely fuel peat and, in the context of the present volume, the discussion will be limited to that portion of the harvest.

Soviet interest in the use of peat fuel, which actually predates the Bolshevik Revolution, derived from the scarcity of fossil fuels in some heavily populated and economically developed parts of the Soviet Union, mostly Central Russia and Belorussia, and the policy of making use of local resources, even if costlier than alternative fuels brought in from the outside. As a northern region with a vast territory, the Soviet Union has large areas of bog and other types of

Table 31. Regional Distribution of Fuel Peat Production
(million metric tons)

Region	1940	1950	1960	1970	1975	1976	1977
USSR total	33.2	36.0	53.6	57.4	53.8	32.7	41.2
RSFSR total	25.6	27.5	36.8	39.4	36.4	17.8	30.0
Northwest	3.37	3.57	5.96				
Leningrad Obl.	2.40	2.53	2.80	2.13	1.90	0.45	2.0
Central Russia	19.3	20.5	26.4	33			
Moscow Obl.	5.5	5.4	5.3	6			
Yaroslavl' Obl.		1.5	3.6	4			
Kalinin Obl.		1.5	3.4	3.2	2.7		
Ivanovo Obl.		2.8	3.5				
Gor'kiy Obl.		2.6	2.4	2.5			
Urals	2.09	2.58	3.51				
Sverdlovsk Obl.	1.64	1.89	2.05	1.86			
Ukrainian SSR	3.54	2.93	4.66	4.08	4.14		
Belorussian SSR	3.36	3.91	8.31	9.24	9.27	8.0	
Estonian SSR	0.28	0.47	0.47	0.97	1.05	0.98	
Latvian SSR	0.21	0.62	1.80	2.14	1.9	1.6	
Lithuanian SSR	0.10	0.51	1.55	1.49	1.01		

Sources: USSR and republic totals from respective statistical yearbooks. 1940, 1950, 1960 regional breakdowns from: *Promyshlennost' SSSR* [Industry of the USSR] (Moscow: Gosstatizdat, 1957), p. 165, and *Promyshlennost' RSFSR* (Moscow: Gosstatizdat, 1961), pp. 98–99.

water-saturated environment in which the partly decomposed plant matter known as peat tends to accumulate. Roughly 60% of the world's peat resources are believed to be in the USSR, whose potential peat area has been estimated at 170 million acres with inferred reserves of 158 billion tons. Detailed exploration has identified about 25 billion tons of measured reserves in the $A+B+C_1$ categories, extending from the Baltic republics and Belorussia in the west to West Siberia. Although the largest reserve share is found in West Siberia (25%) and the Urals (35%), most production is concentrated in Central Russia, which contains only 8% of the reserves, and in Belorussia, with 1.5% of the reserves. Unlike the production of other fossil fuels, of which peat is one of the lowest ranking with a heating value of 2 million kilocalories per ton (30% of standard fuel), the output of fuel peat has been fairly stable since World War II, with an average annual harvest of about 50 million tons (Table 31). Because the peat harvest is dependent on drying in the field, weather conditions tend to affect year-to-year production; the harvest was particularly low, for example, in 1976, a wet year, when Soviet fuel peat production was 32.7 million tons, or 60% of the previous year's level. The decline was particularly pronounced in Central Russia and the

Northwest, the main producing areas of the RSFSR. The 1976 output of the RSFSR dropped to 17.8 million tons (49% of the 1975 level).

The beginnings of peat-fueled electric power in the Soviet Union go back to 1914 when a 15 MW station went into operation next to a peat deposit 50 miles east of Moscow, at the settlement of Elektroperedacha (Russian for "electric transmission"), renamed Elektrogorsk in 1946. In light of the emphasis on local resource use in the early period of Soviet economic development, several peat-fueled stations were built in the 1930s, including heat and power stations serving the municipal needs of such cities as Gor'kiy, Ivanovo, Vladimir as well as Moscow and Leningrad.[31] Peat-fueled stations played a particularly important role during the wartime emergency of the 1940s, when supplies of fossil fuels were disrupted and peat constituted 5 to 6% of all Soviet fuels production. By the middle 1960s, the share of peat had declined to 1.7% in the Soviet Union as a whole, but regionally, in Central Russia, it represented 8.1% of fuel consumption.[32] At that time, there were 77 peat-fueled power stations in operation in the Soviet Union, including 14 condenser-type stations producing only electricity and 63 heat and power stations producing both electricity and steam heat for urban areas. Because of the low efficiency of peat utilization for power generation, most of the stations were small, with 56 (about three-fourths) with installed capacities of less than 50 MW each. One of the largest was the Shatura central station, east of Moscow, in operation since 1925.

Despite the apparent drawbacks of using the low-calorific peat for power generation and the growing interest in agricultural uses, further expansion of peat-fueled power was ordered in the late 1960s because of the shortage of fossil fuels in Central Russia. The Shatura station was expanded by the addition of three 200 MW units to 700 MW, making it the Soviet Union's largest peat-fired power station by 1972, with further growth under way in the late 1970s.[33] Three stations of 600 MW each were to be built in the oblasts of Smolensk, Pskov and Vologda, making use of newly developed peat deposits. The Smolensk station, at Ozernyy, 50 miles northeast of Smolensk, went into operation in 1978 with the first of three planned 200 MW units,[34] and the Cherepovets station (Vologda Oblast) at Kaduy started up the third of three 200 MW units in late 1978. But the Pskov station, in the Dedovichi area, appeared stalled in the late 1970s amid indecision on whether to proceed with such projects. Growing feeling was being expressed by planners, particularly in light of the potential agricultural uses of peat in the rural development of the Non-Chernozem region of north central European Russia, that no more peat-fueled power stations should be built.[35] In West Siberia, a 600 MW heat and power plant based on peat has been working at Tobol'sk.[36]

NOTES

[1] V. A. Shelest. *Regional'nyye energoekonomicheskiye problemy SSSR* [Regional Energy-Economic Problems of the USSR] (Moscow: Nedra, 1975), pp. 112–131; *Optimizatsiya razvitiya i razmeshcheniya ugledobyvayushchey promyshlennosti* [Optimization of the Development and Location of the Coal-Mining Industry] (Novosibirsk: Nauka, 1975), pp. 10–16.

[2] *Ugol'*, 1972, No. 5, and 1976, No. 5.

[3] *Pravda Ukrainy*, Feb. 8, 1972, and Nov. 21, 1973.

[4] F. N. Sukhopara and V. I. Udod. *Problemy razvitiya i razmeshcheniya proizvoditel'nykh sil Donetsko-Pridneprovskogo rayona* [Problems of Development and Location of Productive Forces in the Donets-Dnieper Region] (Moscow: Mysl', 1976), pp. 49–54 and 95–99; *Izvestiya*, Jan. 25, 1974.

[5] *Pravda*, Feb. 10, 1976.

[6] *Pravda Ukrainy*, Nov. 19, 1976.

[7] Total Ukrainian coal production, which includes the L'vov-Volhynian bituminous basin in the western Ukraine and the Aleksandriya lignite mines, declined from 218 million tons in 1976 to 217 million in 1977 and an indicated level of 211 million in 1978; Ukrainian coking coal output dropped from 84.7 million tons in 1975 to 84.3 million in 1976 and 83.1 million in 1977 (see Tables 26 and 28).

[8] Shelest, op. cit., p. 124.

[9] *Ugol'*, 1977, No. 6, and 1978, No. 2.

[10] *Ugol'nyye mestorozhdeniya dlya razrabotki otkrytym sposobom* [Coal Deposits Suitable for Surface Mining] (Moscow: Nedra, 1971), p. 99.

[11] *Pravda*, Aug. 25, 1978.

[12] *Vneshnyaya Torgovlya*, 1976, No. 1.

[13] *Pravda*, Dec. 29, 1975; *Sovetskaya Rossiya*, June 11, 1977.

[14] "Dynamics and structure of coking coal deliveries by economic regions of the USSR in 1960–75," *Koks i Khimiya*, 1978, No. 6, pp. 10–11.

[14a] *Vodnyy Transport*, Dec. 23, 1978.

[15] *Ekonomicheskaya Gazeta*, 1978, Nos. 27 and 29. The first movement of Neryungri steam coal was reported in *Gudok*, Oct. 3, 1978.

[16] *Kazakhstanskaya Pravda*, Oct. 22, 1976; *Planovoye Khozyaystvo*, 1977, No. 12.

[17] *1985 projection from Kazakhstanskaya Pravda*, Dec. 25, 1977; 1990 projection from *Narodnoye Khozyaystvo Kazakhstana*, 1977, No. 12.

[18] For decree on Ekibastuz development, see, *Pravda*, April 29, 1977.

[19] *Kazakhstanskaya Pravda*, Feb. 17, 1978.

[20] *Kazakhstanskaya Pravda*, April 22, 1977 and Dec. 11, 1977.

[21] *Soviet Geography*, November 1976, p. 670.

[22] *Sovetskaya Rossiya*, Jan. 13, 1978; for background on Kansk-Achinsk Basin, see also: K. N. Grigorev. *Kansko-Achinskiy ugol'nyy basseyn* [The Kansk-Achinsk Coal Basin] (Moscow: Nedra, 1968), 183 pp.; G. G. Bruyer et al. *Kansko-Achinskiy basseyn–toplivnaya baza strany* [The Kansk-Achinsk Basin–Fuel Base for the Nation] (Krasnoyarsk, 1972), 84 pp.; *Planovoye Khozyaystvo*, 1974, No. 4, pp. 142–143 (Gosplan decree on the study of Kansk-Achinsk development); Ya. Mazover, "Prospects of the Kansk-Achinsk coal basin," *Planovoye Khozyaystvo*, 1975, No. 6, pp. 65–73.

[23] *Sotsialisticheskaya Industriya*, June 24, 1978.

[24] *Izvestiya*, Nov. 22, 1974; Jan. 5, 1975, and April 24, 1976; *Sotsialisticheskaya Industriya*, Sept. 17, 1976; *Stroitel'naya Gazeta*, Sept. 20, 1978.

[25] *Sovetskaya Rossiya*, March 16, 1976; *Sovetskaya Belorussiya*, April 4, 1976.

[26] *Sotsialisticheskaya Industriya*, June 24, 1978.

[27] *Soviet Geography*, 1976, No. 9, pp. 490–491.

[28] "Dynamics and structure. . . ," op. cit.

[29] Shelest, op. cit., p. 141; *Soviet Geography*, June 1978, pp. 429–430.

[30] *Sovetskaya Estoniya*, Feb. 4, 1976; July 9, 1976; Jan. 4, 1977; Aug. 8, 1977; *Energetik*, 1978, No. 2.

[31] *Torf v narodnom khozyaystve* [Peat in the Soviet Economy], edited by A. M. Matveyev (Moscow: Nedra, 1968), pp. 7–8 and 59–64.

[32] *Tsentral'nyy ekonomicheskiy rayon* [Central Economic Region]. Moscow: Nauka, 1973, p. 128.

[33] *Torfyanaya Promyshlennost'*, 1977, No. 9; *Stroitel'naya Gazeta*, Nov. 7, 1978.

[34] *Soviet Geography*, May 1978, p. 352.

[35] M. Gorlin and A. Yampol'skiy, "Basic trends in the use of the peat resources of the USSR," *Planovoye Khozyaystvo*, 1975, No. 3, pp. 123–129; *Proizvoditel'nyye sily nechernozemnoy zony RSFSR* [Productive Forces of the Non-Chernozem Zone of the RSFSR] (Moscow: Mysl', 1977), pp. 177–178.

[36] *Ekonomicheskaya Gazeta*, Oct. 15, 1960; *Stroitel'naya Gazeta*, Nov. 13, 1970; *Elektricheskiye Stantsii*, 1978, No. 8.

Chapter 5

HYDROELECTRIC POWER

The generation of hydroelectric power, using the energy produced by running water to create electricity, represents a net addition to the energy supply since, unlike fossil-fueled thermal power, it does not consume other forms of energy. Although hydroelectricity represents only a small part of the total energy mix when converted to standard fuel equivalents (see Tables 7 and 8 in Chapter 2), its contribution becomes more significant when considered within the framework of the electric power industry of the Soviet Union.

Hydroelectric projects have a number of favorable characteristics that provide incentives for developing economical waterpower sites. They utilize a renewable resource and they do not contribute to air pollution. In many cases, the development of hydroelectric power makes possible such associated benefits as recreation, urban water supply, irrigation, improved navigation and the enhancement of fisheries. The reservoirs created by hydroelectric dams may also provide sources of cooling water for steam-electric plants. Other factors may inhibit hydroelectric development. Since hydro plants must be located along rivers, there is less flexibility in the selection of sites than for other types of power-generating facilities. Hydro development may come into conflict with other economic activities, precluding the construction of power plants in some areas. An example in the case of the Soviet Union was a large hydro plant once considered for the lower reaches of the Ob' River in West Siberia. The project was abandoned when it became evident that its reservoir would flood a vast region in which some of the Soviet Union's largest reserves of oil and natural gas were being discovered. Hydro power generation may also be affected by drawdowns of reservoirs for irrigation purposes, as happened in the Central Asian

Table 32. Regional Distribution of Hydroelectric Capacity and Production (capacity in thousand megawatts; output in billion kilowatt-hours)

Year	USSR total					
	All power		Hydroelectric power			
	MW	KWH	MW	KWH	Share (%)	
					MW	KWH
1940	11.2	48.6	1.59	5.26	14	11
1950	19.6	91.2	3.22	12.7	16	14
1960	66.7	292	14.8	50.9	22	17
1965	115	507	22.2	81.4	19	16
1970	166	741	31.4	124	19	17
1975	217	1039	40.5	126	19	12
1976	228	1111	43.1	136	19	12
1977	238	1150	45.2	147	19	13
1980 plan	283	1380	53.6	197	19	14

Region	Hydroelectric power								
	1960		1965		1970		1975		1976
	MW	KWH	MW	KWH	MW	KWH	MW	KWH	MW
EUROPEAN USSR[a]	12	38	14.5	51	17	67	21	60	
Northwest	1.4	5.0	1.9	8.8	2.3	9.7	2.5	12	
Baltic	0.2	0.8	0.6	0.7	1.0	2.1	1.4	3.1	
Central Russia	1.0	2.6	1.0	2.4	1.0	2.7	1.0	2.5	
Volga	5.0	15.5	5.0	21.1	6.1	28.1	6.3	20	
Urals	0.6	2.0	1.6	4.9	1.8	4.6	1.8	3.8	
Ukraine-Moldavia	1.8	4.0	2.2	6.3	2.6	11.9	3.8	10.0	
North Caucasus	0.4	1	0.6	1.5	0.9	2.5	1.7	3.5	
Transcaucasia	1.6	5	1.8	6.0	1.9	4.7	2.3	5.3	
ASIAN USSR[a]	2.7	12	7.3	29	13.5	57	18	60	
West Siberia	0.4	1.6	0.4	1.7	0.4	2.1	0.4	2	0.4
East Siberia	0.8	3.7	4.6	18.3	10.0	42.2	12.5	48	13.7
Far East	–		–		0.3	0.6	0.7	1	
Kazakhstan	0.7	2.0	1.1	3.7	1.4	5.6	1.6	3	
Central Asia	0.8	4.6	1.2	4.8	1.4	6.5	2.8	6	

[a]Regional figures do not add up to total USSR figures because isolated hydro stations are not included in regional subtotals.

Sources: Nar. khoz. SSSR for various years; 1960, 1965 and 1970 regional figures from: *Elektroenergeticheskaya baza ekonomicheskikh rayonov SSSR* [Electric Power Basis of Economic Regions of the USSR] (Moscow: Nauka, 1974), pp. 112–113, 125, 134–135, 144, 156, 165, 171, 178, 208, 217; 1970 data also from *Energetika SSSR v 1971-1975 godakh* [Electric Power in the USSR in 1971–75] (Moscow: Energiya, 1972), p. 144; 1975 data from *Elektrifikatsiya SSSR 1967-1977* [Electrification of the USSR] (Moscow: Energiya, 1977), p. 197.

hydro projects of Toktogul and Nurek in the dry years of the mid-1970s. Because of the needs of irrigation, water had to be released from the reservoirs, depriving the power generators of sufficient head and reducing the output of electricity.

Over the years, hydroelectric plants have provided a substantial but variable portion of the Soviet Union's electric power supply. In the early stages of hydroelectric development, when both hydro capacity and steam-electric capacity were relatively small, the hydro plants furnished 11 to 14% of total electrical production. The share rose to 17 to 20% during the period 1957–1963, when the first large hydro stations began to come on stream and the Soviet Union had yet to embark on a policy of building large central steam-electric stations. Since then, with a policy of building both large hydro plants and large thermal stations, the hydro share has settled around 13 to 14% of power output, dropping to 12% in occasional dry years, such as 1975. The hydro share of installed generating capacity, ranging close to 19% in the late 1970s, tends to be higher than the share of power production because most hydro plants are designed to serve peak loads during periods of high electricity demand and therefore operate fewer hours than steam electric generating units, which are adapted to work continuously to meet the so-called base-load demand (Table 32).

A survey of economically potential hydroelectric sites in the Soviet Union has identified sites with a possible electrical power output of 1,095 billion KWH, including 201 billion in the European USSR and 894 billion in the Asian USSR, of which 350 billion are in East Siberia and 294 billion in the Soviet Far East.[1] Early hydroelectric development was concentrated in the European USSR, particularly on the Volga-Kama and Dnieper rivers, where a relatively high percentage of the potential has already been developed (about 60% in the Ukraine's Dnieper drainage basin, and close to 70% in the Volga basin). Future development is therefore likely to be concentrated in the Siberian regions, and that is where much of the current construction of hydroelectric stations is taking place. The hydroelectric capacity of the Volga stations, totaling about 6,000 MW, was surpassed in the late 1960s by the Siberian stations of the large potential Angara-Yenisey drainage basin, where the aggregate installed hydroelectric capacity reached 13,700 MW at the end of 1976, or nearly one-third of the Soviet total.

The buildup of Siberian hydroelectric capacity and the distance separating these stations from the power-consuming centers of the European USSR has resulted in contrasting uses of the hydro capacity in the European and Asian USSR. In the European regions, hydroelectric plants have been used increasingly for peak-load operations. With the growing trend toward the construction of very large steam-electric stations, both fossil-fueled and nuclear, which can operate most efficiently at high plant factors (see note on p. 188), there is increasing need for plants designed specifically for peak-load operation. Because of their ability to start quickly and make rapid changes in power output, hydroelectric plants are well adapted for serving peak loads and providing reserve capacities in case of power failures. Many of the hydro plants in the European USSR operate in this

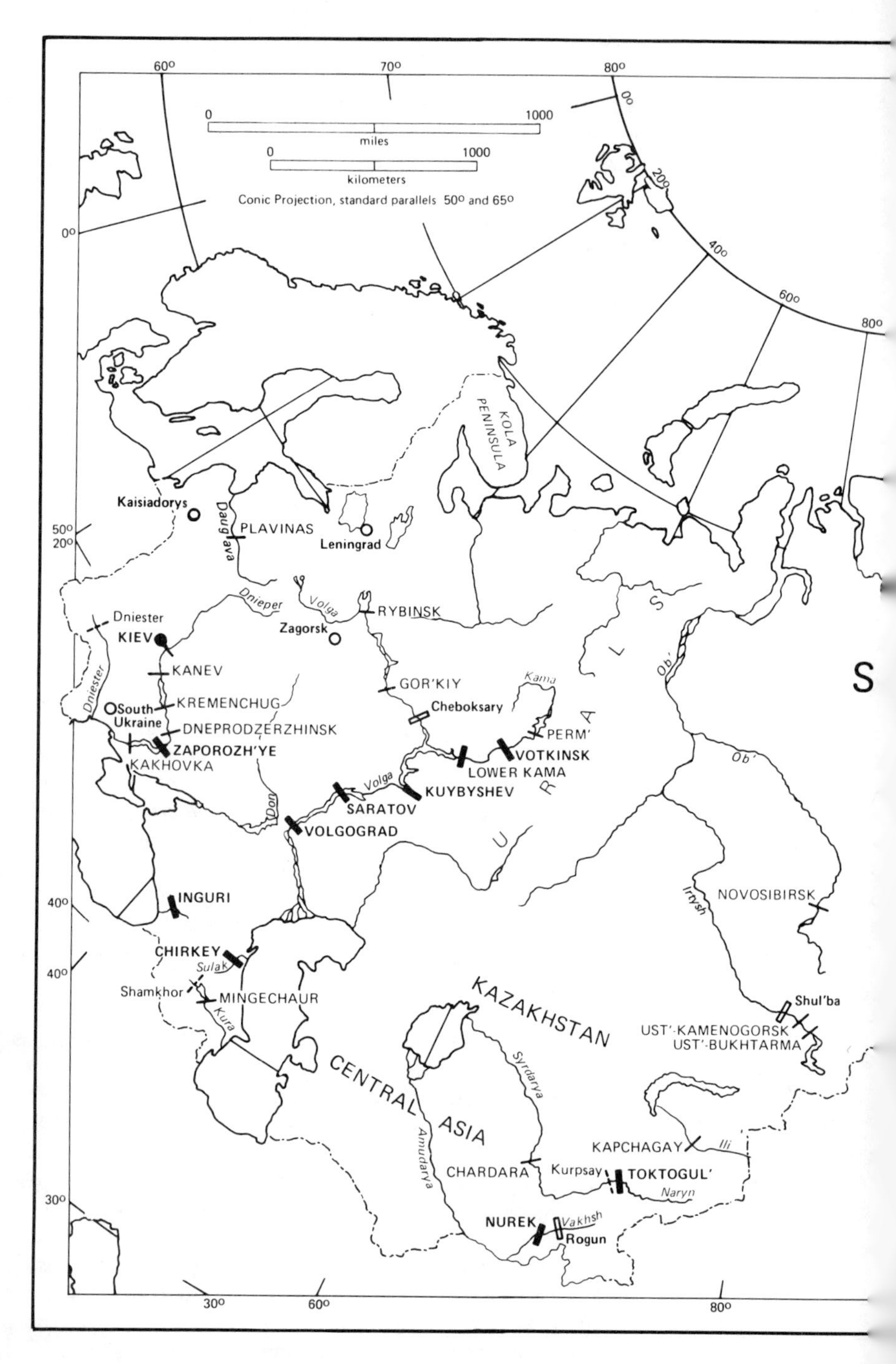

Fig. 7. Principal hydroelectric stations.

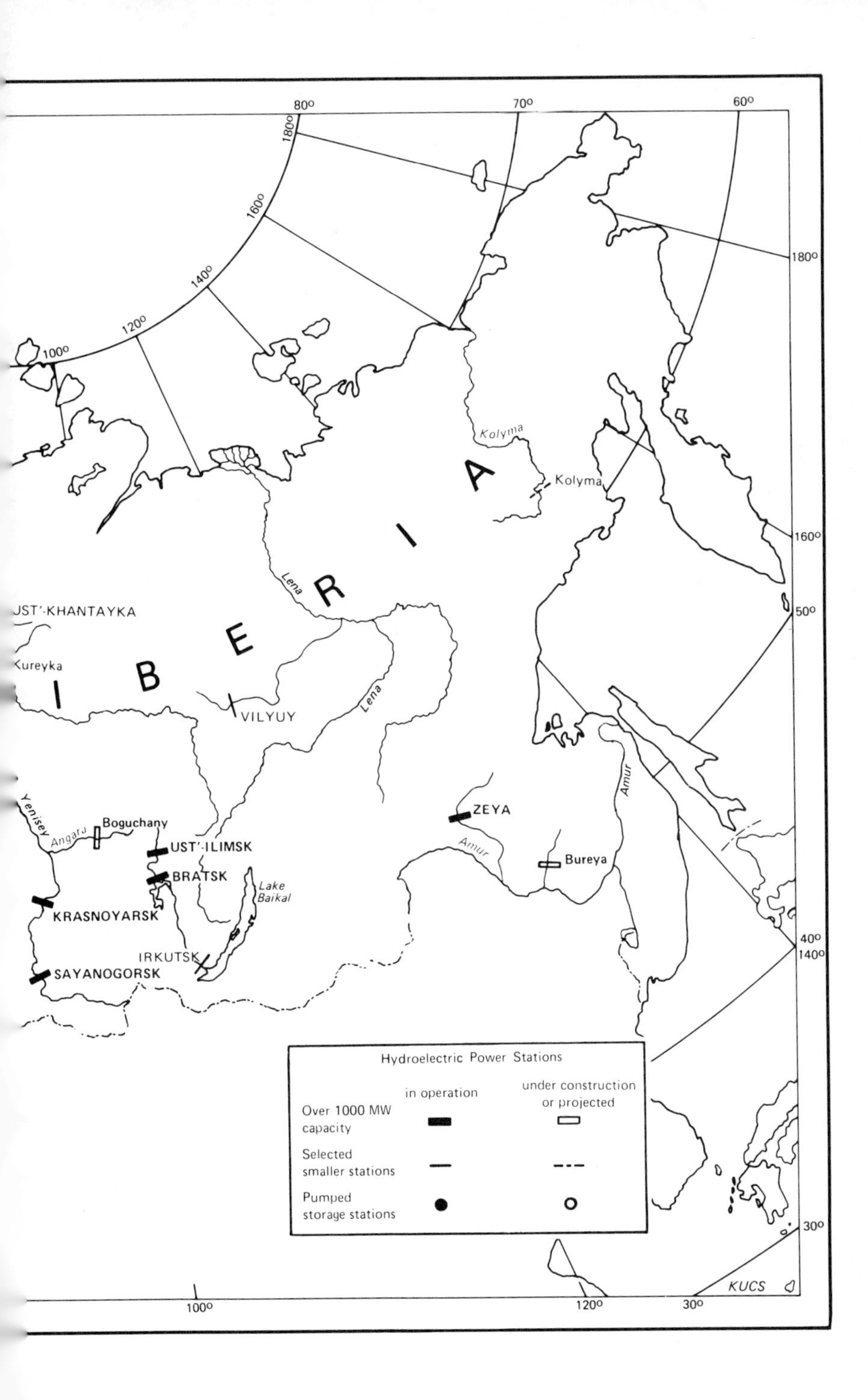

S I B E R I A
UST'-KHANTAYKA
Kureyka
Lena
VILYUY
Kolyma
Yenisey
Angara
Boguchany
UST'-ILIMSK
BRATSK
Lake Baikal
KRASNOYARSK
IRKUTSK
SAYANOGORSK
ZEYA
Amur
Bureya
Hydroelectric Power Stations
in operation
under construction or projected
Over 1000 MW capacity
Selected smaller stations
Pumped storage stations
KUCS

manner. In the Ukraine, for example, a highly industrialized Soviet republic where the construction of large steam-electric and nuclear stations has been given priority, these thermal stations operate about 5,500 hours a year (out of a theoretical maximum of 8,760 hours). These stations serve the base load, meaning the portion of the electric power output that tends to fluctuate relatively little. The Ukrainian hydro stations, located along the Dnieper River, have been operating about 2,500 hours a year, or only about half the time of the steam-electric plant operations, supplying the additional power needed at times of peak demand.[2]

In Siberia, by contrast, hydro capacity represents a larger portion of total installed capacity than in the European USSR; at the end of 1976, hydro capacity in the Siberian power system, covering West and East Siberia, represented 49% of the total installed capacity of 28,800 MW. The large Siberian hydro stations thus serve mainly the base load although they do set some generating units aside for peaking use. The availability of large amounts of waterpower, which tends to be one-sixth of the cost of steam-electric power, has attracted power-intensive industries to Siberia, particularly aluminum reduction plants, which are heavy users of electricity. The differences in use of hydro capacity in the European and Asian regions of the USSR are evident from Table 32, which shows that hydro stations in Siberia in 1975 operated an average of 3,900 hours in 1975 compared with 2,800 hours in the European regions (hours of operation can be derived by dividing output in kilowatt-hours by capacity in kilowatts).

Early Hydro Developments

The development of hydroelectric power in the 1920s and 1930s tended to be in small stations situated in easily developed sites in mountain areas or on streams where rapids provided a natural head for power generation (difference in level between the water surface above and below the dam); few stations exceeded 100 MW in generating capacity. The first stations were clustered in the Northwest, where streams in the Kola Peninsula and Karelia form rapids on the resistant rocks of the Fenno-Scandian Shield; in the Caucasus, both on the north slopes and in the Transcaucasian republics of Armenia and Georgia; in the Altay Mountains of eastern Kazakhstan, where waterpower served lead-zinc mines, and in Central Asia, in the vicinity of the city of Tashkent. Most of these areas were also distinguished by a shortage of fossil fuels for power generation.

From the very beginning, the cheap electricity produced by hydro stations attracted power-intensive industries like aluminum. The Volkhov station in the Northwest, east of Leningrad, was one of the first Soviet hydro plants to be put into operation, in 1926, with an installed capacity of 64 MW (eight units of 8 MW each), and was chosen as the site of the Soviet Union's first aluminum reduction plant, with a capacity of 11,000 tons of metal. The mean annual output of the Volkhov station was 350 million KWH and the power consumption of the aluminum plant, at the rate of 20,000 KWH per ton of aluminum, was 220 million KWH.

The largest of the early Soviet stations, far out of proportion to the others, was the first Dnieper River station at Zaporozh'ye, with a designed capacity of 560 MW. This hydro plant, which started power generation in 1932 and reached its designed capacity in 1939, was one of the world's largest at the time, and attracted power-intensive industries like aluminum, electric steels, ferroalloys and magnesium. During reconstruction after World War II, the Dnieper station was expanded to 650 MW by 1950. Because of the presence of power-intensive industries, the station was designed to meet base-load requirements and, in contrast to the Dnieper stations that were to follow, operated as much as 5,600 hours a year. Like other hydro plants along the Dnieper, this first station was a multipurpose facility, intended to improve navigation in a rapids-strewn segment of the river and to store water for irrigation in the semi-arid southern Ukraine.

Development of Dnieper Hydro Potential

In the 1950s began the further development of the hydroelectric potential of the Dnieper River, with the construction of five additional stations and the expansion of the original Dnieper station for peak-load purposes. These stations were designed to work in combination with the Ukraine's growing steam-electric generating capacity, which was serving base-load requirements. Four of the additional stations were situated upstream from the original Dnieper plant and one downstream. The upstream installations were Dneprodzerzhinsk, with an installed capacity of 352 MW and a start of power production in 1963; Kremenchug, with 625 MW, in 1959; Kanev, with 420 MW, in 1972, and Kiev, with 350 MW, in 1964. The conventional hydro plant near Kiev was associated with a pumped storage facility installed in the first half of the 1970s to enhance peaking capacity. (For separate discussion of pumped storage developments in the Soviet Union, see final section of this chapter.) The downstream hydro plant on the Dnieper River was the Kakhovka station, with 351 MW, put in initial operation in 1955. The Dnieper stations, like those built in the Volga-Kama drainage basin, differed from the early Soviet stations in being built in lowland areas, forming large reservoirs that flooded often valuable arable land and, because of the gentle gradient, created relatively small heads for power generation. By far the largest of the Dnieper reservoirs was the Kakhovka reservoir, with a total storage capacity of 18 billion m^3, including 6.8 billion m^3 of usable capacity; the water accumulated in the Kakhovka reservoir plays a particularly important role in irrigation of adjoining areas of the southern Ukraine, including the Crimea. (For some power-generating characteristics of the Dnieper River hydro stations and other major hydro plants in the Soviet Union, see Table 33; for location map, see Fig. 7.)

Volga River Hydro Development

Even as work began in the early 1950s on the Kakhovka dam, the first of the postwar Dnieper projects, construction also started on two of the largest Volga River hydro stations, near Kuybyshev and near Volgograd, each with more than

Table 33. Characteristics of Principal Hydroelectric Stations

Station name	Location	Installed capacity (MW)	Mean annual output (bill. KWH)	Construction: Start	Construction: First unit	Construction: Last unit
Dnieper Hydro System						
Kiev	Vyshgorod	350 (20x17.5)	0.635	1960	1965	1968
Kanev	Kanev	420 (24x17.5)	0.823	1963	1972	1975
Kremenchug	Svetlovodsk	625 (12x52.1)	1.506	1954	1959	1960
Dneprodzerzhinsk	Dneprodzerzhinsk	352 (8x43.8)	1.25	1956	1963	1964
Dnieper[a]	Zaporozh'ye	650 (9x72.2)	3.64	1944	1947	1950
Expansion	"	828 (8x103.5)	0.50	1969	1974	(1980)
Kakhovka	Novaya Kakhovka	351 (6x58.5)	1.42	1951	1955	1956
Volga–Kama Hydro System						
Ivan'kovo	Dubna	30 (2x15)	0.13	1932	1937	1938
Uglich	Uglich	110 (2x55)	0.24	1937	1940[b]	1947
Rybinsk	Rybinsk	330 (6x55)	1.1	1936	1941	1950
Gor'kiy	Zavolzh'ye	520 (8x65)	1.51	1948	1955	1957
Cheboksary	Novocheboksarsk	1404 (18x78.4)	3.5	1968	(1980)	
Volga (Lenin)	Zhigulevsk	2300 (20x115)	10.9	1950	1955	1957
Saratov	Balakovo	1290 (21x60 +2x45+1x10)	4.5	1955	1967	1970
Volga (22d Party Congress)	Volzhskiy	2530 (22x115)	11.1	1951	1958	1962
Kama	Perm'	504 (24x21)	1.7	1948	1954	1957
Votkinsk	Chaykovskiy	1000 (10x100)	2.3	1955	1961	1963
Lower Kama	Naberezhnyye Chelny	1248 (16x78)	2.7	1963	(1979)	
Others in European USSR						
Plavinas (Daugava R.)	Stucka	825 (10x82.5)	1.5	1961	1965	1966
Kegums (Daugava R.)	Kegums	68	0.39		1939	
Expansion	"	168 (3x56)	0.18	1970	(1979)	
Riga (Daugava R.)	Salaspils	384 (6x64)	0.7	1966	1974	1975
Narva (Narva R.)	Narva	125 (3x42.7)	0.7	1950	1955	1955
Tsimlyansk (Don R.)	Volgodonsk	200 (4x50)	0.7	1948	1952	1954
Dniester	Novodnestrovsk	702 (6x117)	0.83	1975	(1983)	
Caucasus						
Chirkey (Sulak R.)	Dubki (Dagestan)	1000 (4x250)	2.47	1963	1974	1976
Miatli (Sulak R.)	below Dubki	220		1971		
Irganay (Sulak R.)	above Dubki	800		1977		
Mingechaur (Kura R.)	Mingechaur	360 (6x60)	1.4	1946	1953	1954
Shamkhor (Kura R.)	Kyur	350 (4x87.5)	0.83	1975	(1981)	
Inguri	Dzhvari (Georgia)	1300 (5x260	5.4	1961	1978	
Angara–Yenisey Hydro System						
Irkutsk	Irkutsk	660 (8x82.5)	4.1	1950	1956	1958
Bratsk	Bratsk	4500 (18x250)[c]	22.6	1954	1961	1966
Ust'-Ilimsk	Ust'-Ilimsk	4320 (18x240)	21.9	1962	1974	1977[d]
Boguchany	Kodinskiy Zaimka	4100 (12x340)	17.8	1975	(1983)	
Sayan-Shushenskoye	Sayanogorsk	6400 (10x640)	23.5	1963	1978	
Krasnoyarsk	Divnogorsk	6000 (12x500)	20.4	1956	1967	1971
Others in Siberia						
Novosibirsk	Novosibirsk	400 (7x57.2)	1.68	1950	1957	1959
Ust'-Khantayka	Snezhnogorsk	441 (7x63)		1963	1970	1972
Kureyka	Svetlogorsk	500		1975	(1981)	
Vilyuy	Chernyshevskiy	308 (4x77)	0.9	1960	1967	1969
Expansion	"	340 (4x85)	0.9	1970	1975	1976
Zeya	Zeya	1290 (6x215)	4.9	1964	1975	
Bureya	Talakan	2000 (7x285)	10	1976	(1983)	
Kolyma	Sinegor'ye	750		1970	(1980)	

Table 33. (cont'd.)

Station name	Location	Installed capacity (MW)	Mean annual output (bill. KWH)	Construction		
				Start	First unit	Last unit
		Kazakhstan				
Bukhtarma (Irtysh R.)	Serebryansk	675 (9x75)	2.2	1953	1960	1966
Ust'-Kamenogorsk (Irtysh R.)	Ust'-Kamenogorsk	331 (4x82.8)	1.58	1939	1952	1959
Shul'ba (Irtysh R.)	Shul'binsk	1350 (6x225)	2	1976	(1981)	
Kapchagay (Ili R.)	Kapchagay	434 (4x108.5)	1.16	1965	1970	1971
		Central Asia				
Chardara (Syrdarya)	Chardara	100 (4x25)	0.54	1958	1965	1966
Charvak (Chirchik R.)	Charvak	600 (4x150)	2.0	1963	1970	1972
Toktogul (Naryn R.)	Kara-Kul'	1200 (4x300)	4.4	1962	1975	1975[e]
Kurpsay (Naryn R.)	below Kara-Kul	800 (4x200)		1975	(1981)	
Nurek (Vakhsh R.)	Nurek	2700 (9x300)	11.4	1961	1972	(1979)[f]
Rogun (Vakhsh R.)	Obigarm	3600 (6x600)	13	1976		

Notes:

[a]The Dnieper station at Zaporozh'ye was rated at 560 MW before World War II, when its first unit was installed in 1932 and the last in 1939; it was upgraded in the course of postwar reconstruction.

[b]The Uglich station was rebuilt in 1944, and the first unit restarted in 1945.

[c]The Bratsk generating units, originally rated at 225 MW, were upgraded to 250 MW during the 1970s.

[d]The Ust'-Ilimsk station, though designed for 18 units, was declared completed in December 1977 after installation of the 15th unit, but it was later announced that the three others would also be installed.

[e]The Toktogul generating units were largely idle in 1975–77 because of reservoir drawdowns for irrigation. Power generation began in 1978.

[f]The first three Nurek units, installed in the winter 1972–73, were largely idle in 1973 and 1974 because of low reservoir levels and delays in construction of the Tursunzade aluminum plant, the principal power consumer. Power generation began in 1976, and the installation of additional generating units resumed late that year.

Sources: Gidroenergetika i kompleksnoye ispol'zovaniye vodnykh resursov SSSR [Hydroelectricity and Multipurpose Use of Water Resources in the USSR] (Moscow: Energiya, 1970), pp. 312–317; *Elektrifikatsiya SSSR, 1967–1977* (Moscow: Energiya, 1977), p. 203, and scattered sources.

2,000 MW of installed capacity. In contrast to most hydro stations in the European USSR, the two plants were designed for high load factors, with around 4,500 hours of operation a year, primarily for the Moscow electric power system. This system, with rapidly growing electricity requirements in industry and households, was already getting the benefit of hydro power transmission from nearby stations on the upper reaches of the Volga, particularly the 330-MW Rybinsk station, completed in the late 1940s northeast of Moscow.

The construction of the two new stations on the middle and lower reaches of the Volga posed the problem of transmitting large blocks of electricity over distances of up to 650 miles to the Moscow area. The basis of a Soviet high-voltage grid was laid with the completion of transmission lines from the Kuybyshev station to Moscow in 1956 and from the Volgograd station to Moscow in 1959 as well as a third line running eastward from Kuybyshev to the Urals in 1959.[3] These early high-voltage lines, some of which first operated at 400 KV, were upgraded in 1961–62 to 500 KV, which became the standard voltage for Soviet intersystem connections, totaling 11,700 miles by the end of 1975. In addition to transmitting electricity over long-distance lines to the

Moscow system and to the Urals, the two large Volga stations also provided the basis for local power-intensive industries. In the case of the 2,300 MW Kuybyshev station, they included synthetic rubber, nitrogen and phosphorus based chemicals; at the 2,530 MW Volgograd station, they were synthetic rubber and fibers in the new city of Volzhskiy adjoining the power station as well as an aluminum reduction plant with a capacity of 200,000 tons in Volgograd itself.

The hydro capacity on the Volga River and its main tributary, the Kama, was further supplemented in the 1960s with the construction of the Saratov station (at Balakovo) with an installed generating capacity of 1,290 MW, and in the 1970s by the Lower Kama station (at the truck manufacturing city of Naberezhnyye Chelny); the first units of the Lower Kama station, with a designed capacity of 1,248 MW, were to start operations in 1979. The last major Volga station now under construction, the 1,404 MW plant in the Cheboksary area, is expected to be completed in the early 1980s. The original designs for the development of the Volga's waterpower potential envisaged two additional stations, the Perevoloki station near Kuybyshev, and the Lower Volga station below Volgograd. The Perevoloki station, with a tentative capacity of 2,400 MW, was reported being designed in the late 1960s,[4] but the Lower Volga project is expected to be shelved because it would flood the fertile floodplain formed by the lower reaches of the Volga and its arm, the Akhtuba River.

The Large Siberian Hydro Stations

The second half of the 1950s saw the start of the development of the vast hydroelectric resources of Siberia. In contrast to the hydro stations of the European USSR, whose peaking functions were to be combined increasingly with the base-load service of large thermal power stations, the hydro plants of Siberia were to contribute much of the continuous power supply of the region and work at high load factors. One of the first Siberian stations to be built was the Novosibirsk plant on the Ob' River, with an installed capacity of 400 MW completed by 1959. This was a modest station by Siberian standards, and it remained the only hydro development in West Siberia after tentative proposals for a huge station on the lower reaches of the Ob' River had been shelved.[5] West Siberia was to rely mainly on fossil-fueled thermal stations burning either the high-quality bituminous coal of the Kuznetsk Basin or the oil and gas of the newly developed fields of the Ob' basin.

Waterpower became a more important form of energy in East Siberia, where the Yenisey River and its main tributary, the Angara, offered the largest hydro potential of any drainage basin in the Soviet Union. The Angara is particularly well suited because it flows out of Lake Baikal, a deep lake that forms a natural storage reservoir regulating the stream flow for downstream hydro stations. Furthermore a steep gradient created a series of excellent sites with a considerable head for power generation. East Siberian hydro development began modestly with the construction of the 660 MW Irkutsk station, which was built at the same time as the Novosibirsk project in West Siberia. The Irkutsk station set the pattern for the hydro developments of the Angara-Yenisey system by attracting the first of a series of large aluminum reduction plants. But it also displayed a shortcoming of planning that was to be repeated in other Siberian projects–failure to coordinate the completion of hydro-generating capacities with the completion of the principal power consumers, the aluminum plants. The Irkutsk station, on the southeast outskirts of the city, reached its designed capacity in 1958; the aluminum plant, at Shelekhov, 10 miles southwest of the city, did not start operations until 1962.[6] It attained its designed capacity of about 240,000 tons only in the early 1970s.

Hydro development on the Angara River assumed "Siberian" magnitudes with the construction of the Bratsk station, farther downstream. Work on the 400-foot high dam began in 1954, and the first generator was in place in 1961. After a series of powerhouse design changes, planners appeared to have finally settled on 20 units of 225 MW each, for an aggregate generating capacity of 4,500 MW.[7] But when the No. 18 unit was installed in 1966, the planners called a halt. During the 1970s, each generating unit was upgraded from 225 to 250 KW, raising the station's combined capacity from 4,050 MW to 4,500 MW. The storage capacity provided by Lake Baikal and by the Bratsk dam's own huge reservoir were to insure a high plant factor, allowing at least 4,500 hours a year of operation even in dry years. However, the power utilization problems of Siberian hydro projects were to plague Bratsk in its early years. The Bratsk aluminum plant, with a designed capacity of 500,000 tons, did not begin production until 1966 when all Bratsk turbines had already been installed; the aluminum plant reached about one-half of its designed capacity in 1970 and was completed only in 1976, or a full 10 years after its power supply source had been built. In the first years of operation, the Bratsk station was generating only about 25 to 30% of its potential, transmitting its output over two 220 KV lines running 360 miles to Irkutsk. In 1963 these transmission lines were raised to 500 KV and, more importantly, a 500 KV line was completed westward to Krasnoyarsk, opening up a new market for Bratsk power. Utilization of the Bratsk generating potential reached 50 to 60%, or about 2,000 MW of the 3,600 MW installed by that time.[8] With the start of production of the aluminum plant in 1966 and the Soviet Union's largest woodpulp mill the following year, the utilization of hydro capacity improved. Over the years, an average of 40% has been consumed locally, 40 has been transmitted to the Irkutsk area and 20% has moved westward toward Krasnoyarsk.[9] The 500 KV lines based on Bratsk were to form the nucleus of a high-voltage Central Siberian grid that extended from Irkutsk westward to the Novosibirsk–Barnaul area by 1978, when it was interconnected through the North Kazakhstan grid with the unified power system of the European USSR.[10]

Bratsk, which lies 115 miles north of the Trans-Siberian Railroad, represented the first major northward penetration of the Siberian forest with a large hydro station and associated industries. In the early 1960s, workers and equipment were moved another 115 miles or so to the north to begin construction on the next Angara project, at Ust'Ilimsk. This station was designed virtually as a twin of the Bratsk plant, with 18 units of 240 MW each, or an aggregate capacity of 4,320 MW. As a striking departure from the rule in the development of the

Angara-Yenisey hydroelectric potential, no aluminum plant was envisaged for Ust'-Ilimsk. The only local industry was to be a 500,000-ton woodpulp mill, built through the joint efforts of the Comecon countries and scheduled for operation in 1979-80. The absence of a power-consuming industry like aluminum meant that most of the electricity generated by the Ust'-Ilimsk station, of the order of 21.9 billion KWH at full capacity, would have to be fed into the Central Siberian power grid. The first 500 KV link was in place when the first Ust'-Ilimsk turbines went into operation in 1974; a second line was completed in 1976, and a third was reported under construction in 1977.[11] The problem of overcapacity that had plagued the Bratsk project was also evident in the Soviet planners' apparent hesitancy over the ultimate parameters of the Ust'-Ilimsk station. Although the station design had called for 18 generating units, it was declared to have been completed and to have reached its designed capacity when the No. 15 unit was installed in October 1977.[12] Some planners evidently felt that hydro capacity was exceeding potential utilization. However, in early 1978, it was announced that the three remaining units would be installed after all.[13]

Both the Bratsk and the Ust'-Ilimsk hydro projects were the first such large industrial enterprises undertaken by the Russians so far north off the beaten path represented by the Trans-Siberian Railroad. In a pattern that was to be repeated elsewhere in undeveloped pioneering regions, the hydro stations and their associated economic activities gave rise to urban centers of considerable magnitude. Bratsk, industrially more diversified, has become the larger of the two, with a population that rose from 43,000 in 1959 (during the construction stage) to 155,000 in 1970 and 209,000 in early 1978. It has become the largest urban center in the heart of Siberia north of the Trans-Siberian mainline. Ust'-Ilimsk, of more recent creation and more limited industrial development, grew in population from 21,000 in 1970 (during the construction stage) to 53,000 in early 1977.

In the methodical step-by-step fashion that has characterized hydro development along the Angara, construction workers began moving their base of operations to the next station site of Boguchany as the last generating units were being installed at Ust'-Ilimsk. The first cross-country truck convoys traveling on seasonal winter roads delivered supplies from Bratsk to the Boguchany site in the winter of 1975-76, and work began in 1977 on a 160-mile 220 KV line to deliver Bratsk power to the Boguchany construction site. Although rail access had been considered essential in the development of both the Bratsk and Ust'-Ilimsk sites, there appeared to be controversy over the need to extend an existing timber railroad running north from the Trans-Siberian so that it would also serve the Boguchany project.[14] As an indication of progress, an urban settlement was formally inaugurated in April 1978 at the hydro site at Kodinskiy, approximately 70 miles east of the old Siberian village of Boguchany. Project designs called for a dam 300 feet high and 1.6 miles long, mostly earthfill except for a concrete middle section, and the start of power generation, in what may once again be an optimistic schedule, has been planned for 1983.[15] Tentative powerhouse designs call for 12 units totaling 4,100 MW.

While development thus proceeded on the Angara River, a set of stations of even larger capacities was being built on the Yenisey River. While the Angara projects ranged around 4,000 MW of station capacity, and unit capacities of about 250 KW, the Yenisey stations were designed for about 6,000 MW each, with generating units of 500 to 600 KW. The two sets of stations also differed in utilization patterns. The Angara stations were distinguished by high plant factors designed to meet base-load requirements, and operated as much as 5,000 hours a year. Hydro capacity in Irkutsk Oblast, which is served by the Angara stations, accounted for three-fourths of the total generating capacity of the Irkutsk power system at the end of 1975.[16] The hydro stations on the Yenisey were designed to work to a greater extent in conjunction with steam-electric plants of the adjoining Kuznetsk Basin and ultimately perhaps with the projected lignite-burning mine-mouth stations of the Kansk-Achinsk Basin. The Yenisey stations were therefore designed to perform at least some peaking functions, and were to operate an average of 3,500 hours a year (see the ratio of power output to capacity in Table 33).

The first Yenisey project was the 6,000 KW Krasnoyarsk station, on which construction began in 1956 in the western outskirts of Krasnoyarsk, where the new town of Divnogorsk arose 20 miles from the city. The first unit generated power in 1967, and the last in 1971. In this case, an aluminum plant, with a designed capacity of 450,000 tons, was not built next to the station because of the lack of industrial terrain, but on the northeast outskirts of Krasnoyarsk city, in the suburb of Zelenaya Roshcha. Contrary to the usual experience, the aluminum plant actually started production early, in 1964, before the Krasnoyarsk hydro station began to supply power, and used electricity both from the Nazarovo steam-electric station in the Kansk-Achinsk lignite basin and from the Bratsk hydro plant (over the newly completed 500 KV line). The Krasnoyarsk aluminum plant was approaching its designed capacity in 1978.[17] In view of the importance of navigation on the upper reaches of the Yenisey River, the Krasnoyarsk project was provided with an unusual ship elevator that went into operation in 1975.[18]

The dam construction workers had been rebased in the 1960s farther upstream to the site of the Sayan-Shushenskoye station, named for the Sayan Mountains and the nearby village of Shushenskoye, where Lenin once lived in Czarist exile, from 1897 to 1900. Work on the 6,400 MW project began in 1963, and the first unit was put in operation in late 1978. In an effort to obtain early production from the huge power project, the first generating unit began operation when the dam had reached a height of only 200 feet, one-fourth of the ultimate designed height of nearly 800 feet. The low initial head was likely to be used for peaking purposes. At any rate, the main base-load power consumer, a 500,000-ton aluminum plant, was once again far behind schedule. In contrast to the Krasnoyarsk project, which was built next to an established industrial center, the Sayan station is giving rise to an entirely new, diversified industrial complex. In addition to aluminum production, which will be located at the new city of Sayanogorsk just below the hydro station, the Sayan complex will include a car-building and container-manufacturing plant, already in operation on the northern outskirts of Abakan; and an electrical equipment manufacturing complex at Zelenyy Bor, on the southern outskirts of Minusinsk. Titanium, ferroalloys and phosphorus-based chemicals have also been mentioned among power-intensive industries that may be attracted to the area.[19]

Looking far into the future, Soviet planners envisage three more hydro stations along the Yenisey River: the Middle Yenisey station, with a tentative generating capacity of 6,400 MW, near the confluence of the Yenisey and Angara rivers; the Osinovo station (6,100 MW), near the mouth of the Podkamennaya Tunguska (Stony Tunguska), and the 5,000 MW Igarka station, on the lower Yenisey. While the Middle Yenisey hydro plant is a realistic possibility for the 1980s,[20] after the completion of the Sayan project, the other stations are a more distant prospect. Their construction is tied in with the whole question of the future economic development of the northern portion of Krasnoyarsk Kray, specifically the lower Yenisey valley.[21]

A number of Siberian hydro stations are associated with isolated economic developments outside the Angara-Yenisey system (Table 33). The Ust'-Khantayka station, put into operation in the early 1970s with 441 MW, supplies electricity to the growing nonferrous complex in the Noril'sk area of northern Siberia, and a second hydro station on the Kureyka River is under development. The Vilyuy station, built originally in the 1960s and expanded in the middle 1970s, serves the diamond-mining industry of western Yakutia. The electric power needs of the economic development zone along the Baikal-Amur Mainline are expected to be met in part by the Zeya hydro plant, where the first 215 MW unit of a designed capacity of 1,290 MW, began to produce power in 1975, and by the Bureya hydro plant, where construction was getting under way in the late 1970s. The Zeya and Bureya dams would also perform important flood-control functions in the Amur basin. The remote Kolyma hydro station, under construction in the 1970s, will improve the power supply of gold and tungsten mining operations in extreme northeastern Siberia.

Hydro Development in Central Asia

Until the 1970s, Central Asia was not a significant factor in Soviet hydro development. A number of small stations had been built since the 1930s, some as large as 100 MW in capacity, but total hydro capacity was only 1,400 MW, compared with 5,800 MW of steam-electric capacity. Only in the mountainous Tadzhik SSR did hydro capacity exceed thermal capacity, with 70% of the total Tadzhik power-generating capacity.

The start of construction in the early 1960s of two large hydro stations–Nurek on the Vakhsh River in the Tadzhik SSR and Toktogul on the Naryn River in the Kirghiz SSR–was intended to develop the Central Asian waterpower potential on a large scale. The Nurek station, in particular, was designed to form the power-supply base for a projected new industrial complex in southern Tadzhikistan that was to include aluminum reduction and power-intensive chemical industries. The two stations were to be connected by a 500 KV grid with large gas-burning thermal electric stations in Uzbekistan to form a Central Asian high-voltage system extending from Alma-Ata and Frunze in the east through the Tashkent area to Turkmenia in the west. Nurek, was designed for a relatively high load factor, running about 4,200 hours a year, because of the projected nearby location of power-intensive industries; Toktogul, with an average of 3,600 hours, was intended to perform more of a peaking function in conjunction with the base-load service of the steam-electric stations. Nurek placed its first three 300 MW units in service in the winter of 1972-73, and Toktogul installed all four of its 300 MW units in 1975, but unforeseen circumstances delayed the full operation of both stations.

In the case of Toktogul, the interests of cotton irrigation outweighed the electric power interests in a series of dry years (1975-77), requiring reservoir drawdowns so that the water level could not rise to provide the head needed for power generation. The installed capacity of 1,200 MW at Toktogul, representing one-half of the total power generating capacity of the Kirghiz SSR, was thus forced to remain idle for three whole years.[22] It was only by the beginning of 1978 that the reservoir filled sufficiently to allow the start of power generation, and electricity began flowing through a 500 KV line that had been ready since 1975 through the Fergana Valley to the gas-fueled Syrdarya steam-electric station at Shirin, at the western valley exit.[23] As a result of the infusion of Toktogul hydropower, electricity generation in the Kirghiz SSR jumped from 4.9 billion KWH in 1977 to more than 6 billion in 1978. Pending the completion of a 500 KV link across the Tien Shan mountains from the Toktogul station to the Frunze area, where most of the Kirghiz consumption is concentrated, Toktogul power is being transmitted into Uzbekistan's 500 KV grid. When the transmission line to Frunze is completed, Toktogul will not only be able to meet Kirghizia's own needs, but turn the republic from a net importer to a net exporter of electricity.

The Toktogul project is being followed by the construction of the 800 MW Kurpsay hydro station, about 20 miles farther downstream on the Naryn River, where the first 200 MW units had been planned for the end of the 10th five-year plan (1976-80). The next project on the Naryn, which offers a number of favorable sites, is expected to be the 2,000 MW Kambarata station, just upstream from the Toktogul reservoir.[24] In the case of the 1,000-foot high Kambarata dam, Soviet designers have been experimenting with a time-saving construction method, in which rock fill from the valley wall is to be laid down across the river in a gigantic directed blast.[24] The technique was reported to have been successfully modeled in a test explosion carried out in February 1975 on the Burlykiya, a small nearby tributary of the Naryn.

In Tadzhikistan, the 2,700 MW Nurek project on the Vakhsh River acquired its first three 300 MW generating units in the winter of 1972-73. By that time, the earthfill dam, the highest of that type in the world, had reached only one-third of its designed height of 1,000 feet, and the first generating units were able to work only at reduced capacity because of the low head. In any event, the main consumers of Nurek power—the Tursunzade aluminum plant and the Yavan caustic—chlorine complex—were not ready, and the station had not been connected with the Central Asian 500 KV grid. In addition there were drawdowns of irrigation water from the Nurek reservoir during the dry years of the middle 1970s. As a result, the 900 MW of installed Nurek capacity, representing one-half of the total power-generating capacity of the Tadzhik SSR, operated only to a limited extent for about three years. The first five sections of the aluminum plant finally were put into production in 1975-76 and a 500 KV line from Nurek to Tursunzade was completed in August 1976, enabling the hydro station to deliver its power output directly to the aluminum plant. At the same time, the installation of generating units resumed in late 1976. The Nurek hydro station was finally coming into its own. In 1978, when the second major power consumer, the Yavan chemical complex, also began operations with a calcium hypochlorite unit, the Nurek station generated close to 5 billion KWH of electricity, nearly one-half of its projected long-term average output of 11.4 billion.

As in the Siberian hydro projects, workers began to be transferred from the Nurek station in 1976 to start preliminary groundbreaking on the site of the next large Vakhsh River station, the Rogun project, just above the Nurek reservoir. Also a rockfill dam, the Rogun project, with 1,130 feet, will be even higher than Nurek, and is designed to include six 600 MW generating units, for a total capacity of 3,600 MW.[25]

While the large hydro stations of Siberia and Central Asia are probably of the greatest interest in a discussion of the Soviet energy supply, mention should also be made of two additional areas with a significant hydroelectric potential—the upper Irtysh River in eastern Kazakhstan and the Caucasus (Table 33).

On the Irtysh River, two stations had been built at an earlier stage of Soviet hydro development to improve the power supply of a crucial mineral industries region that accounts for substantial shares of Soviet nonferrous metals (zinc, lead and coproducts) as well as titanium and magnesium. The 331 MW station of Ust'-Kamenogorsk, the center of the mineral industry, was put in operation in the 1950s, and was followed by the 675 MW Bukhtarma station, farther upstream, in the first half of the 1960s. These hydro products were designed mainly for power production, and their generation regimes and reservoir drawdowns were geared to the needs of the electric power industry, to the detriment of irrigation, navigation and other water needs in the lower reaches of the Irtysh River. The need for a reregulation project farther downstream gave rise to plans for construction of the Shul'ba station, with a designed capacity of 1,350 MW. Unlike the two other Irtysh hydro plants, the Shul'ba station will only incidentally generate electric power—its average use is projected at 1,300 to 1,500 hours a year. Its main purpose will be the storage of stream flow for release as needed by downstream users. Construction of the project began in 1976, and completion is expected in the early 1980s.

In the Caucasus, the most significant developments have taken place on the Sulak River in the Dagestan ASSR, the Kura River in Azerbaijan, and the Inguri River in western Georgia (Table 33). The Sulak hydro series, including the 1,000 MW Chirkey plant, is designed to supply peaking capacity for the North Caucasus system, which relies mainly on thermal power. Peaking needs also weigh heavily in the case of the hydro station in Azerbaijan, where oil-burning steam-electric plants prevail. In Georgia, however, the large Inguri station, with 1,640 MW of installed capacity, is projected as a more significant producer, with more than 4,000 hours of operation a year. Previous hydro stations in Georgia, with small reservoirs, have been suitable only for peak loads. The Inguri project, with the world's highest concrete arch dam, at 890 feet, has been under construction for what may well have been a record period. Work on the project began in 1961, and the generation of first power was originally planned for 1970 before completion dates were continuously put off. In the meantime, the original cost estimate of 213 million rubles more than tripled, to 731 million rubles. The first units of the main 1,300 MW powerhouse finally started operation in late 1978. An additional generating capacity of 340 MW had been installed previously on downstream derivation canals, in the early 1970s.[26]

An unusual episode in Soviet hydroelectric development occurred in another Transcaucasian republic, Armenia, where a whole series of hydro stations, totalling 550 KW in capacity, has been virtually shut down because of unforeseen environmental damage. The stations were built on the Razdan River, the outlet of Lake Sevan, and by 1960 provided nearly all of Armenia's electric power output of 2.7 billion KWH. However, the enhanced outflow from Lake Sevan for power generation was found to cause a drop in the lake level and environmental deterioration, such as a lowering of the watertable and increased erosion, along the shoreline. It was therefore decided to shift the basis of Armenian power supply from the installed hydro capacity to projected gas-burning steam-electric stations, and to replenish the waters of Lake Sevan by building a 30-mile tunnel from the nearby Arpa River to the lake. Outflow from the lake was to be limited to irrigation needs. By the middle 1970s, 90% of Armenia's electricity was produced by the newly built thermal stations, and hydro capacity was used only incidentally as a byproduct of irrigation needs.[27] The republic's power supply was further enhanced in 1976 with the start of power generation at a nuclear plant, and the Arpa-Sevan derivation tunnel was expected to start carrying water to the depleted lake in 1979, after nearly two decades of work.

Pumped Storage Developments

The growing need for electric power plants designed for peak-load operation and the scarcity of suitable hydroelectric sites in the European USSR, where most of the electricity is consumed, have focused interest in the Soviet Union on the potentialities of pumped storage stations. Such power projects, operating essentially as large storage batteries, use off-peak surplus energy to pump water from a lower to an upper reservoir, where it is stored and subsequently released through the generating plant for peak-load use. Specially designed reversible turbines serve both as pumping and generating units. Pumped storage plants, like conventional hydroelectric stations, are well adapted for peaking and reserve use because of their ability to start up quickly and make rapid changes in power output.

The Soviet Union, feeling itself secure in what were thought to be ample supplies of conventional energy, was a latecomer in the pumped storage technique, which had gained wide acceptance in Western Europe, the United

States and Japan. The Soviet hydroelectric design agency Gidroproyekt has identified about a hundred possible sites, of which 15 have been chosen for design studies. These priority sites are situated near peak-load centers and are said to have a combined generating potential of 15,000 to 20,000 MW.[28]

The first Soviet pumped storage installation, considered experimental by planners, was built in conjunction with the 350 MW Kiev hydroelectric station on the Dnieper River, which reached its designed capacity in 1968. Taking advantage of a suitable site for an upper reservoir on the high right bank of the Dnieper north of Kiev, the Soviet Union added a 225 MW pumped storage facility in the early 1970s, for a combined installed capacity of 575 MW in the Kiev project. This experimental project was said to have proved itself by enhancing the maneuverability of the power system at times of abrupt increases in load.

In 1976, the Russians began the construction of the first large pumped storage installation to serve the needs of the Moscow power system, with its heavy load fluctuations. This station, situated 10 miles north of Zagorsk, will have a generating capacity of 1,200 MW, consisting of six 200 MW reversible generating-pumping units. A suitable site, with a level difference of about 300 feet between lower and upper reservoirs, has been identified on the Kun'ya River (a left tributary of the Dubna) near the village of Bogorodskoye. The station, tentatively planned for power production in 1980, would feed its output into the Moscow system over two 500 KV transmission lines.[30]

Design studies appear to be well advanced on at least three other pumped storage stations. They are the 1,600 MW Kaišiadorys station in Lithuania; a 1,300 MW station in eastern Leningrad Oblast, both serving the Northwest power system, and a 2,000 MW station in the Dniester River valley, possibly in the Dubossary area, to serve the Southern power system of the Ukraine and Moldavia. The Kaišiadorys installation would be situated near the village of Kruonis, at the confluence of the Streva and Nemunas (Niemen) rivers, using the reservoir of the Kaunas hydro station (built in the late 1950s) as the lower basin and adding an upper reservoir to create a head of 300 feet. The projected station in eastern Leningrad Oblast is to be located on the lower course of the Shapsha River, at the village of Ratigora, 30 miles southeast of Lodeynoye Pole.[32] It will be equipped with eight reversible units of 165 MW each.

An unusual pumped storage project is planned to be part of a combined nuclear–hydroelectric–pumped storage complex in the Ukraine, on the Southern Bug River, with a combined capacity of 6,200 KW. It would consist of the 4,000 South Ukrainian nuclear station, currently under construction; the 1,800 MW Tashlyk hydro station, on a tributary of the Southern Bug, and the Konstantinovka combined hydro and pumped storage station, with 400 MW capacity, on the Southern Bug. The clustering of these interconnected operations, which would yield about 30 billion KWH at full capacity, would make it possible to use the Tashlyk reservoir as a cooling pond for the heated water discharged by the nuclear station; at the same time the heating of water would facilitate winter operation of the pumped storage system.[33]

NOTES

[1] *Energetika SSSR v 1971-1975 godakh* [The Electric Power Industry of the USSR in 1971-75] (Moscow: Energiya, 1972), pp. 144-147; *Energetika SSSR v 1976-1980 godakh* (Moscow: Energiya, 1977), p. 129.

[2] *Narodnoye khozyaystvo Ukrainskoy SSR. Yubileynyy statisticheskiy yezhegodnik* [The Economy of the Ukrainian SSR. Anniversary Statistical Yearbook] (Kiev: Tekhnika, 1977), p. 112.

[3] *Elektrifikatsiya SSSR*, edited by P. S. Neporozhnyy (Moscow: Energiya, 1970), p. 334.

[4] *Sovetskaya Rossiya*, July 17, 1968.

[5] For a discussion of the pros and cons of the Lower Ob' hydro project, see: *Kompleksnoye osvoyeniye vodnykh resursov Obskogo basseyna* [Multipurpose Development of the Water Resources of the Ob' Basin] (Novosibirsk: Nauka, 1970), 255 pp.

[6] Theodore Shabad and Victor L. Mote, *Gateway to Siberian Resources (the BAM)* (New York: Halsted Press, 1977), pp. 28-29.

[7] *Sovetskaya Rossiya*, Aug. 19, 1959.

[8] *Voprosy Ekonomiki*, 1966, No. 8, p. 55.

[9] *Sovetskaya Rossiya*, Sept. 1, 1975.

[10] *Sotsialisticheskaya Industriya*, July 15, 1978, reported the completion of the last 180 km link between Barnaul and Novosibirsk to link up the 500 KV grids of Central Siberia and North Kazakhstan–European USSR.

[11] *Pravda Ukrainy*, Feb. 13, 1977.

[12] *Pravda*, Oct. 20, 1977; *Soviet Geography*, January 1978, p. 67.

[13] *Stroitel'naya Gazeta*, Jan. 25, 1978; *Soviet Geography*, April 1978, p. 285.

[14] *Lesnaya Promyshlennost'*, Jan. 8, 1976, and June 24, 1976.

[15] *Turkmenskaya Iskra*, Jan. 8, 1976; *Stroitel'naya Gazeta*, March 13, 1977.

[16] *Soviet Geography*, February 1977, p. 133.

[17] *Stroitel'naya Gazeta*, Jan. 18, 1978, and July 16, 1978.

[18] *Sotsialisticheskaya Industriya*, July 29, 1975.

[19] V. V. Sokolikova. *Sayanskiy narodnokhozyaystvennyy kompleks* [The Sayan Economic Complex] (Moscow: Mysl', 1974), 207 pp.

[20] Design studies for the Middle Yenisey hydro station were reported under way in 1976 (*Sotsialisticheskaya Industriya*, April 21, 1976).

[21] V. M. Myakinenkov. "Prospects of development of production and settlement in the northern Yenisey region," *Soviet Geography*, November 1975, pp. 579-583; also *Problemy rasseleniya v rayonakh Severa* [Problems of Settlement in the Regions of the North], edited by V. M. Myakinenkov (Leningrad: Stroyizdat, 1977), 224 pp., particularly pp. 86-224.

[22] *Izvestiya*, Dec. 17, 1977.

[23] *Soviet Geography*, April 1976, p. 285.

[24] *Sovetskaya Kirgiziya*, Dec. 22, 1974; *Gidrotekhnicheskoye Stroitel'stvo*, 1977, No. 5, pp. 15-16, 18-20.

[25] *Izvestiya*, Sept. 28, 1976; *Kommunist Tadzhikistana*, Sept. 28, 1976, and Dec. 22, 1976.

[26] *Soviet Geography*, March 1973, pp. 663, 664; March 1976, 212; April 1977; pp. 273-274.

[27] *Soviet Geography*, March 1970, pp. 218-221; March 1977, pp. 206-207.

[28] *Gidroenergetika i kompleksnoye ispol'zovaniye vodnykh resursov SSSR* [Hydroelectricity and the Multipurpose Use of Water Resources in the USSR], edited by P. S. Neporozhnyy (Moscow: Energiya, 1970), pp. 250-251.

[29] *Elektrifikatsiya SSSR, 1967-1977.* Moscow: Energiya, 1977, p. 217.

[30] *Soviet Geography*, November 1972, p. 648; *Sovetskaya Rossiya*, April 16, 1976.
[31] *Sovetskaya Litva*, Sept. 27, 1974, and Dec. 25, 1975.
[32] *Sovetskaya Rossiya*, April 24, 1978.
[33] *Elektrifikatsiya SSSR*, 1967–1977, op. cit., p. 218; *Gidrotekhnicheskoye Stroitel'stvo*, 1978, No. 5.

Chapter 6

NUCLEAR POWER*

On June 27, 1954, the USSR received electricity from the first nuclear electric power reactor. In the ensuing quarter century, the Soviet Union has assigned a high priority to all forms of nuclear research as a necessary part of maintaining its role as a superpower. Soviet planners expect to rely heavily on nuclear electricity to provide a growing percentage of their electrical needs for the future, and the USSR is moving resolutely ahead in the development of all major forms of atomic power generation.

Commercial Reactor Development through 1978

At the conclusion of the 9th Five-Year Plan, ending in 1975, the Soviet Union had installed 6,200 MW(e) of nuclear generating capacity.[1] This represented about 3% of the total electrical generating capacity available at that time (Table 34). The rate of growth of nuclear power was greater in the first half of the 1970s than either thermal (fossil fuel) or hydroelectric power. In part this rapid rate of growth can be explained statistically from having started from a smaller base, but future additions to installed nuclear capacity are expected to increase at a similar rapid rate.

The pace of Soviet nuclear energy development during the 1970s was greater than in previous years. In the United States, development during the 1960s and

*The discussion on nuclear power is by Philip R. Pryde of the Department of Geography, San Diego State University.

Table 34. Installed Nuclear Electrical Generating Capacity (thou. megawatts)

Installed capacity	1965	1970	1975	1980 plan
Total capacity[a]	115.0	166.1	217.5	284
Nuclear capacity:				
Utility stations[b]	0.31	0.88	4.7	18.4
Total civilian stations[c]	0.91	1.6	5.5	
Highest reported capacity[d]			6.2	19.4

Notes and Sources:

[a]Total capacity of all types of Soviet power stations is from: *Nar. khoz. SSSR*, various issues.

[b]Soviet reporting of total nuclear capacity has been ambiguous because of the existence of a secret "Siberian" facility, believed to produce materials for weapons programs. Commonly, Soviet reporting of nuclear capacity is limited to utility stations, which are under the administration of the Ministry of Electric Power and are connected to regional power grids and systems. This category of stations excludes special-purpose stations, such as the "Siberian" station, various small experimental stations, and a dual-purpose station at Shevchenko (Kazakhstan) used both for power generation and for water desalination. Data for utility stations are from: *Elektricheskiye Stantsii*, 1977, No. 12, p. 6, for 1965; and *Energetika SSSR v 1976-1980 godakh* (Moscow: Energiya, 1977), pp. 11, 61, 111, 220. The utility series is consistent with summary data on capacity additions (ibid, p. 220): 3,800 MW added in 1971-75, and 13,800 planned to be added in 1976-80.

[c]Based on the table of individually reported capacity additions (Table 38).

[d]A preliminary 1975 figure (about 6,000 MW) was reported in *Elektricheskiye Stantsii*, 1977, No. 1, p. 3; a subsequent 1975 figure (6,200 MW) was reported in *Elektricheskiye Stantsii*, 1977, No. 6, p. 3, which also gives the planned 1980 capacity of 19,400 MW, suggesting planned additions totaling 13,200 MW. These figures apparently include the weapons-oriented Siberian facility, variously described in official Soviet sources as having "reached," "exceeded" and "substantially exceeded" 600 MW. For some recent Soviet sources giving all three formulations, see: A. M. Petros'yants, *Atomnaya energetika* (Moscow: Nauka, 1976), p. 21 ("has now reached 600 MW"); *Atomnaya nauka i tekhnika v SSSR* (Moscow: Atomizdat, 1977), p. 30 ("exceeded 600 MW by 1964"); A. M. Petros'yants, *Sovremennyye problemy atomnoy nauki i tekhniki v SSSR* (Moscow: Atomizdat, 1976), p. 123 ("now substantially exceeds 600 MW").

early 1970s was very rapid, with a decided slowing in the mid-1970s because of economic and environmental factors. In the Soviet Union, the pace of construction was fairly slow in the 1960, but accelerated markedly during the 1970s, especially the second half (Table 35).

The first Soviet reactor, which was started up at the Obninsk research center in 1954, was a small water-cooled, graphite moderated reactor with an output of 5 MW(e) or 30 MW(th). Since 1954, Obninsk has remained essentially a research center; there are no large-scale commercial reactors at this location. Several other cities have research reactors, the largest of which is probably the 70 MW reactor at Dimitrovgrad (formerly Melekess).

During the late 1950s, a large facility, officially designated simply "Siberian," was built by the Russians. Little information is available on this station, probably because it is associated with the production of fissile material for weapons. United States lists of foreign reactors have given the location of the "Siberian" plant as "Troitsk." The only significant known place by that name is an industrial center in the Urals, 75 miles south of Chelyabinsk, that is not generally considered to be part of Siberia. Soviet sources have not identified the location of the Siberian plant, whose reactors are of the water-cooled, graphite moderated, channel type. The first 100-MW unit was put in operation at the Siberian plant in 1958, and an installed capacity of 600 MW was reported to have been achieved by 1964. The present number of units and total installed capacity are uncertain (see Note d, Table 34).

Until 1972 there were only two major commercial atomic power complexes in the USSR (aside from the Siberian facility). These were located at Beloyarskiy in the Urals, with two reactors, and at Novovoronezhskiy, in Central Russia, where there were four. At Beloyarskiy, a town 30 miles east of Sverdlovsk, there are two boiling water-superheat, graphite-moderated, channel-type reactors now on line. A third reactor is now under construction at this site; as it is of the breeder type it will be discussed below with other breeder facilities. Some steam from these plants is used for heating in nearby communities, and a reservoir has been

Table 35. Nuclear Power Production

Year	Installed capacity (end of year)			Power generation		
	Total (thou. MW)	Nuclear[a] (thou. MW)	Nuclear[a] (%)	Total (bill. KWH)	Nuclear[a] (bill. KWH)	Nuclear[a] (%)
1960	66.7	–		292	–	
1965	115	0.31	0.27	507	1.4	0.28
1970	166	0.88	0.53	741	3.5	0.47
1975 plan	228	8.1	3.5	1070		
1975[b]	217.5	4.7	2.2	1039	20.2	1.94
1976	228.3	5.7	2.0	1111		
1977	238	7.1	3.0	1150	34	3.0
1978		8.1		1195		
1980 plan	284	18.4	6.5	1380	80	5.8

Notes:

[a]Nuclear capacity and power production figures are for utility stations only (see Table 34).

[b]Actual.

Sources: Energetika SSSR v 1971-1975 godakh (Moscow: Energiya, 1972), pp. 92, 217; *Energetika SSSR v 1976-1980 godakh* (Moscow: Energiya, 1977), pp. 11, 61, 220; *Elektricheskiye Stantsii*, 1977, No. 12, p. 6.

Table 36. Principal Characteristics of Water-Moderated, Water-Cooled Reactors

Characteristics	Novovoronezhskiy-3 VVER-440 (1971)	Novovoronezhskiy-5 VVER-1000 (1979)
Power		
MW (electrical)	440	1,000
MW (thermal)	1,375	3,000
Fuel		
type	Uranium dioxide pellets	Uranium dioxide pellets
enrichment	3.5%	3.3–4.4%
cladding	zirconium, 1% niobium subassembly cladding is zirconium, 2½% niobium	same
average burnup, MW days/ton	28,600	26,000–40,000
Reactor efficiency	32%	33%
Coolant		
type	water	water
inlet/outlet, °C	269/300	289/324
Moderator	water	water
Steam pressure at turbine, atmosphere	44	60
Safety considerations	37 control rods of boron steel.	109 control rods of boron steel.

Source: I. D. Morokhov et al., *Atomnoy energetike XX let* (Moscow: Atomizdat, 1974), p. 68.

constructed for cooling purposes. Apart from the breeder reactor now under construction, no additional units are apparently envisioned for this site in the foreseeable future.[3]

Novovoronezhskiy, located about 30 miles south of the city of Voronezh, on the Don River, is one of the largest atomic-energy producing complexes in the country, with four reactors now on line having a combined capacity of 1,455 MW(e). All four are of the pressurized water type, using water both for cooling and for moderating. The first unit became operational in 1964, with an initial output of 210 MW(e), and the second unit, of 365 MW(e), achieved criticality in 1969. Both of the first two units use water withdrawn from the Don River for condenser cooling. The average resulting increase in river temperature is not known, but the allowable downstream standards are a 5°C increase in the summer and 3°C in the winter. The Don at Novovoronezhskiy is not large, except in spring floods. The flow of the river at this point is normally well under 5,000 cubic feet per second, and has been as low as 600 cfs.[4]

The third and fourth units at Novovoronezhskiy are identical 440 MW(e) water-cooled, water-moderated reactors. They are referred to as VVER-type reactors (for the Russian words for "water-water power reactor"–in the West they are referred to as PWR or "pressurized water reactors"), and VVER reactors have been designated as one of two standard types now being manufactured in the Soviet Union. Seven cooling towers have been built in conjunction with Units 3 and 4. Unit 3 went on line in 1971, and Unit 4 in December 1972. A fifth pressurized-water reactor of 1,000 MW(e) capacity was to go on line in 1979, bringing total output at this location to about 2,500 MW(e). Unit 5 will have a 600-hectare cooling pond constructed for it. Design characteristics of the two standard types of water-cooled, water-moderated reactors are presented in Table 36.

VVER-440 reactors of the Novovoronezhskiy type have been constructed or are planned at three other sites in the USSR. One reactor of this type has been built near the town of Metsamor, about 25 miles west of Yerevan in Armenia, with a second unit scheduled for 1979. (Because of specific cooling conditions, the Armenian units are rated at 405 to 410 MW(e).) Two others have been

Table 37. Principal Characteristics of Graphite-Moderated Reactors

Characteristics	Beloyarskiy-2 (1968)	Leningrad-1 (RBMK-1000) (1973)
Power		
MW (electrical)	200	1,000
MW (thermal)	530	3,200
Fuel		
type	998 fuel channels (uranium alloyed with molybdenum)	1693 channels (no nuclear superheated channels)[a]
enrichment	2% and 3%	1.8%
average burnup, MW days/ton	12–16,000	18,500
Coolant		
type	water	water
inlet/outlet, °C	300/330	270/?
Moderator	graphite	graphite
Steam pressure at turbine,atmospheres	76	70
temperature, °C	500	284[a]
Safety considerations	78 control rods	?

[a]The next generation reactors of channel type, RBMKP-2000 or RBMKP-2400, will again utilize nuclear superheating as in Beloyarskiy-2.

Source: I. D. Morokhov, et al., *Atomnoy energetike XX let* (Moscow: Atomizdat, 1974), pp. 82–95.

Table 38. Nuclear Power Reactors in the USSR[a]

Station name and reactor number	Station site	Reactor designation[b]	First power output	Commercial MW (f)
Obninsk	Obninsk	AM-1	1954	5[d]
Siberian	Troitsk (?)	Graphite	1958[c]	600[c]
Obninsk	Obninsk	BR-5	1959	5[d]
Beloyarskiy-1	Zarechnyy	AMB-1	1964	100
Novovoronezhskiy-1	Novovoronezhskiy	VVER	1964	210[e]
Dimitrovgrad	Dimitrovgrad	VK-50	1965	50[d, e]
Beloyarskiy-2	Zarechnyy	AMB-2	1967	200
Novovoronezhskiy-2	Novovoronezhskiy	VVER	1969	365
Dimitrovgrad	Dimitrovgrad	BOR-60	1969	12[d]
Novovoronezhskiy-3	Novovoronezhskiy	VVER	1971	440
Novovoronezhskiy-4	Novovoronezhskiy	VVER	1972	440
Shevchenko	Shevchenko	BN-350	1973	150[f]
Bilibino-1, 2, 3	Bilibino	VK-12	1973–75	36
Kola-1	Polyarnyye Zori	VVER	1973	440
Leningrad-1	Sosnovyy Bor	RBMK	1973	1000
Kola-2	Polyarnyye Zori	VVER	1974	440
Leningrad-2	Sosnovyy Bor	RBMK	1975	1000
Bilibino-4	Bilibino	VK-12	1976	12
Kursk-1	Kurchatov	RBMK	1976	1000
Armenia-1	Metsamor	VVER	1976	405
Chernobyl'-1	Pripyat'	RBMK	1977	1000
Chernobyl'-2	Pripyat'	RBMK	1978	1000
Kursk-2	Kurchatov	RBMK	1979*	1000
Novovoronezhskiy-5	Novovoronezhskiy	VVER	(1979)	1000
Leningrad-3	Sosnovyy Bor	RBMK	(1979)	1000
Rovno-1	Kiznetsovsk	VVER	(1979)	440
Beloyarskiy-3	Zarechnyy	BN-600	(1979)	600
Armenia-2	Metsamor	VVER	(1979)	410
Kursk-3	Kurchatov	RBMK	(1980)	1000
Leningrad-4	Sosnovyy Bor	RBMK	(1980)	1000
Rovno-2	Kuznetsovsk	VVER	(1980)	440
South Ukraine-1	Konstantinovka	VVER	(1980)	1000
Smolensk-1	Desnogorsk	RBMK	(1980)	1000
Kola-3	Polyarnyye Zori	VVER	(1980)	440
Kola-4	Polyarnyye Zori	VVER	(1980)	440
Kalinin-1, 2, 3, 4	Udomlya	VVER	(after 1980)	1000 each
West Ukraine-1, 2, 3, 4	Khmel'nitskiy	VVER	(after 1980)	1000 each
South Ukraine-2, 3, 4	Konstantinovka	VVER	(after 1980)	1000 each
Kursk-4	Kurchatov	RBMK	(after 1980)	1000
Chernobyl'-3, 4	Pripyat'	RBMK	(after 1980)	1000 each
Smolensk-2	Desnogorsk	RBMK	(after 1980)	1000
Ignalina-1, 2	Sneckus	RBMK	(after 1980)	1500 each

*Achieved criticality in December 1978.

Notes:

[a]Site preparations were also reported under way in the late 1970s for stations in the Crimea (at Aktash, on the Kerch' Peninsula), near Zaporozh'ye, Rostov (at Tsimlyansk) and near Saratov. Possible future sites have been reported under consideration near Yaroslavl' and Gor'kiy on the Volga River; Chelyabinsk in the Urals; Krasnodar and Chogray in the North Caucasus; in Azerbaijan (at Sangachaly) and in Georgia; Tuzly (near Odessa), Otashev (near Kiev), Rozhnyatov (in the Carpathians) and Olkhovatka (near Kremenchug), in the Ukraine; Pavilosta (Latvia) and Segozero (Karelia) (Fig. 8, p. 160).

[b]Reactors designated AM, AMB and RBMK are varieties of graphite-moderated, water-cooled reactors; VK are boiling-water reactors; VVER are pressurized-water reactors; BR, BOR and BN are breeders.

[c]The first 100 MW unit at the "Siberian" station was installed in 1958, with a total capacity of 600 MW reached or exceeded by 1964. Further unannounced expansion has apparently taken place (see Note d, Table 34).

[d]Obninsk and Dimitrovgrad (called Melekess until 1972) are reactor research and development centers, and their reactors are mainly experimental.

[e]The VK-50 reactor at Dimitrovgrad was upgraded to 65 MW(e) in 1973–74; the Novovoronezhskiy-1 reactor was first upgraded to 240 MW and then, in a brief experiment, to 280 MW(e) in January 1969.

[f]Of the total electrical capacity of 350 MW of the BN-350 reactor at Shevchenko, 150 MW is used for electrical output and 200 MW is used to raise steam for a water desalination unit.

Sources: Power Reactors in Member States, 1977 edition (Vienna: International Atomic Energy Agency, 1977), p. 29; *Energetik*, 1978, No. 3, p. 33, and sources listed in Note 9.

---

constructed on the south shore of Lake Imandra on the Kola Peninsula with two more scheduled. Another pair of pressurized water reactors of this type are under construction near Rovno in the western Ukraine with the first due in 1979.

The channel-type (sometimes referred to as "pressure tube") reactors at Beloyarskiy have been used to prove the design concept for the second of the standard types of reactors now being built in the Soviet Union. This is a larger channel, uranium-graphite reactor termed the RBMK-1000 (the initials stand for Russian words meaning "large-capacity channel reactor"). The characteristics of the second Beloyarskiy unit and the newer RBMK-1000 units are compared in Table 37. The earliest Soviet channel reactor was the Siberian facility, in operation since 1958. The first station using the new standard 1000 MW reactors of the channel type has been constructed at Sosnovyy Bor, about 45 miles west of Leningrad, and consists of two units, each of 1000 MW(e), with two additional units scheduled. The ultimate 4,000 MW capacity at this site will make it the largest nuclear generating facility in the USSR. Other channel reactor stations containing 1,000 MW(e) units each are under construction near the cities of Kursk, Chernobyl' (north of Kiev) and Smolensk. At the Kursk station, the first of four scheduled units started up in 1976; at Chernobyl', the first of four scheduled units went into operation in 1977 and the second in 1978, and at Smolensk, the first of two planned units was due to start up in 1980 (Table 38).

The Soviet Union apparently has high expectations for the RBMK type of reactor, since most of the new generating capacity that was to be built in 1976–1980 was planned to be of this model. Soviet scientists cite as one of its main advantages the possibility of larger station size, and RBMK units of 1,500 MW each are planned for the Ignalina station in Lithuania. A future generation of channel reactors, with electrical output of 2,000 MW or even 2,400 MW, is being referred to as the RBMKP-2000 or RBMKP 2400, with the P in the abbreviation signifying the use of superheated steam.[5] Other advantages noted are that the large number of individual fuel channels (1,693 in the RBMK-1000; 1,920 in the projected RBMK 2400) reduces hazards associated with breakdowns in individual channels, and that one channel may be replaced without shutting down the entire station. Their main disadvantages at present are that channel reactors require greater space and construction costs, and operate at lower thermal-to-electrical efficiency (about 31%). However, Soviet researchers believe these problems can be overcome and that, if desired, superheated steam capabilities can be added as well.

In comparing the Soviet graphite-moderated reactors to reactor types used in the United States and West Europe, one difference in particular is apparent. Gas-cooled reactors using carbon dioxide or helium are an advanced design concept in the Western world, but there have been no indications that Soviet nuclear engineers are employing this type of technology for their commercial units. They have, however, aided in the construction of a gas-cooled reactor in Czechoslovakia. The Soviet Union also has not shown interest in commercial heavy-water reactors although they have carried out research on this application.

Finally, a small atomic power plant has also been constructed in a remote area of Siberia at Bilibino, which is in the Chukchi Autonomous Okrug, about 125 miles southeast of the mouth of the Kolyma River. Output is 48 MW(e), obtained from four boiling-water, graphite-moderated reactors of 12 MW(e) each. Criteria for this facility, which serves an isolated gold-mining region, were simplicity and reliability. The fourth unit was completed in 1976, and the station replaced a floating 20 MW coal-fired plant. In 1978, the nuclear plant also began to provide steam heat for buildings as well as electrical power.

The Soviet Union's two breeder reactors, one in operation and one under construction, will be discussed below in a separate section.

Nuclear Power Planned for 1980 and Beyond

By 1975 the Soviet Union's installed nuclear capacity was 6,200 MW(e) (see Note d, Table 34), and by the end of 1978, 9,600 MW, including 8,100 MW in utility stations (Table 35). The target figure for the 10th five-year plan (1976–80) was 19,400 MW. The original planned increase in the capacity of utility stations during the five-year period was 13,200 MW, of which only 3,400 had been installed during the first three years.* The 6,200 MW(e) installed in

*An article in *Ekonomicheskaya Gazeta*, 1978, No. 7, indicated a figure of only 11,700 MW(e) as the capacity to be added through 1980, suggesting that a downward revision of the original plan figures had possibly taken place.

1975 represented 2.8% of the total electrical generating capacity of 217,500 MW. The stated goal for 1980 of 19,400 MW would represent just under 7% of the electrical industry total for that year.[6] This meant that the 13,200 MW to be installed during the five years of the plan period was to represent 19.8% of all new generating capacity constructed in that period.

Despite the rapid rate of growth planned for the nuclear electric industry, it will not expand uniformly across the country in terms of the geographic distribution of new stations. With only one minor exception (Bilibino), there are no atomic power stations built or projected anywhere east of the Urals. All the new facilities started during the 10th five-year plan were located in the European third of the USSR.

This concentration of the industry west of the Urals can be explained by the fact that, in general, Siberia is an energy-rich portion of the country, whereas many parts of European USSR are not. West Siberia is rich in coal, oil, and natural gas, and East Siberia has plentiful coal and hydroelectric resources. Central Asia has adequate reserves of natural gas and a rapidly expanding hydroelectric base, although in terms of fossil fuels is not nearly as well endowed as Siberia. The European USSR has the Donets coal reserves and the Volga-Urals oil fields, but the coal is deep and becoming expensive to extract and the oil fields are believed to be past peak production, so that a new energy source is seen as vital for the future. Apparently nuclear power plants represent the source that has been selected. These large nuclear facilities are expected to perform the same function as large fossil-fuel plants in providing base-load electrical needs. Atomic power plants in the Soviet Union had an overall utilization coefficient of .71 in 1974.[7]

Of the planned addition of 13,200 MW to the total capacity of nuclear electrical stations scheduled during the 10th five-year plan, 3,417 MW was added in 1976–80. This included the first 1,000 MW unit at Kursk, the first two units at Chernobyl' (2,000 MW) and the first (405 MW) of the two units at the station in Armenia, plus the last of the four small units at Bilibino in northeast Siberia.

The 1979 plan called for the power startup of Kursk-2 (which had achieved criticality in late 1978), Novovoronezhskiy-5 and Leningrad-3, all of 1,000 MW capacity; Rovno-1 (440 MW), the second Armenian reactor (410 MW) and the breeder that had long been under construction at Beloyarskiy (BN-600).

This would leave about 5,300 MW of nuclear capacity to be installed in 1980 if the five-year plan were to be fulfilled, which appears unlikely. These remaining stations were to have included four 1,000 MW units: Kursk-3, Leningrad-4 as well as the first reactor of the South Ukrainian station and the first of the Smolensk station. Others were 440 MW reactors, including the second at the Rovno station, and two additional units at the Kola station. All these sites are shown in Fig. 8.

A number of other facilities are undergoing initial site preparation or early stages of construction. These stations include Ignalina (in Lithuania), Aktash (on the Crimea's Kerch Peninsula), Zaporozh'ye, Tsimlyansk, and the West Ukraine station near Khmel'nitskiy. Work had reportedly begun at the Ignalina, Aktash and Tsimlyansk sites by the end of 1977, with initial site preparation at

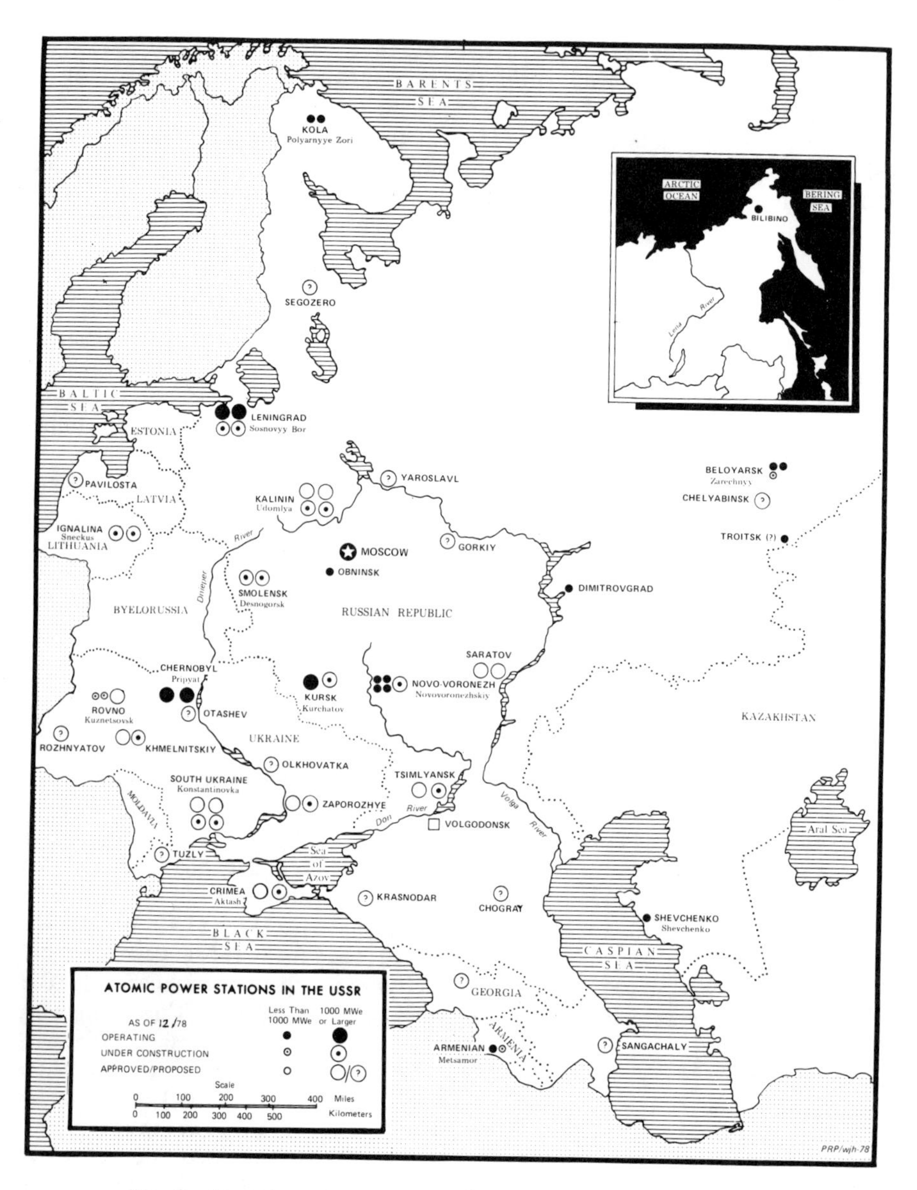

Fig. 8. Atomic power stations of USSR in December, 1978.

Zaporozh'ye and Khmelnitskiy scheduled to begin during 1978.

Soviet sources suggest that additional sites are under consideration. These include stations near Yaroslavl' and Gorkiy in the Volga valley; Chelyabinsk in the Urals; Krasnodar and Chogray (Stavropol' Kray) in the North Caucasus; in Azerbaijan (at Sangachaly) and in Georgia; Tuzly (near Odessa), Otashev (near Kiev), Rozhnyatov (in the Carpathians) and Olkhovatka (near Kremenchug), in the Ukraine; Pavilosta (Latvia) and Segozero (Karelia). In view of the growing trend toward larger nuclear generating units, it is likely that such stations will be equipped with reactors of at least 1,000 MW each and possibly 1,500 MW or the projected 2,400 MW size.

In summary, even if the 1976-80 plan were not fulfilled, as appeared likely, there was no question that a more rapid pace of development was being realized in the late 1970s and that it would continue into the 1980s.

One other significant construction project of the 10th five-year plan in the field of nuclear engineering needs to be mentioned. It is not a power station, but rather a huge manufacturing complex for serial production of reactor vessels to be used in the stations. Known as Atommash (an acronym for "atomic machinery"), it was being built in the late 1970s along the Don River near the city of Volgodonsk. Its river-access location will allow large reactor units to be easily transported within the European USSR. The plant will produce units of 1000 MW each. To accommodate an assembly line for units of this magnitude, a site of 1,600 acres has been selected, on which a main assembly plant is being built that is 150 feet high, one-quarter mile wide, and one-half mile long.[10] The first stage of the plant, with a manufacturing capacity of 3,000 MW of nuclear reactors, went into operation in late 1978.

In addition, a new nuclear engineering college, known as the Atomic Power Institute, was being created at the research complex at Obninsk. It is planned that up to 3,500 students could be accommodated for training there.

Breeder and Fusion Reactors

Research in advanced atomic electrical generating systems, such as breeder and fusion reactors, has been vigorously pursued in the USSR for about three decades. Soviet scientists have been interested in breeder reactors since 1949 and now have about 20 years of experience in their construction and operation. The advantage of breeder reactors is that their design allows tham to produce more fissionable material than they consume. Several breeder reactors used for research purposes have been constructed at Obninsk, using the prefix BR. The BR-5, first operated in 1959, has undergone periodic modifications and now has a power rating of 10 MW(th). A more recent experimental breeder reactor is the BOR-60 at Dimitrovgrad. This unit went critical with sodium in December 1969, and had a planned output of 60 MW(th) and 12 MW(e).

The Soviet Union's first full-sized commercial breeder reactor (the BN-350) achieved criticality in November 1972, and is rated at 350 MW(e). It is located at the town of Shevchenko, on the Mangyshlak Peninsula on the east side of the Caspian Sea. This is a dual-purpose facility, with part of its energy output going to operate a desalinization plant. The desalinization plant, which operated on steam from an oil-fired power plant before the BN-350 was completed, is a multi-stage flash distillation facility capable of producing 120,000 m^3 of fresh water per day. This facility was about 200 MW(e), or over half, of all the power produced by the BN-350, and thus the Shevchenko breeder's electrical output is usually given at 150 MW(e).[11]

Construction has long been under way on a larger breeder reactor (the BN-600) at Beloyarskiy, adjacent to the two existing boiling water units at the same site, but in a separate compound. This 600 MW(e) reactor was scheduled for 1979, as part of the 10th five-year plan, after having failed to meet an earlier deadline by 1975. The BN-600 is a departure from earlier breeders in one way; it is a tank-type design with all components of the primary system fitting compactly into one vessel. Previous breeders, including the BN-350, placed the heat exchangers and pumps outside the core vessel itself, and the units were connected by double contained piping that could be valved off. The BN-600 is expected to demonstrate the feasibility of using this less expensive and more compact tank design concept, which is intended to produce increased power with a higher fuel burnup and a longer time between refueling. A comparison of technical data for the BN-600 and the BN-350 is presented in Table 39.

With Shevchenko on line, the Soviet Union operated the largest breeder power plant in the world and with the completion of Beloyarskiy-3 will maintain its position as the leader in this regard. These units differ somewhat from design practices in the United States. The most fundamental difference in concept is that Soviet breeders are designed only for credible malfunctions such as site electric power failure. The Russians do not consider propagation of nuclear reactions due to failed fuel elements to be a credible occurrence. Failed fuel elements are known to occur but they have not been shown to cause uncontrollable power surges. This basic assumption means that there have been limited core detection instrumentation, no fuel meltdown backup system, and no containment structures other than the steel lined reactor cavity itself. (The Novovoronezhskiy-5 reactor was the first Soviet reactor with a containment dome.) As is the case with most nuclear reactors in the Soviet Union, the reactor building for breeders is a conventional factory-type design with no special containment characteristics. The breeder design emphasizes reliability of all components, and includes extensive site power backup systems, automatic operation procedures in case of an emergency, and independently activated safety rods in the core of the reactors.

Soviet references mention the design of an 800 MW sodium-cooled breeder (an improved version of the BN-600) and ultimately a 1,600 MW reactor as the basis for future development of the breeder stage of the nuclear power industry. Work is also going forward, at the Belorussian Nuclear Power Institute in Minsk, with the design of a 300 MW gas-cooled breeder demonstration reactor using nitrogen tetroxide as a coolant. The demonstration facility is viewed as a possible step toward the further development of such gas-cooled reactors with electrical outputs of 1,000 to 1,500 MW.[12] It was envisioned in the early 1970s that the solving of all the problems associated with breeder reactors would allow them to

Table 39. Characteristics of Soviet Breeder Reactors

Characteristics	Shevchenko BN-350 (1972)	Beloyarskiy BN-600 (1979 plan)
Power		
MW (electrical)	350	600
MW (thermal)	1,000	1,430
Fuel		
type	UO_2; PuO_2 to be introduced at a later time	same
enrichment	inner core–17% outer core–26%	inner core–21% outer core–28%
average burnup, MW days/ton	38,000 or 5%	10%
refueling	every 50 days, 1/6 core replaced	150 days. Pu doubling time is 10 years.
core sub-assemblies	199	397
Coolant		
type	liquid sodium	liquid sodium
inlet/outlet, °C	300/500	380/550
Steam pressure at turbine, atmospheres	50	130
temperature, °C	430	500
Safety considerations	7 coarse, 2 fine and 3 independent control rods made of B_4C	total of 26 safety control rods of B_4C

Source: Soviet Power Reactors–1970, (Washington: Atomic Energy Commission, August 1970), and I. D. Morokhov et al., *Atomnoy energetika XX let* (Moscow: Atomizdat, 1974), p. 118.

be used as the main form of increased electrical power production in the European part of the USSR after 1985, but with the delays in completing the BN-600, this may be optimistic.

The Soviet Union is also recognized as carrying out some of the world's most advanced research on controlled thermonuclear fusion. Preliminary investigations in this area were begun in the USSR in 1950. As has been the experience with fusion elsewhere, the problems have turned out to be more complex than expected and no sudden break throughs are expected.

The two overriding problems of fusion research are to prepare a high temperature plasma and then to maintain that thermal energy level for a sufficient period of time. Almost all methods to create and maintain a plasma with such characteristics depend on the use of a powerful magnetic field. Soviet scientists have developed two types of systems for closed (toroidal) magnetic fields—the Tokamak devices and stellerators. Open magnetic systems sometimes called "magnetic mirrors" have also been investigated. Newer research topics include the interaction of plasma with high frequency electromagnetic fields and the use of laser pulses to heat matter to ultra high temperatures. The USSR has also become increasingly interested in investigating laser fusion techniques for so-called "hybrid reactors". These are sometimes called "fusion assisted breeders", and are devices in which a fusion reactor is inside a hollow fission reactor so that neutrons from the inner portion can be used to breed plutonium and/or produce power from fission in the outer part. The USSR participated in the first international conference on hybrid reactors, held in California in 1976.

The main center for fusion research in the Soviet Union is at the Kurchatov Institute in Moscow, where more than 200 scientists are employed in fusion research alone. Also in Moscow is the smaller Lebedev Institute, involved in research with stellarators and laser-pulse heating. At the Kharkov Physical-Technical Institute is the large stellarator "Uragan." Two fusion research centers are located in Leningrad, the Ioffe Physical-Technical Institute and the Yefremov Institute of Electro-Physical Apparatus. Other experimental work relating to controlled fusion is carried out at the Nuclear Physics Institute in Novosibirsk and at the Physical-Technical Institute of the Georgian Academy of Sciences at Sukhumi.[13]

The Soviet Union and the United States both view controlled fusion as a highly desirable form of electrical energy generation for the future, due to the abundance of fuel (deuterium), relative absence of harmful residuals, and freedom from the risk of "nuclear accidents." However, American scientists are not optimistic of sustaining fusion reactions in a form suitable for generating electricity before 1990, and there have been no indications that the achievement of sustained, commercial fusion power in the Soviet Union is less than some decades away.

Siting Considerations

One of the more difficult problems in any national program of commercial atomic power development is the selection of appropriate sites for the reactor facilities. The degree of difficulty encountered in siting atomic power stations will be in part a function of how potentially dangerous officials perceive such stations to be. The Soviet Union, having adopted a public position that such plants are relatively safe, is somewhat less rigid with regard to siting criteria than is the case in the United States: "The successes in the area of atomic power plant safety as a whole have been so great that the selection of locations for atomic power plants at the present time is not limited by safety requirements, being determined only by technical and economic factors."[14]

More specific siting criteria have been enumerated in a few Soviet studies and monographs. A paper prepared for a 1971 conference of the International Atomic Energy Agency listed the following siting considerations:

(1) Sites should be open and well ventilated, and should be downwind from populated areas.

(2) Sites should be constructed on a thick layer of impermeable rock that would retard the accidental spread of radioactive liquids.

(3) Sites should be on level or slightly sloping land not subject to inundation or waterlogging.

(4) All portions of the station below ground level should be above the groundwater level. It is desirable to have favorable conditions near the station for injecting liquid radioactive wastes into the ground.

(5) It is mandatory that sanitary-protective zones be established around the station. These may contain administrative and service facilities but not residences.[15]

The article does not suggest the widths of such zones, but does state that determining maximum probable emergency situations for various types of atomic power stations is an important consideration in locating them near cities.

An earlier study (1965) lists 10 siting considerations that, in addition to the above, include proximity to cooling water, holding ponds for radioactive waters, proximity to fuel-manufacturing plants, proximity to roads or rail lines, and availability of an independent electrical power supply.[16] Interestingly, neither work specifically mentioned earthquake safety, a major consideration in the United States. Soviet specialists apparently feel that locating plants in geographic regions of low earthquake probability precludes the need for detailed seismic studies. However, the Metsamor station in Armenia is in an area that has experienced moderate earthquakes in the past. It is also noteworthy that this facility was under construction much longer than its "sister" station at Polyarnyye Zori on the Kola Peninsula.

Prominent Soviet authorities have spoken favorably of the possibility of siting nuclear power stations near urban centers: "The successful use of the first nuclear power stations and their freedom from radiation hazards creates confidence in the possible siting of atomic thermal power stations close to large cities."[17]

It has been stated that such stations must be at least 6 miles from populated places, and 20 miles from major towns.[18] To date, none has been built closer than about 25 miles from major urban centers. However, an American atomic energy exchange delegation to the USSR reported: "The Soviets locate their power plants in the general location of load demands and at the location of the coolant. They do not require an exclusion area around their reactors. Townsites are developed conveniently for personnel access to the industrial site or activity. Housing for atomic plant workers is usually located close to the plant."[19]

Depending on what is meant by "close," this could suggest that either criterion (5) above is not followed in practice or the protection zone must be relatively small. Plants constructed more recently, such as those at Kursk and Smolensk, have been located no closer than 22 miles from the major city, but some more recently proposed, such as the South Ukraine and Zaporozh'ye facilities, may be within 12 miles of major populated areas. Further, it is known that the Soviet Union is interested in using waste heat from atomic power plants for communal heating, and is doing so in at least two locations (Beloyarskiy and Bilibino). Also, there are plans to heat greenhouses with the thermal effluent from the Kola and Bilibino atomic stations.[20]

Reactor Safety and Waste Disposal

The probability of a major accident at a commercial nuclear reactor site is perceived as less of a concern in the USSR that it is in the United States. The Soviet approach to reactor safety emphasizes excellence of design, reliability of equipment, and careful operating procedures to prevent releases of radioactivity to the environment. Special containment structures (domes) are generally not thought to be justified because of the improbability of any serious accident, and such domes are therefore viewed as costly and superfluous precautions. The first containment shell in the Soviet Union was constructed at the Novovoronezhskiy-5 site.[21] Loss of coolant in the core is also deemed to be an unlikely event, and thus the reactors do not have a special emergency core-cooling system. Soviet writers question the philosophy of designing redundant systems, for "an excess of such backup systems, where the need or the reliability is not clearly assured, introduces operational complexity and reduces overall safety."[22] In addition, the RBMK channel-type reactors favored in the Soviet Union are inherently safer in design than other types of light water reactors.

Soviet scientists acknowledged that some types of accidents might release radiation accumulated in the cooling medium, or possibly even some of the radioactivity from unsealed fuel cans, but such releases were not projected as exceeding the daily permissible releases from power stations ($\leqslant 10^3$ to 10^4 curies). In cases of radioactive contamination of the area immediately around the reactor itself, all ventilation openings in the factory-type building could be sealed, and closed-cycle air decontamination started.[23] However, since there are no pressure suppressant chambers, this slow sealing process could not prevent sudden releases of radioactivity into the atmosphere in the case of a serious accident.

Assurances of the impossibility of catastrophic nuclear accidents are repeatedly stated in Soviet sources. For example: "There is every reason to consider such [nuclear power] plants no more dangerous than conventional power plants. Radiation injury to the population is practically impossible, and any presumable emergency situation in nuclear power stations with water-cooled, water-moderated power reactors cannot be of a catastrophic nature."[24]

Despite such assertions, some public concern about nuclear power plant safety appears to persist in the Soviet Union. The extent of such concern is not known, but in a book intended for widespread public distribution the editors state that they hope the section on radiation safety will "help to overcome the popular misconceptions concerning the danger of atomic plants for the population." The same source further assures the public that "Soviet specialists have developed a system of measures to eliminate the possibility of leakage of nuclear radiation from reactors and other installations to their surroundings."[25]

Despite the precautions taken, reports circulated in 1976 and 1977 of an explosion of radioactive wastes (the exact cause of the accident is a matter of dispute) in 1958 in the Kyshtym-Kasli area of the southern Urals, causing extensive damage to the landscape and possibly loss of life. Additionally, a steam line explosion occurred at the Shevchenko breeder facility (BR-350) in 1974.

Table 40. Maximum Permissible Radiation Doses, USSR (rem/year)

Category	Groups of critical organs			
	I Whole body, gonads, bone marrow	II Muscles, fatty tissue, liver, kidney, spleen lungs, eye, other organs except I, III, IV	III Bone tissue, thyroid, skin	IV Wrists, hands, forearms, ankles, feet
Staff	5[a]	15	30	75
General population in a monitored exposure area	0.5	1.5	3	7.5
General population	0.17[b]			

Notes:

[a]5 rem/year for 30 years is the maximum dose. At any age X, the maximum cumulative dose should not exceed 5(X-18); 18 is the minimum age for employment in the atomic industry.

[b]Maximum exposure for the sum of internal and external doses. All values exclude background and medical doses.

Sources: Data from I. K. Dibobes et al., "Radiation Safety Standards," *Soviet Atomic Energy*, Vol. 28, No. 6 (June 1970), p. 594.

The radiation safety standards in effect in the Soviet Union were drawn up by the National Commission for Radiation Protection and confirmed by the Ministry of Public Health in August 1969. These values (known as NRB-69) were slightly modified in 1976 in a set of standards called NRB-76. They are based mainly on the previously published and widely accepted standards of the International Commission on Radiological Protection.[26] The protection standards presume the absence of any threshold dose and may change as more becomes known about the effects of radiation. Some radiation protection standards now in effect are shown in Table 40.

The Soviet standards have the force of law, and each nuclear power plant must establish an independently operated radiation safety department within its organization that is responsible for radiation monitoring and practices leading to personnel safety. In addition, the State Sanitary Inspectorate under the Ministry of Public Health has responsibility for ensuring that radiation safety standards are met.

Generalized results of external monitoring programs have been published for some locations. One of the most extensively studied areas has been around the

Novovoronezhskiy power complex. Tests have included Don River water, river bottom sludge, surrounding soil, and vegetation up to 25 miles away. These and similar studies at Beloyarskiy and Dimitrovgrad have not resulted in any reported radiation levels above background. As a result, Soviet specialists have always maintained the position that "all these data permit the conclusion to be drawn that the problem of handling radioactive waste at Soviet nuclear power stations has been solved successfully."[27]

What are the procedures by which these wastes are successfully managed? In many cases Soviet technical practices in this regard do not differ significantly from those of other countries. There are several categories of waste products from nuclear power plants that must be treated and disposed of in a safe manner. Solid and gaseous wastes involve substances that are only weakly radioactive and/or have short half-lives, and these usually present no substantial disposal problems. Certain Soviet disposal practices (burial of solid wastes, delayed release of gaseous emissions) appear to be similar to those used in the West.

Some information has been made available concerning the levels of gaseous releases at certain specific locations. Beloyarskiy averages 250 curies/day from stack emissions, of which 80% is argon-41 with a half-life of 110 minutes. At Novovoronezhskiy the release of inert radioactive gases is given as 250 curies/day. At both Novovoronezhskiy and Beloyarskiy, the iodine-131 emission is in the range 0.005–0.05 curies/day and it does not appear to be accumulating in the environment.[28]

Liquid wastes from nuclear reactors come from cooling water leaks, water from basins holding spent fuel elements, and wash water from contaminated clothing. If only slightly active ($\leqslant 10^{-5}$ curies/liter) the wastes can be purified by coagulation, settling, filtration, ion exchange, electrodialysis, or distillation and released into local disposal systems. Higher level wastes, including concentrated solutions containing up to 10^{-1} curies/liter, are placed in special holding tanks located at each atomic research center and nuclear power station. Each large city that uses radioactive materials has at least one such centralized burial (holding) site for wastes from all research laboratories, hospitals, and industrial and power plant facilities in that area. Highly active liquid wastes, to several tens of curies/liter, also may be temporarily stored in these onsite holding tanks, or may be transferred to special centralized stations for better isolation.

The basic Soviet approach to the disposal of high-level nuclear wastes is to solidify them in concrete, bitumens, or glass-type (vitreous) solids, and then to encase them in concrete and stainless steel containers. They are then buried at or near the surface, so that they will be accessible for inspection and future storage modifications (if necessary), and in case future technologies would enable them to be processed into useful materials.[29]

An alternative approach to the long-term storage of high level wastes is injection into deep geologic formations. An experimental program of this type has been carried out at Dimitrovgrad. The site started operation in 1966 and in that year received 40,000 m^3 of low and medium activity liquid wastes injected at depths of up to 1,550 meters. In 1971 it was reported that a total of 290,000

m^3 of wastes had been injected.[30] At that time no change in the radiation environment around the surface of the bore holes had been noted. The share of high-level wastes receiving deep burial at the present time is unknown.

Various other long-range disposal schemes have been investigated. One that was studied in the early 1960s was disposal into the Black Sea, but the exchange rate between water layers proved to be faster than expected, and the idea was abandoned. Furthermore, a large proportion of the strontium-90 present came into the sea by means of river-borne isotopes.[31] This indicates that accumulation could result in the sea if levels of radiation in the Don River, which provides the cooling water for Novovoronezhskiy, were allowed to rise above background levels, even if there were no direct disposals of wastes into the Black Sea. Another disposal scheme that has been discarded is ocean dumping; the USSR is a signer of the 1972 Convention on Prevention of Pollution of the Sea by Discharge of Wastes and Other Materials.[32]

There is some basis for concluding that the problem of the long-term storage of high-level radioactive wastes has not been any more conclusively resolved in the USSR than it has in the United States. In the fairly comprehensive Ioirysh article (footnote 18), most topics associated with radiation protection from atomic power plants are brought up and discussed. Interestingly, there is no mention of the disposal of high-level wastes. Also, in the comprehensive book edited by Morokhov (footnote 3), there is only one short paragraph on monitoring the area around liquid high level waste burial grounds against possible leaks, although other aspects of radiation safety are covered in detail. The Soviet Union is usually willing to publicize areas where it feels it has formulated better scientific practices than have other nations.

Other Peaceful Uses of Nuclear Energy

In addition to commercial power reactors, nuclear reactions and controlled radioactivity can be used in other ways in a nation's economy. One of the most common applications is the use of radioactive isotopes for industrial, research, and medicinal purposes. The Soviet program is similar to that pursued in the United States, and specific literature on it can easily be found by the interested reader.[33]

Also, like the United States, the Soviet Union has an increasing interest in exporting nuclear power plants. The first two to be constructed outside the USSR were the Nord station at Lubmin near Greifswald in East Germany (1973) and the Kozloduy plant in Bulgaria (1974). The first unit of a Finnish facility at Loviisa opened in 1977 and represents the first Soviet-built atomic plant outside the Warsaw Pact nations. Other plants are under construction in Hungary, Rumania, and Czechoslovakia, and one is being planned for Poland.[34] Almost all of the exported Soviet reactors are of the VVER-440 type, similar to Novovoronezhskiy-3 and 4.

An area where the Soviet Union has conducted a broader variety of experiments is in the use of underground nuclear explosions to create reservoirs, dig canals, stimulate recovery of mineral resources (especially oil and gas), create underground storage chambers, and in at least one instance to put out a fire in a natural gas field. Atomic blast tests were carried out in the mid-1970s, for example, to test the feasibility of digging a water diversion canal from the Pechora to the Kama rivers by nuclear explosives (a concept since rejected). In other cases, reservoirs, now used for recreational purposes, were created by underground nuclear devices.[35]

Another unique Soviet effort is the development of atomic-powered icebreakers, which have been used for about two decades to maintain shipping channels along the Northern Sea Route through the Arctic Ocean. The *Lenin* has operated since 1958; it was modernized in the late 1960s, and operates on three pressurized water reactors that can develop 44,000 horsepower. More recently launched are the larger and more powerful 75,000 HP *Arktika* (1974) and *Sibir'* (1977). In 1977, the *Arktika* became the first surface vessel to sail to the North Pole, and in 1978 the *Sibir'* reconnoitered a new northerly shipping route through the Arctic Ocean.[36]

Summary

The USSR places a great deal of emphasis on its commercial nuclear power industry, and considerable expansion during the 10th five-year plan and on into the 1980s can be expected. Standardized VVER and RBMK reactors of 1000 MW(e) will be used initially, with larger units of 1500 and 2000 MW(e) being projected for the future. Although past experience would suggest that the 1980 target of 19,400 MW(e) of installed nuclear capacity is not likely to be met, a tremendous increase in operating capacity will surely occur in the remainder of the 10th and through the subsequent five-year plans. Continued advanced research on breeder and fusion reactors will take place, and with the completion of the Atommash plant an increased emphasis on export of nuclear technology may occur, as well.

GEOGRAPHY OF URANIUM RESOURCES*

In contrast to progress in the nuclear energy industry, the Soviet Union has published little on the geographical distribution of the mining and processing of uranium, the basic nuclear fuel. Nor are there any Western sources that discuss the spatial aspects of the Soviet uranium industry. No data occur in a nuclear energy report issued in 1977 by the Central Intelligence Agency or in the periodic world assessments of uranium supply and demand prepared jointly by the Nuclear Energy Agency of the Organization for Economic Cooperation and Development and the International Atomic Energy Agency.[37] According to the Soviet geological literature,[38] which tends to discuss types of uranium deposits without relating them to specific places, the Soviet Union has identified and explored a wide variety of domestic uranium deposits, in many parts of the

*This section is by Theodore Shabad.

country, and is assured of a supply for the foreseeable future. In addition, the Russians have been collaborating with geologists in allied countries in exploration for uranium and are known to be importing uranium from Eastern Europe, mainly from Czechoslovakia and East Germany.

Soviet uranium deposits are said to fall into two main classes: ore deposits found in igneous-rock environments, mainly of the vein type, and deposits in sedimentary formations, in which uranium deposition stems from infiltrating water solutions. Vein deposits occur both in the Precambrian metamorphic rocks of shields and in fold belts, where the ore is associated with granites or with volcanic rocks. The principal sedimentary sites are either in the Mesozoic sediments of foreland troughs adjoining areas of recent tectonic activity or in intermontane basins. The sedimentary deposits are mainly in sandstones, but may also occur in lignites, clays and limestones. Both vein-type deposits and sedimentary formations are said to be of equal importance in the localization of Soviet uranium resources.

Many of the deposits found in the Soviet Union have no analogs abroad. They include uranium-molybdenum associations of the stockworks type, with myriads of intersecting veinlets, and associations of uranium with albitite, with titanium or with iron ore in Precambrian metamorphic rocks. Another distinctive type of deposit is a uranium-phosphate association in clays with fish-bone detritus. The coproducts of uranium deposits are often recovered together with uranium, yielding molybdenum, iron, rare earths and phosphatic fertilizers. The recovery of these coproducts usually improves the economics of the mineral operations and makes it economical to work ores with a low uranium content.

Three mining methods are employed in the uranium industry of the Soviet Union. Rich ores found in hard rock at depths to 600–700 feet are mined by underground methods, a technique that is still said to account for the greater part of Soviet uranium production. Lower-grade deposits covering large areas near the surface are mined by open-pit methods. Since the middle 1960s, the Russians have also been making increasing use of in-situ leaching (using strong sulfuric acid) to exploit low-grade ores that cannot be economically extracted either by open-pit or by underground methods. In the early phase of the Soviet uranium industry, treatment mills at the mines concentrated the ore by traditional leaching methods used in other branches of the nonferrous metals industry. In an effort to achieve cost savings in the labor-consuming recovery of uranium from leach solutions, the Russians, like uranium producers elsewhere in the world, developed more efficient processes using ion exchange or solvent extraction that "pull" the uranium from the solution. Uranium concentrate must be converted to a form suitable for enrichment, in which the nuclear fuel, uranium-235 (which makes up only 0.71% of natural uranium), is separated from the nonfissionable isotope, uranium-238. Since the enrichment process most commonly used is gaseous diffusion, the uranium concentrate must be converted to a volatile compound, such as uranium hexafluoride (UF_6), so that it can be processed in a gaseous state. There are relatively few facilities outside the Soviet Union converting uranium concentrate to hexafluoride—two each in the United States and France, and one each in Canada and Britain. There is no published information on the number or location of Soviet conversion facilities or of gaseous diffusion plants, which are known to have huge power requirements.

There are also no published figures on Soviet uranium production. Total Western production in the middle 1970s was about 20,000 tons of uranium concentrate (in terms of U_3O_8, the oxide produced by ore-treatment mills). The leading producers were the United States, with about 9,000 tons; Canada, with 4,700 tons; South Africa, with 2,600 tons; France, with 1,700 tons, and Niger, with 1,200 tons.[39] In terms of its nuclear fuel needs, estimated by the writer at about 2,000 to 2,500 tons in the late 1970s, as well as military requirements and reserves, the Soviet Union's uranium production may be of the magnitude of Canadian output.

In the absence of a systematic published discussion of the geography of Soviet uranium mining, a tentative treatment is offered here on the basis of indirect evidence. The association of an urban settlement with uranium mining and processing can sometimes be inferred from the political-administrative status of the inhabited place in the Soviet hierarchy of civil divisions, for example, when an urban settlement is endowed with high administrative status such as oblast-level city, while reporting a relatively small population that would not normally entitle it to such status. Uranium-related activities may also be inferred when Soviet publications refer to a populated place simply as an unspecified mining site or avoid detailed discussion of its economic base or pass it over altogether in silence while providing specifics on nearby localities. It goes without saying that the following discussion of presumed uranium centers is highly circumstantial and should in no way be considered complete.

One of the most important uranium centers in the Soviet Union–and one of the few that have been publicly identified as such[40]–is the city of Zheltyye Vody in the Ukraine, at the northern end of the Krivoy Rog iron-ore basin. This is a Precambrian site in which uranium is associated with iron quartzites. Known since before World War II and operated as a small iron-ore mine, the site was occupied by the urban settlement of Zheltaya Reka (Yellow River), with a population of 6,522 in the 1939 census. With the start of the nuclear era after World War II, it developed rapidly after the start of uranium production in 1948. It was raised to the status of oblast city under the new name of Zheltyye Vody (Yellow Waters) in 1957, and reported a population of 32,406 in the 1959 census. Its continued vitality was evident in its population growth, which rose to 40,334 in the 1970 census and 52,000 in early 1977. Zheltyye Vody had a satellite operation farther south within the Krivoy Rog iron-ore basin at Terny, another prewar iron-mining town, with 2,874 people in 1939. It, too, grew after World War II, was placed under the immediate jurisdiction of Zheltyye Vody, and reached a population of 13,213 in 1959. In the 1960s, with uranium apparently becoming less significant in Terny mining operations, the town was selected as the site of one of the new iron-quartzite mining and concentrating complexes in the Krivoy Rog basin. The complex, known as the North Concentrator, opened in 1964, and five years later Terny was incorporated within the city limits of Krivoy Rog.[41] The administrative shift appeared to reflect the severing of a linkage between Terny and Zheltyye Vody, which continued to function primarily as a uranium center. The large size of Zheltyye Vody may suggest not only the significance of its mining and milling operations, but possibly also the presence of further processing facilities such as hexafluoride conversion.

In Estonia, the town of Sillamäe, on the Baltic Sea west of Narva, has been long identified as a uranium center, associated with uranium-phosphate ores in clays with fish-bone detritus. In addition to phosphate, this type of uranium deposit also yields rare-earth elements, mainly of the cerium subgroup. Sillamäe, which is situated on the coast of the Estonian oil-shale basin, arose as an urban settlement in 1949 and was raised to the status of republic-level city, as indication of its economic significance, in 1957. However, its narrow economic base, limited to uranium mining and milling, has kept the town's population to a low level; it was reported as 8,200 in the 1959 census, 13,500 in the 1970 census, and 15,400 in early 1975.[42] Its growth rate during that period (88%) was among the highest in Estonia. The unusual character of Sillamäe is also evident in its relatively large municipal area of 4 square miles, which is 40% of the area of Tartu, a city of 100,000 population.

The town of Lermontov, in the Caucasus, appears to be a uranium mining center, possibly of the uranium-molybdenum type associated with volcanic rocks. The town is situated in the well known Caucasus mineral springs district of Pyatigorsk, distinguished by a cluster of inselbergs of the laccolith type. Lermontov is situated at the west foot of Beshtau, the highest of these mountains, about six miles northwest of Pyatigorsk. The existence and significance of the uranium extraction site became evident when the urban settlement associated with mine and mill was raised to the status of oblast-level city in 1956; it was officially known as Lermontovskiy until August 1967, when it dropped the adjectival ending to become simply Lermontov (in honor of the early 19th-century Russian poet whose career is associated with the Pyatigorsk area). Typically for a uranium center, Lermontov, though of high administrative rank, has a modest population; it was reported as 10,820 in 1959, 15,063 in 1970 and 17,000 at the beginning of 1974. This uranium center, too, has an unusually large municipal territory, of 11 square miles, the same as Pyatigorsk, which has a population of 105,000.[43]

In the Urals, some smaller mining places extracting unspecified minerals are believed to be associated with uranium. The most important is Vishnevogorsk, an urban settlement 8 miles northwest of Kasli, in northern Chelyabinsk Oblast. Vishnevogorsk is associated with one of a series of extensive nepheline syenite intrusions that yield a wide array of rare minerals, including radioactive ores. It was raised to the status of urban settlement in 1949, and had a population of 9,100 in the 1959 census, nearly one half the size of the nearby town of Kasli. It is in this area that an unspecified nuclear mishap is said to have occurred in 1958, causing damage to the environment from radioactive wastes. Other mining settlements believed to be associated with the extraction of unspecified rare metals from intrusive bodies are Novogornyy, established in 1954 about 10 miles southeast of Kyshtym, and Yuzhnyy, which lies 40 miles east of the iron and steel city of Magnitogorsk and became an urban settlement in 1961.

In Siberia, at least two major localities appear to be associated with uranium production–Vikhorevka in Irkutsk Oblast and Krasnokamensk in Chita Oblast. Vikhorevka, in contrast to some other uranium-related urban places, has achieved only the status of rayon-level city, one rung below oblast-level city in the Soviet urban hierarchy, but its population development has been significant. Situated 25 miles west of Bratsk, it was formally given the status of urban settlement in 1957, 10 years after it was reached by a railroad from Tayshet on the Trans-Siberian mainline. In the 1959 census, Vikhorevka was reported to have a population of 10,781, making it the largest urban place on the railroad segment between Tayshet and Bratsk. Moreover, though overshadowed by the rapidly expanding urban center of Bratsk, Vikhorevka continued to grow. It was raised to the status of rayon city in 1966 and reported a 1970 census population of 17,901, a 66% growth in the intercensal period, one of the largest percentage increases in Irkutsk Oblast. Vikhorevka lies in the Siberian shield and, if it is indeed involved in uranium production, it may be mining titanium-uranium ores associated with ultrametamorphic Archean rocks in deep-seated fractures of the type that traverses the area.

The case of Krasnokamensk in southeast Chita Oblast, 20 miles from the Soviet Union's Argun' River border with China, is unusual in that it was raised to the high status of oblast-level city in 1969 without having been previously reported as an urban place of lower rank. In the 1970 census, Krasnokamensk had a population in the range of 12,000–15,000, and by the beginning of 1974 it was credited with an estimated population of 34,000,[44] making it the second largest urban place of Chita Oblast. Its economic significance became further evident in March 1977, when it was made the administrative seat of a separate Krasnokamensk Rayon.[45] Southeast Chita Oblast is a highly diversified mineralized zone in which some of the Soviet Union's principal deposits of fluorspar and pegmatite-related minerals such as tantalum, columbium, lithium and beryllium occur. If indeed Krasnokamensk's sudden appearance in the southeast Chita steppe is related to uranium mining, it may involve one of the Soviet Union's uranium-fluorspar associations found in basin-shaped volcanic depressions formed during volcanic activity of the Mesozoic era. The operation may be one referred to in 1967 as an unidentified mine in the steppe of Transbaikalia, with the Soviet Union's first fluorspar pellet plant, producing a higher-grade fluxing agent for use in basic oxygen steel furnaces.[46]

Among mining places in Kazakhstan that have not been identified with a particular mineral activity and are therefore presumed to be uranium-related is the city of Stepnogorsk, another oddity in the Soviet published literature. Although referred to occasionally in the daily press and in books that appear in the Soviet Union, Stepnogorsk is not being shown on Soviet maps[47] and does not appear in the Soviet administrative-territorial guides that identify the nation's minor and major civil divisions. From the available evidence, Stepnogorsk appears to have arisen in the middle 1960s on the Selety River, 80 miles northeast of the city of Tselinograd. A reservoir on the Selety, built to provide a crucial water supply in this semi-arid region, began filling in 1965.[48] A railroad branch running from Yermentau (on the Tselinograd–Pavlodar railroad) to the north for a distance of 137 miles through Turgay, Stepnogorsk and Aksu to a railhead known as Aysary first appeared in Soviet railroad guides and rail maps in 1965-66. The 1970 census, without specifically identifying Stepnogorsk, listed a population of 25,956 as a residual.[49] Election district lists in 1971 and in 1975, again without naming Stepnogorsk, provided two districts (City district and Construction district) for the town, with nominal populations of 25,000-30,000 each. However, during the 1975 election campaign, the Soviet press reported rallies being held in Stepnogorsk, identifying it as the seat of the Tselinnyy (Virgin Lands) Mining and Chemical Complex.[50] The city, still omitted from Soviet maps and reference books, was again in the news in early 1978 when the first stage of large bearings plant, manufacturing journal bearings for railroad freight cars, started operations.[51] The manufacturing plant was evidently intended to diversify the economic base of Stepnogorsk. Any uranium activity at Stepnogorsk might be associated with deep-seated fractures at the northern end of an ore-bearing belt that cuts across Kazakhstan from north to south along roughly Long. 73-74°E. It is possible that the Stepnogorsk operation involves in-situ leaching, which is said to have started commercially in the Soviet Union about the time that Stepnogorsk is presumed to have arisen.

Two other Kazakhstan mining operations, about which less is known, are presumed to be associated with uranium mining farther south along the north-south ore-bearing belt in the areas of Koktas and Kiyakhty, near the western end of Lake Balkhash.[52] The urban settlement of Koktas was founded in December 1960 just north of the border separating Dzhambul Oblast from Dzhezkazgan Oblast. It is situated 10 miles from the town of Saryshagan, on the north-south Trans-Kazakhstan railroad. The unusual character of Koktas was indicated by its being placed directly under oblast jurisdiction even though it was only a low-ranking urban settlement. After having been listed in administrative-territorial guides in the early 1960s, Koktas was dropped from published reference books and maps. In recent elections, a separate election district was allocated to it, suggesting a nominal population of 25,000-30,000. The other area is identified with Kiyakhty, a rail station situated 110 miles farther south in Dzhambul Oblast, where another unspecified election district was set up. The importance of Kiyakhty, not publicly identified as an urban area, is suggested by the fact that it has regularly scheduled air service from Frunze. Nearby mining activities may be associated with two urban settlements: Aksuyek, 30 miles east, and Akbakay, 80 miles northwest of Kiyakhty rail station.

Uranium mining and processing have long been known as a significant activity in Central Asia, involving the Kirghiz, Tadzhik and Uzbek republics. Most of the mines are located around the margins of the Fergana Valley and other intermontane depressions in the Tien Shan mountain system. Several of the uranium deposits are associated with coal-bearing sediments.

In central Kirghizia, occupying a small mountain basin at 8,000 feet in the Tien Shan, is the uranium center of Min-Kush, which was established as an urban settlement in 1953. Its uranium operation is associated with the extraction of lignite, which is strip-mined and is burned in a local 10 MW power plant; lignite production has been ranging around 100,000 to 120,000 tons a year.[53]

Min-Kush, which has been described as having a well developed urban appearance despite its low hierarchical standing, reported a 1959 census figure of 10,907. In the 1970 census, it showed unusual decline to 6,218, but was again reported at 12,000 in 1975.[54] In an evident effort to diversify the economic base and provide female employment in the heavily male-oriented mining center, a light manufacturing plant producing ball-point pens and office copying machines began operations in 1971.[55]

Mill concentrates from Min-Kush are transported by truck over the 11,762-foot Tyuz-Ashuu Pass to a processing plant at Kara-Balta, 130 miles to the north, in the densely settled and economically developed Chu Valley of northern Kirghizia. This plant, which probably involves a further step in the manufacture of nuclear fuel, such as conversion to hexafluoride, also dates from the 1950s. It was originally established in conjunction with the urban settlement of Kosh-Tegirmen, whose formation was announced in late 1956. It had a population of 7,921 in 1959 and about 10,000–12,000 in 1970. In 1975, it was merged with the adjacent agricultural processing town of Kara-Balty (beet sugar, flour) to form the republic-level city of Kara-Balta, with a population estimate of 45,000.[56]

A smaller North Kirghiz operation apparently also feeding uranium concentrate to the Kara-Balta processing plant is the mining settlement of Kadzhi-Say on the south shore of Lake Issyk-Kul'. Kadzhi-Say, one of the earliest Soviet uranium mining sites, was established as an urban settlement in 1947 after a lignite deposit in the local Soguty intermontane basin had proved to be uraniferous. It reported a 1959 population of 5,922 and apparently remained in that range in 1970. A shaft mine, the Tsentral'naya (Central) mine, yields lignite for local consumption in a small power plant (about 100,000 tons). As in the case of Min-Kush, the economic base was diversified at Kadzhi-Say with the opening (in 1965) of a manufacturing plant producing semiconductor products such as transistors and rectifiers.[57]

Another source of uranium may be a cluster of mining settlements at the eastern end of the Chu Valley, including two older lead mines (Ak-Tyuz and Bordunskiy) and the newer Orlovka, which was given urban status in 1969. Orlovka is the seat of the so-called Kirghiz Mining and Metallurgical Plant, among whose products are yttrium and other elements of the yttrium group of rare-earth metals.[58] Some of the yttrium-bearing minerals are known to contain significant quantities of uranium and thorium. Also focused on the Kara-Balta processing plant may be another uranium mine to the west, just across the border of the Kazakh SSR, at Granitogorsk, an urban settlement founded in 1954.

On the margins of the Fergana Valley are also some of the oldest uranium sources of the Soviet Union, including the uranium-vanadium associations of Tyuya-Muyun (also spelled Tuya-Moyun), 15 miles southwest of the city of Osh, and the town of Mayli-Say, on the northeast margins of the Fergana Valley. Tyuya-Muyun was an early Soviet source of radioactive materials, mainly radium, before the advent of the uranium-based nuclear era. Continuing activity in the Osh–Tyuya-Muyun area is believed associated with a lignite strip mine, about 12 milcs southwest of Osh. The strip mine, called Almalyk, began operations in 1962 and yields about 200,000 tons of lignite a year. No separate urban area has been publicly identified with the operation, which lies outside the Osh city limits.[59]

The Mayli-Say site had been known to contain uranium-vanadium ores even before World War II, and thus became one of the earliest uranium mines to be developed. An urban settlement was first established in 1947, and was raised to the high status of oblast-level city in 1956, with a population of 24,711 in the 1959 census. The uranium reserves became depleted by the early 1960s and the town's population declined to around 15,000. In the middle of the decade, alternative sources of employment began to be provided with the construction of manufacturing plants oriented toward the electrical industry. A plant producing silicone rubber and other electric insulating materials began operations in 1965 in the southern suburb of Kok-Tash, and a large light-bulb factory followed in late 1967. The bulb factory turns out one-sixth of Soviet production.[60]

Two other uranium sites on the northern margins of the Fergana Valley are within the Kirghiz SSR. One is the urban settlement of Kyzyl-Dzhar, which was founded in 1952, about 14 miles southwest of the coal town of Tashkumyr. The other is Sumsar, also founded in 1952 and situated about 25 miles northwest of the city of Namangan, in the Uzbek section of the valley.

Farther west, in the Uzbek section, lies another former uranium mine, Uygursay, that became exhausted at an early stage and acquired light manufacturing to take up the employment slack. Uygursay (near Pap) was originally established as an urban settlement in 1947, indicating the start of uranium-related activities. But reserves were evidently among the smallest of any of the early mines, and mining appeared to have ceased in the middle 1950s. A rubber-goods fabricating plant was built on the site of the former uranium mill and went into operation around 1960.[61]

An important uranium mining and processing complex is located at the entrance of the Fergana Valley within the territory of the Tadzhik SSR. Here, on the south slopes of the Kurama Mountains, the uranium-vanadium mine of Taboshar was a source of radioactive minerals even before World War II, having been founded as an urban settlement in 1937. It grew in the 1940s, with the increasing need for uranium, and had a population of 11,283 in the 1959 census.[62] For 15 years Taboshar was not referred to in published Soviet reference books or shown on maps, but it re-emerged in 1974 together with the name of the city of Chkalovsk, a presumed uranium-processing center.[63] Published information about Chkalovsk disclosed that it had been established originally in 1944 as an urban settlement, 10 miles southeast of Leninabad, on the site of the village of Kostakoz. It vanished from public reference in 1956 on being raised to the status of oblast-level city, and had inferred census populations of 17,082 in 1959 and 24,239 in 1970.[64] A photograph published after its re-emergence in public print showed it to be a modern town with high-rise buildings bordering on an embankment along the Great Fergana Canal.[65] The city's main industrial establishment, presumed to involve advanced nuclear-fuel processing such as conversion to hexafluoride, has been referred to in published sources simply as the "construction plant" (*stroitel'nyy kombinat*).[66]

Administratively, Chkalovsk has two uranium mining and milling centers under its jurisdiction. One is Taboshar, 30 miles by road to the north; the other is a more recent development, the uranium-fluorspar mine of Naugarzan, founded in 1964, about 65 miles northeast of Chkalovsk. The Naugarzan mine, situated high in the Kurama Mountains at 6,000 feet elevation, was actually developed from the Uzbek side of the mountain range, and a three-mile tunnel was built just below the crest to carry fluorspar ore to a mill at Toytepa in the lower Angren valley. The Naugarzan mine and tunnel began operations in 1960–61.[67] In view of the apparent uranium association with fluorspar, Naugarzan then dropped out of public print, and its elevation to urban status in 1964 was not disclosed until it re-emerged 10 years later in conjunction with Chkalovsk.

To the north, in Uzbek SSR, there has long been uranium mining in the Chatkal Mountains east of Tashkent. Administratively the senior place among these uranium sites is the town of Yangiabad, overlooking the Angren valley, 10 miles northwest of the coal mine of Angren. Despite a small population–9,517 in the 1959 census and 8,560 in the 1970 census–Yangiabad has been an oblast-level city since the 1950s within Tashkent Oblast, where other cities of this rank are industrial centers of at least 50,000 inhabitants. Yangiabad has under his jurisdiction another uranium mining settlement, Krasnogorskiy, which was established in 1955 about 22 miles west of Yangiabad. A third mine, Chavlisay, which was located near Krasnogorskiy, operated only until 1960, when its urban status was abolished. Also under the jurisdiction of Yangiabad, is the Angren valley settlement of Chigirik, just south of Toytepa. Chigirik was the site of the Toytepa fluorspar mill in the 1950s, and was merged into a single urban area with adjacent Toytepa in 1960. But in 1963, Chigirik was again separated administratively and placed under the jurisdiction of distant Yangiabad, suggesting a uranium related activity such as some processing stage; Chigirik had a population of 2,486 in the 1970 census, making it the smallest urban place within Tashkent Oblast.

A major mineral extracting and processing center presumed to be associated with uranium is the town of Uchkuduk ("three wells") in the Kyzyl Kum desert of western Uzbekistan. Its urban development began in 1961, when it was established as an urban settlement on the site of the desert outpost of Aytym, at the south foot of the mountain range Bukan-Tau. The importance of the Uchkuduk development was suggested by the construction of a 175-mile railroad from Navoi, on the Central Asian rail system. Uchkuduk retained a low urban status in the first stage of its development, although the base town of Navoi was raised to oblast-level city in 1958, when it had a population of only about 5,000 (5,443 in the 1959 census). The presumed uranium center of Uchkuduk arose in a gold-bearing area explored in the late 1950s in the central Kyzyl Kum, and may be associated, in particular, with the Kokpatas gold mine. The Kyzyl Kum gold occurrences have been compared by some Soviet authors to the South African type, in which the Witwatersrand deposits are known to be important sources of uranium.[68] While other gold developments in the Kyzyl Kum, for example, the development of the Zarafshan-Muruntau district, received wide publicity in the Soviet press in the middle 1960s, no information was published about Uchkuduk. It reported a population of 18,393 in the 1970 census, was shifted from the jurisdiction of Navoi to that of Zarafshan in 1972, and was finally raised in urban status to that of oblast-level city in its own right in April 1978. Uchkuduk has been described in the Soviet press variously as an industrial town and as being based on an unspecified mineral discovery.[69] It is the seat of the so-called Navoi Mining and Metallurgical Complex.[70] Uchkuduk was one of several resource-oriented developments during the 1960s that accounted for a major influx of ethnic Russians into Bukhara Oblast, making this the only major area in Uzbekistan and in Central Asia as a whole in which the percentage of ethnic Russians in the total population increased between the censuses of 1959 and 1970. Navoi, after having served as a base town for the development of Uchkudukon went on to become an industrial center in its own right, with the first units of a large gas-fueled power plant opening in 1961, a nitrogen fertilizer complex in 1965, and a cement plant in 1977. It had a population of 87,000 in early 1977.

NOTES

[1]*Elektricheskiye Stantsii*, January 1976, pp. 206.

[2]*Nuclear Engineering International*, April 1977 (Supplement), pp. 22 and 56.

[3]For more information on the types of reactors installed at Beloyarskiy and Novovoronezhskiy, see A. M. Petros'yants et al., "Prospects of the development of nuclear power in the USSR," *Soviet Atomic Energy*, Vol. 31, No. 4 (October, 1971), pp. 1067–68, or I. D. Morokhov et al., *Atomnoy energetike XX let* (Moscow: Atomizdat, 1974), pp. 64–93.

[4]L. K. Davydov, *Gidrografiya SSSR–II* (Leningrad, 1955), p. 132, and *Soviet Power Reactors–1970*, U.S.A.E.C., 1970 (WASH-1175), p. 6.

[5]A. M. Petros'yants et al., "The Leningrad nuclear power station and the outlook for channel-type BWR's," *Soviet Atomic Energy*, Vol. 31, No. 4 (October, 1971), p. 1086. The RBMKP-2000 reactor is discussed in N. A. Dollezhal' and I. Ya. Yemel'yanov, "Experience in the construction of large power reactors in the USSR," *Soviet Atomic Energy*, Vol. 40, No. 2 (February, 1976), pp. 137–146. The RBMKP-2400 reactor, apparently a subsequent design, is described in *Atomnaya nauka i tekhnika v SSSR* (Moscow: Atomizdat, 1977), pp. 38–41.

[6]*Elektricheskiye Stantsii*, loc. sit.

[7]*Soviet Atomic Energy*, Vol. 40, No. 2 (February 1976), p. 219.

[8]*Pravda*, January 2, 1978, p. 1. These units were originally scheduled for 1978, but startup was delayed to 1979.

[9]Chuvilkin, O. D., "Atomic energy in the USSR in the 10th five-year plan," *Geografiya v Shkole*, 1976, No. 5, pp. 4–8, and Shelest, V. A., *Regional'nyye energoekonomicheskiye problemy SSSR* (Moscow: Nauka, 1975), pp. 285–293.

[10]"Why the Russians go all-out for nuclear power," *Business Week*, August 2, 1976, pp. 52–53.

[11]The BN-350, and other Soviet breeder reactors, are reported on in some detail in *Soviet Power Reactors–1970* (Report of the U.S.A. nuclear power reactor delegation visit to the U.S.S.R., June 15–July 1, 1970), USAEC, 1970 (WASH-1175).

[12]*Nuclear Engineering International*, October 1971, p. 825, and *Nuclear News*, March 1978, p. 63; also *Atomnaya nauka i tekhnika v SSSR* (Moscow: Atomizdat, 1977), pp. 55 and 68–70.

[13]For more technical details on the Soviet fusion research program, consult L. A. Artsimovich, "Research in controlled thermonuclear fusion in the USSR," *Soviet Atomic Energy*, Vol. 31, No. 4 (October, 1971), pp. 1119–1128, and *World Survey of Major Facilities in Controlled Fusion* (Vienna: I.A.E.A., 1970).

[14]A. N. Komarovskiy, *Design of Nuclear Plants* (Moscow: Atomizdat, 1965); translated by the Israel Program for Scientific Translations (1968), p. 159 of translation.

[15]A. S. Belitskiy et al., in *Environmental Aspects of Nuclear Power Stations* (Vienna: IAEA, 1971), pp. 179–181. These criteria are repeated in the 1973 Ioirysh article cited in note 18.

[16]Komarovskiy, op. cit., pp. 132–134.

[17]Petros'yants et al., "Prospects . . . ," p. 1074.

[18]A. I. Ioirysh, "Legal regulation of environmental effects of atomic energy," *Sovetskoye gosudarstvo i pravo*, 1973, No. 6, translated in *Soviet Law and Government*, Vol. XII, No. 4 (Spring 1974), p. 50.

[19]*Soviet Power Reactors–1970*, op. cit., p. 11.

[20]*Pravda*, July 30, 1977, p. 2.

[21]J. Lewin, "The Russian approach to nuclear reactor safety," *Nuclear Safety*, Vol. 18, No. 4 (July–August, 1977), pp. 438–450. For more information on radiation safety at plant sites, gaseous emission, results of monitoring programs, etc., see I. D. Morokhov et al., *Atomnoy energetike XX let* (Moscow: Atomizdat, 1974), pp. 172–185; and N. G. Gusev, "Radiation safety at nuclear power stations," *Soviet Atomic Energy*, Vol. 41, No. 4 (October 1976), pp. 890–896.

[22]"The USSR nuclear power program: Report of the U.S. reactor delegation," *Nuclear News*, October 1970, p. 20.

[23]L. M. Luzanova et al., in *Handling of Radiation Accidents* (Vienna: I.A.E.A., 1969), p. 371.

[24]V. F. Ostashenko et al., "Some safety problems in nuclear power plants with water-cooled, water-moderated power reactors," *Soviet Atomic Energy*, Vol. 30, No. 2 (February 1971), p. 157.

[25]K. I. Shchelkin, *Sovetskaya atomnaya nauka i tekhnika* (Moscow, 1967); translated by the Joint Publications Research Service as JPRS-45331, p. 126 of translation.

[26]A. I. Ioirysh, op. cit., p. 48. The 1976 modifications are reported in *Atomnaya nauka i tekhnika v SSSR*, op. cit., pp. 190–191.

[27]B. S. Kolychev, et al., "Handling of radioactive waste at Soviet nuclear power stations," *Soviet Atomic Energy*, Vol. 26, No. 5 (May 1969), p. 469.

[28]*Soviet Power Reactors–1970*, op. cit., p. 6, and Belitskiy, op. cit., pp. 178–179.

[29]Interviews with Dr. Vadim Artemkin of the State Committee for the Utilization of Atomic Energy and a member of the 1977 USSR Los Angeles Exhibition staff, San Diego, Nov. 22, 1977, and Los Angeles, Nov. 26, 1977.

[30]V. F. Bagretsov et al., "Operation of the facility for deep burial of liquid radioactive waste," *Soviet Atomic Energy*, Vol. 31, No. 2 (August 1971), p. 847.

[31]International Atomic Energy Agency, *Disposal of Radioactive Wastes into Seas, Oceans and Surface Waters* (Vienna: I.A.E.A., 1966), p. 373.

[32]For a more detailed discussion of Soviet research on possible waste disposal techniques, see Joint Publications Research Service, *Disposal of Radioactive Wastes* (USSR), (JPRS-58764), dated 17 April 1973.

[33]Nonelectrical uses of nuclear technology in the Soviet economy are discussed in A. K. Kruglov, "Atomic science and technology in the national economy of the USSR," *Soviet Atomic Energy*, Vol. 40, No. 2 (February 1976), pp. 123–136.

[34]*Nuclear Engineering International*, Vol. 22, No. 58 (April 1977), pp. 56–62.

[35]Kruglov, op. cit.; see also K. Parker, "Engineering with nuclear explosives–Achievements and prospects," *Journal of the Institution of Nuclear Engineers*, Vol. 12, No. 6 (Nov.-Dec. 1971), p. 159.

[36]For an account of the *Arktika* voyage, see *Polar Geography*, Vol. I, No. 4, October-December 1977, pp. 325–326; for the *Sibir'* trip, see *Polar Geography*, Vol. II, No. 3, July-September 1977, pp. 219–220.

[37]U.S. Central Intelligence Agency. *Nuclear Energy* (ER 77-10468), August 1977, and Organization for Economic Cooperation and Development. *Uranium–Resources, Production and Demand*, December 1975.

[38]For discussions of the geology of Soviet uranium deposits, without reference to specific geographical locations, see: *Atomnaya nauka i tekhnika v SSSR*, op. cit., pp. 120–134; *Mestorozhdeniya urana i redkikh metallov* [Deposits of Uranium and Rare Metals] (Moscow: Atomizdat, 1976), 288 pp.; A. B. Kadzhan, *Osnovy razvedki mestorozhdeniy redkikh i radioaktivnykh metallov* [Fundamentals of Exploration for Rare and Radioactive Metals] (Moscow: Vysshaya Shkola, 1966), pp. 50–74.

[39]C.I.A., *Nuclear Energy*, op. cit., p. 27.

[40]*Pravda Ukrainy*, Nov. 1, 1967; *Pravda*, Dec. 11, 1967; *Vokrug Sveta*, 1967, No. 10; S. A. Pogodin and E. P. Libman, *Kak dobyli sovetskiy radiy* [How the Soviet Radium was Mined] (Moscow: Atomizdat, 1971), p. 167.

[41]Theodore Shabad, *Basic Industrial Resources of the USSR* (New York: Columbia University Press, 1969), p. 189; *Soviet Geography*, September 1969, pp. 424–425.

[42]*Nõukogude Eesti* [Soviet Estonia] (Tallinn: Valgus, 1975), p. 42; Shabad, op. cit., p. 213.

[43]*Stavropol'skiy kray* (Stavropol', 1961), p. 339.

[44]Krasnokamensk does not appear among lists of places in the published 1970 census (15,000 and over), but it is included in the 12,000–20,000 class size on 1970 census maps in: V. V. Vorob'yev, *Naseleniye Vostochnoy Sibiri* [Population of East Siberia] (Novosibirsk: Nauka, 1977], Figs. 14, 15, 21; the 1974 population estimate is from: *SSSR. Administrativno-territorial'noye deleniye soyuznykh respublik–1974* [USSR. Administrative-Territorial Divisions of the Union Republics–1974] (Moscow: 1974), p. 642.

[45]*Soviet Geography*, January 1978, p. 45.

[46]*Sovetskaya Rossiya*, April 20, 1967.

[47]The only publicly available map that shows Stepnogorsk is the National Geographic map of the Soviet Union, February 1976.

[48]A. B. Avakyan and V. A. Sharapov. *Vodokhranilishcha gidrostantsiy SSSR* [Reservoirs of Hydro Stations in the USSR] (Moscow: Energiya, 1977), 3d edition, p. 383.

[49]*Soviet Geography*, December 1975, p. 691–692.

[50]*Kazakhstanskaya Pravda*, May 5, May 14 and May 31, 1975.

[51]*Soviet Geography*, May 1978, p. 354.

[52]*Soviet Geography*, December 1975, p. 692; *Sovetskaya Kirgiziya*, Dec. 20, 1978, p. 4.

[53]F. T. Kasharin et al. *Uzgenskiy kamennougol'nyy basseyn* [The Uzgen Coal Basin] (Frunze: Ilim, 1975), Table 2, facing p. 40; *Problemy razvitiya i razmeshcheniya proizvoditel'nykh sil Kirgizskoy SSR. Ekonomicheskoye rayonirovaniye Kirgizskoy SSR* [Problems of Development and Location of Productive Forces in the Kirghiz SSR. Economic Regionalization of the Kirghiz SSR] (Frunze: Ilim, 1976), p. 105.

[54]*Sovetskaya Kirgiziya*, July 24, 1975.

[55]*Sovetskaya Kirgiziya*, April 14, 1971.

[56]The history of the Kosh-Tegirmen nuclear fuels plant can be traced in part through the career of its long-time director, Vasiliy N. Mindrul, who was associated with the uranium industry since its inception. He represented Min-Kush in the 1967 election, and the

Kosh-Tegirmen plant in the 1971 and 1975 elections, when it was identified as a manufacturing plant for "experimental elements" or simply "construction articles." On his death in August 1978, Mindrul was said to have served as director of an "important industrial establishment" of the Kirghiz SSR. He was succeeded by Anatoliy P. Yezhov.

[57]*Sovetskaya Kirgiziya*, July 9, 1967; Jan. 11, 1968; Feb. 3, 1972.

[58]*Sovetskaya Kirgiziya*, Oct. 30, 1977.

[59]*Sovetskaya Kirgiziya*, Aug. 25, 1968; Dec. 28, 1968; April 12, 1970; Aug. 13, 1975.

[60]*Pravda*, May 15, 1970; *Sovetskaya Kirgiziya*, March 8, 1975.

[61]*Pravda Vostoka*, Aug. 3, 1957.

[62]The 1959 figure was not explicitly reported, and was obtained as a residual.

[63]*Soviet Geography*, September 1975, p. 472.

[64]Chkalovsk census figures were derived as residuals.

[65]*Kommunist Tadzhikistana*, April 5, 1975.

[66]*Kommunist Tadzhikistana*, May 30, 1975. Like its Kirghiz counterpart, the Kosh-Tegirmen plant, the Chkalovsk establishment has had a long-time director, Vladimir Ya. Oplanchuk.

[67]*Kommunist Tadzhikistana*, Dec. 18, 1959; *Sovetskaya Latviya*, Oct. 23, 1960; *Pravda Vostoka*, Aug. 3, 1961.

[68]*Mineral'no-syr'yevyye resursy Uzbekistana* [Mineral Raw Material Resources of Uzbekistan] (Tashkent, FAN, 1976), Part 1, p. 97.

[69]*Mineral'no* . . . , op. cit., p. 45; *Pravda Vostoka*, Dec. 27, 1969.

[70]The history of the Uchkuduk plant can be traced through its directors. In the 1960s, Zarab P. Zarapetyan was identified as director of an unspecified industrial establishment; in the 1970s, Anatoliy A. Petrov was identified as the director of Navoi Mining and Metallurgical Complex.

Chapter 7

THE ROLE OF THE ELECTRIC POWER INDUSTRY

Electrification and economic modernization have long proceeded hand in hand. As the most versatile and the most conveniently manipulated type of energy, electric power has been instrumental in the rapid rise of labor productivity. The growth of electrification greatly enhanced the efficiency of factory operations formerly constrained by the limitations of mechanical power transfer and plant layout imposed by the coal-burning factory steam engine. To the service and agricultural sectors, too, electrification has helped to impart a flexibility and, in some instances, scale economies they did not previously enjoy. Darmstadter argues persuasively that, in the United States, electrification was strongly responsible for the increased rate of GNP growth per unit of energy input despite the large, though declining, conversion losses in the generation of electric power.[1] In the USSR, the crucial role of electrification found its lapidary expression in Lenin's famous equation, "Communism is Soviet power plus electrification of the whole country." Soviet writers often group the electric power, modern engineering and chemical industries into a special category, as branches forming the basis of technological progress.

Chapters 5 and 6 dealt with primary electricity—the small but important and (at least in the case of nuclear power) rapidly growing portion of the industry that contributes to aggregate energy supply. Most of electric power generation, by burning up fossil fuels, adds nothing to such supplies and is a form of utilization, a transformation of conventional fuel resources. Still, the electric power industry requires comprehensive and separate treatment because of its special characteristics and increasing role in the total energy system.

First, the continuous nature of production and the virtual impossibility of storage exacerbate the reserve problem. Unlike inventories, which are eventually used up and are, at any rate, small compared with total material inputs, reserves in power generation consist of idle, thus wasted, capacity, which on a yearly basis forms a large share of total equipment time. Technological developments to cope with this problem by centralization and increasing system size bring a corresponding weakness of instability to the entire system. Second, partly as a result of nonstorability, fixed investment in the electric power industry is large both absolutely and relatively. In the USSR, the value of this industry's productive fixed capital has for long far exceeded that in all fuel industries combined, and much of the investment in the latter branches is made expressly to supply power plants.[2] In nine of the 15 union republics, productive fixed assets in the electric power industry exceed in value those of all other industrial branches, and on the national scale they take second place only to those in machine building–metalworking.[3] Per unit of labor, electric power is by far the most capital-intensive industry, and per ruble of output, the coefficient of fixed capital in this industry reaches three-fourths of the capital coefficient in all fuel industries combined.[4] Finally, electricity is not only clean and convenient but also a highly specialized form of energy, irreplaceable in a growing number of technological processes regardless of its cost relative to other forms of energy. Its price elasticity, like that of gasoline, is far below those of other energy forms. And in any developed country, the disruption of power supply probably results in more immediate and widespread consequences than than of any other energy form.

SYSTEM FORMATION AND THE PROBLEM OF A DUAL ECONOMY

Resource utilization in the electric power industry is inseparable from the problems of scale and system formation. The industry is extremely sensitive to scale economics, while the nonstorability of the product and the violent fluctuation in demand necessitate careful planning and allocation of reserve capacities. By the interconnection of plants into coordinated systems and the interties of the systems themselves, idle capacity can be reduced and reliability increased.

Unlike the numerous networks in the United States, which grew up piece-meal and in a somewhat haphazard manner, Soviet plans for electrification envisaged from the beginning the eventual creation of a unified power grid over the whole country. The movement toward extending and unifying systems has been purposeful, systematic, and rapid. By World War II, three large centralized systems had grown up–Central, Southern, and Urals–and, together with the smaller Leningrad system, produced roughly three-fifths of all electricity in the country.[5] A centralized European grid was finally forged in the late 1950s to which the Leningrad, Baltic and Transcaucasian networks are now tied in. The large-scale power development in Central Siberia (beginning, except for the Kuzbas, only some 15 years ago) has also consciously striven toward a unified

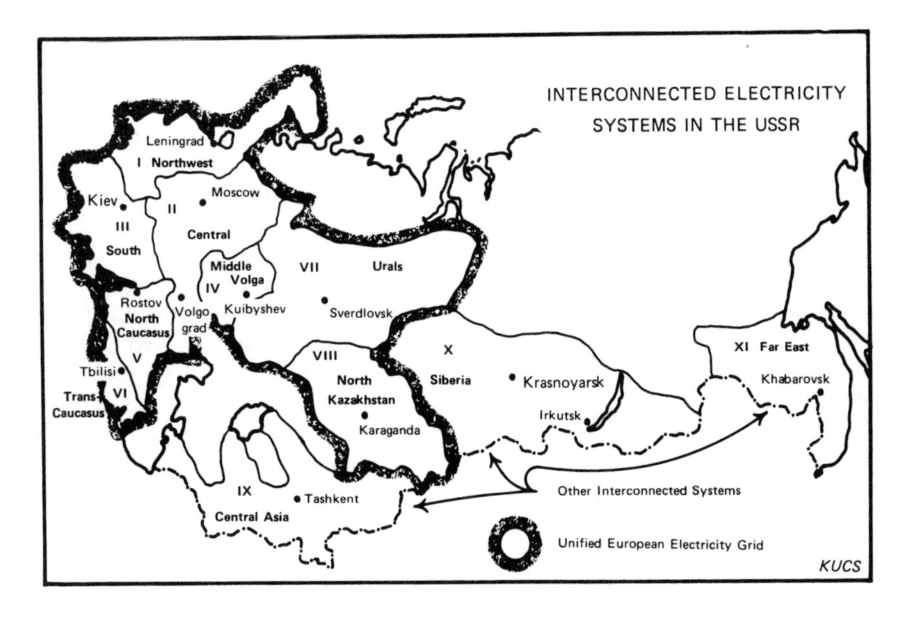

Fig. 9. Interconnected electricity systems in the USSR.

network for the region. A unified Siberian grid extends from East Kazakhstan to Ulan-Ude, east of Lake Baikal, and it is the second largest system in the country (Fig. 9). In 1978 it was linked by a 500 KV line with the European grid; however, significant power transfers will still have to await the planned construction of extra high voltage interconnections.

This conscious policy toward a unified power system has had several diametrically opposed implications. On the one hand, it has increased efficiency, reliability and reduced idle capacity. On the other, it was instrumental in the development of a "dual economy" within the electric power industry in that the factors of production had different marginal productivities and rates of return in different segments of that industry until recently. The pattern had been obvious down to the late 1960s; only in the last few years did the split in the Soviet power economy become negligible.

The Vanishing World of Small Scale Electricity

The preoccupation with the creation of a supergrid, an excellent example of Soviet "gigantomania," had given rise to the parallel growth of small-scale electricity, especially in the countryside. For many decades, the Ministry of Electric Power had concerned itself only with supplying power to large cities and industrial centers and had an interest in efficiency only within this domain. Thus, it built only large central-electric stations and, until recently, only high-voltage links. Until 1953, the stations under the Ministry of Electric Power were actually prohibited from supplying state and collective farms and some other categories of consumers, and the latter remained low on the Ministry's priority scale until the late 1960s.[6]

The low priority sectors were left almost entirely to their own devices and had to improvise vital power supplies from their own meager resources. The result has been the spread of small, unconnected local stations all over the country, even in the large cities. In 1962, there were some 150,000 of them in the RSFSR and the Ukraine alone compared with fewer then 500 central stations.[7] Of all Soviet power plants, numbering over 212,000 in 1965, 98.5% had capacities below 500 KW (Table 41). These stations operated irregularly (fewer than 2,000 hours compared with 4,734 for all stations),[8] with dismal fuel efficiency, and required a large number of workers.[9] The problem was especially acute regarding agricultural thermal stations, whose average capacity in 1967 reached only 54 KW and most of which have been at work a mere 9 to 12% of the time.[10] As late as the end of 1966, some 110,000 of them still provided three-tenths of all electricity consumed in Soviet villages.[11]

Table 41. Size Distribution of Soviet Electric Power Stations, 1950, 1960 and 1965

Totals	1950	1960	1965
Number	56,780	173,930	212,000
Capacity (MW)	19,614	66,721	115,033
Output (mill. KWH)	91,226	292,274	506,672

	Capacity range in KW (%)						
By year	<500	500 to 999	1,000 to 9,999	10,000 to 49,999	50,000 to 99,999	100,000 and over	Total
Number							
1950	96,40	1.47	1.69	0.29	0.07	0.08	100.0
1960	98.23	0.57	0.92	0.16	0.04	0.08	100.0
1965	98.52	0.40	0.80	0.14	0.04	0.10	100.0
Capacity							
1950	12.8	2.9	14.2	18.2	14.2	37.1	100.0
1960	14.2	1.0	6.9	8.6	8.0	61.3	100.0
1965	11.7	0.6	5.1	6.8	5.4	70.4	100.0
Output							
1950	3.8	1.5	10.6	20.3	18.2	45.6	100.0
1960	4.2	0.5	5.0	9.7	9.7	70.9	100.0
1965	4.8	0.2	2.6	5.8	5.5	81.1	100.0

Source: Adopted from Vilenskiy, *Po leninskomu puti sploshnoy elektrifikatsii* (Moscow: Nauka, 1969), pp. 130–131 and 135–36.

Such a distribution of reserve capacity and generally suboptimal use of production factors (i.e., human and natural resources) were closely linked to the preference scale of the Soviet planned economy. It may even be argued that from the viewpoint of central planners, such a distribution was logical until quite recently. Although the capital cost of such rural pygmies far exceeded the investment needed to link a collective farm to a central station within a 75-mile radius,[12] neither the construction nor the high operating costs of rural electric plants had to come out of planners' resources. Links to the centralized network most certainly did. The isolated rural plants have used low-quality local fuels, such as peat and agricultural wastes, and surplus agricultural labor characterized most of the Soviet country side until not long ago. However large, only part of the savings captured here could have been devoted to objectives central planners have been intent in maximizing. In addition, because of highly irregular demand, agricultural consumers tend to depress the efficiency indices of centralized systems. It is not surprising, therefore, that Soviet planners, bent on reducing reserves in high-priority sectors, and in fields from which the savings could be most directly devoted to their own objectives, so long contrived to avoid the burden of rural electrification and of low-priority sectors in general.

The shrinking labor supplies and the need to modernize on an ever expanding front had largely eliminated the utility of such a "dual economy" to the Soviet system's directors by the end of the Khrushchev era. Since the early or middle 1960s a rather determined effort has been made to connect collective and state farms to the grids. The output of electric power from agricultural stations in both the USSR and the Russian Republic, for example, continued to grow until 1965 but then declined drastically.[13] By the late 1960s over four-fifths and by 1976 almost 99% of all rural power consumption was provided from unified networks.[14] No data on the number of rural stations have been available since 1968, but their total capacity declined far less than their output since 1965 (44% compared with 85%).[15] This seems to indicate that many collective farms still may not feel entirely secure about their power supply from the unified grids and prefer to keep these tiny plants for security. There is evidence that in the Asian USSR a large number of small, isolated stations were still in existence at the beginning of the 1970s both in populous and in remote regions,[16] although centralization must have been nearing completion since then.

Unified Systems

Unified systems allow the construction of larger and more economical generating units than would be possible with isolated operations. Interconnections also provide large capacity savings from time zones, random load diversity and from the exchange of power among regions. Because of its northerly location and minimal use of air conditioning, the USSR lacks regions with a clear summer peak, but there are 11 time zones and several waterpower zones with different hydraulic characteristics.

The sizes of the Soviet interconnected systems are impressive indeed. At the beginning of 1976, 11 large energy grids produced 994 billion KWH (Table 42),

96% of the Soviet total.[17] Eight of these grids formed the unified European grid with transmission lines of 220 KV and over, covering 6.5 million km^2, with a population of over 200 million people.[18] More than 950 stations are tied into this huge system, which by 1977 reached a total capacity of 160,000 MW and produced 837 billion KWH of power.[19] The unified Siberian grid, stretching from Omsk and Barnaul in the west to Ulan-Ude, east of Lake Baikal, is the second largest system, with 27,400 MW capacity and 140 billion KWH of output at the end of 1975 (Table 42).

The sheer size of these systems permits electricity to be generated from ever larger power stations, thus reaping substantial scale economies even where local consumption is insufficient. And these stations can operate at high availability (or plant factor*) even where the local load curve** is highly irregular. The number of hydro and thermal plants with installed capacities of over 1,000 MW reached 66 by the end of 1977, and they accounted for close to half of total Soviet capacity.[20] The number of such stations was planned to grow to 76 by 1980 when they were to contribute 53% of total capacity.[21] (For lists of major hydro, nuclear and thermal stations, see Tables 33, 38 and 43.)

The expansion of system size through high-voltage links resulted in a significant smoothing of the load curve. Soviet engineers and economists estimated an ultimate capacity saving of up to 4,400 MW from the interconnection of the individual systems of the European USSR.[22] In 1965, 1970 and 1975, peak shavings of 1,200, 1,685 and 1,185 MW capacity were achieved respectively through the shunting of electricity among subsystems of the grid. Each equivalent to a large power plant, they represented 2.7, 2.0 and 1.0% savings from the maximum load of the unified European system.[23] Despite the reduction of the work week, the relative and absolute increase of power supply to sectors with fluctuating consumption patterns and low load factors (e.g., agriculture, the household-municipal economy, etc.), the availability of all Soviet power stations improved from 50% to 55.6% between 1960 and the end of 1976.[24]

Because of the dominance of industry, particularly heavy industry, in electricity consumption, power demand in the USSR is more regular than in the

*The plant factor equals actual yearly use/maximum possible use, where maximum use is the number of hours of continuous operation theoretically possible, i.e., 8,760 per year. The term "capacity factor" is generally applied to electric systems; its meaning is the same as that of plant factor, which is commonly employed with respect to individual plants. Availability is also a frequently used term for the same concept.

**The term "load curve" refers to the regularity of power demand and, when graphed, shows the fluctuation of electricity consumption for the period it is drawn. The load factor gives a quantitative measurement of that fluctuation for the period under consideration. It is computed as average load/peak load. The closer the index is to unity, the more regular is the demand for power. Since electricity cannot be stored, low load factors result in much idle capacity, mitigated by the possibility of the exchange of power among neighboring systems with different load curves. In addition to the idle capacity resulting from low load factors, power systems must maintain emergency reserve capacity for outages and plant overhauls.

Table 42. Characteristics of Soviet Electric Systems (1975)

Unified system	Total capacity (mill. KW)	Percent hydro	Maximum load (mill. KW)	Annual production (bill. KWH)	Load factor[a]	Capacity factor[b] or availability
Unified European Grid	153.097	14.3	126.0	781.0	.71	.58
of which:						
Central Russia	29.830	12.4	29.1	156.1	.61	.60
Middle Volga	12.754	28.7	10.6	60.8	.66	.54
Urals	25.439	6.9	21.4	148.8	.79	.67
Northwest	23.009	16.8	15.8	96.0	.69	.78
South (Ukraine)	38.445	9.9	29.9	206.3	.79	.61
North Caucasus	8.363	20.4	6.8	38.9	.65	.53
Transcaucasia	8.018	28.9	5.9	35.3	.68	.50
North Kazakhstan	7.240	14.5	6.0	38.8	.74	.61
Central Siberia	27.354	47.3	20.3	140.1	.75	.59
Central Asia–Southern Kazakhstan	11.695	28.4	7.8	49.6	.73	.48
Far East	5.866	11.8	4.5	23.3	.59	.40
All systems of USSR	198.012	19.6	–	994.0	–	.57

Notes:

[a]Load factor = average annual load/peak load × 8760 (i.e., number of hours in a year).

[b]Capacity factor = average annual load/maximum possible load.

Average annual load = yearly production from the system.

Maximum possible load = installed capacity × 8760 (i.e., the number of hours of continuous operation theoretically possible).

Source: Taken and computed from *Energetika SSSR v 1976–1980 godakh* (Moscow: Energiya, 1977), p. 165.

United States. The unified European grid as a whole reached a load factor of .71 in 1975. With the exception of one system, load factors among the various Soviet systems ranged from .65 to .79; these are factors that system managers in the United States would be happy to accept. The highest load factors are observable for the Urals, the South (Ukraine) and Central Siberia, where industries with continuous processes dominate. Still, load variation is a serious problem for Soviet (as for Western) power engineers. With increased electrification of nonindustrial sectors, the peaking problem will intensify in the future, especially in the European USSR.

Despite the high load factors and the generally high level of electric power technology, the Russians have long experienced a difficulty in covering the peak

Table 43. Soviet Thermal Power Stations of More Than 1,000 MW (as of the beginning of 1978)

Region and station name	Town–location	Fuel	Capacity (MW)
NORTHWEST			
Kirishi	Kirishi	oil	2070
BALTIC			
Baltic	Narva	shale	1435
Estonian	Narva	shale	1610
Lithuanian	Elektrenai	oil, gas	1800
BELORUSSIA			
Lukoml'	Novolukoml'	oil	2400
CENTRAL RUSSIA			
Konakovo	Konakovo	oil, gas	2400
Kostroma	Volgorechensk	oil, gas	2400
Kashira	Kashira	lignite	2018
Cherepet'	Suvorov	lignite	1500
No. 22 Heat and Power	Moscow	oil, gas	1250
Ryazan'	Novomichurinsk	lignite	1200
No. 23 Heat and Power	Moscow	oil, gas	1150
No. 21 Heat and Power	Moscow	oil, gas	1100
VOLGA			
Zainsk	Zainsk	oil, gas	2400
Karmanovo	Neftekamsk	oil	1800
Lower Kama Heat and Power	Naberezhnyye Chelny	oil	1005
UKRAINE			
Zaporozh'ye	Energodar	coal, oil	3600
Uglegorsk	Svetlodarskoye	coal, oil	3600
Krivoy Rog No. 2	Zelenodol'sk	coal	3000
Burshtyn	Burshtyn	coal, oil	2400
Gottwald (ex-Zmiyev)	Komsomol'skoye	coal, gas	2400
Pridneprovsk	Dnepropetrovsk	coal, gas	2400
Voroshilovgrad	Schast'ye	coal, gas	2300
Starobeshevo	Novyy Svet	coal	2300
Slavyansk	Slavyansk	coal, gas	2100
Ladyzhin	Ladyzhin	coal	1800
Tripol'ye	Ukrainka	coal	1800
Kurakhovo	Kurakhovo	coal	1460
MOLDAVIA			
Moldavian	Dnestrovsk	coal	2020
NORTH CAUCASUS			
Novocherkassk	Novocherkassk	coal, oil, gas	2400
Nevinnomyssk	Nevinnomyssk	oil, gas	1370
Stavropol'	Solnechnodol'sk	gas	1200
TRANSCAUCASIA			
Tbilisi	Tbilisi	oil, gas	1250
Razdan	Razdan	oil, gas	1110
Ali-Bayramly	Ali-Bayramly	oil, gas	1100

Table 43. (cont'd)

Region and station name	Town–location	Fuel	Capacity (MW)
URALS			
Troitsk	Troitsk	coal	2500
Reftinskiy	Reftinskiy	coal	2300
Iriklinskiy	Energetik	oil	1800
Verkhniy Tagil	Verkhniy Tagil	coal, oil, gas	1625
Sredneural'sk	Sredneural'sk	gas	1198
Yuzhnoural'sk	Yuzhnoural'sk	coal	1000
WEST SIBERIA			
Surgut	Surgut	gas	1494
Tom'-Usa	Mezhdurechensk	coal	1300
Belovo	Belovo	coal	1200
EAST SIBERIA			
Nazarovo	Nazarovo	lignite, oil	1400
Krasnoyarsk	Krasnoyarsk	lignite, oil	1200
No. 10 Heat and Power	Irkutsk	coal	1150
Krasnoyarsk Heat and Power	Krasnoyarsk	lignite, oil	1005
KAZAKHSTAN			
Yermak	Yermak	coal	2400
Dzhambul	Dzhambul	gas	1230
CENTRAL ASIA			
Syrdarya	Shirin	gas	2100
Tashkent	Tashkent	gas	1930

Sources: *Energetik*, 1978, No. 7 (inside front cover); I. A. Yerofeyev, *Elektroenergetika SSSR v novoy pyatiletke (1971-1975)* [Electric Power of the USSR in the New Five-Year Plan (1971–75)] (Moscow: Prosveshcheniye, 1972), pp. 24–25.

and intermediate loads.* For the sharp, short-time peaks, the Russians have not yet developed adequate gas-turbine technology, pump-storage facilities and hydro units for peaking purposes. At the same time, the increasing importance of large generating units of 300 MW or more, which require several hours to start up and a long time to shut down and which cannot operate at low capacity, aggravates the problem of coping efficiently with semi-peak fluctuations (principally daytime versus nighttime demand).[25] In the mid-1970s, poorly maneuverable units (i.e., fit primarily for the base load), such as heat and power stations,

*To minimize interruptions or voltage reductions in electrical service, utilities must be able to meet the peak demand, which may be twice the low demand point. There are major fluctuations in weekly and seasonal load curves. Planning generating capacity for varying load demand is divided into three general categories: (1) base loaded, to operate at a high, and essentially constant load factor, generally on the order of 70%; (2) intermediate loaded, to respond to a varying load demand, with an annual plant capacity factor around 40%; (3) peak loaded, to meet peak demand and shut down after the peak demand period ends, with an annual plant capacity factor of about 10%.

nuclear plants and whole blocks of condenser stations, accounted for 61 to 64% of all generating capacity in the important Central and Northwest systems and are expected to account for 71 to 81% by 1980. In the big Southern system of the Ukraine, the corresponding share is expected to rise from 78 to 87%.[26] Given these problems, there is a view that the entire present and most of the potential hydro capacity west of the Urals and the Caspian Sea should be used for peaking purposes.[27]

A closer examination of Table 42 permits further inferences about the operation and efficiency of the Soviet power systems. In general, they are used more intensively (i.e., at higher capacity factors, or availability) than the United States counterparts. The availability of all Soviet systems in 1975 was .57. The unified European grid and that of Central Siberia, which jointly account for 87% of all Soviet power and 93% of all power from grids, had capacity factors of .58 and .59, respectively, with half the systems in the European grid exceeding .60. By way of comparison, United States utilities operated with capacity factors well below .50 (.43 in 1975), with correspondingly larger idle equipment time.[28] But Soviet systems are faced with more regular demand, thus higher load factors, and carry less reserve capacity in relation to the peak load than United States utilities. Equalizing the load factors and reserves for the two countries results in almost identical capacity factors, suggesting that Soviet power system managers are no more efficient in utilizing equipment than their American colleagues.[29]

An excess of installed capacity relative to maximum load, especially when coupled with high load factor (i.e., regular demand), indicates unused equipment and/or inefficient operation. At least in some regions it may also suggest that large steam turbines are forced to cope with semi-peak loads, a task for which they are technologically unsuited; and for systems with substantial hydro shares, it may indicate low water regimes. On the other hand, too close an agreement between installed capacity and maximum load indicates vulnerability in times of emergency. According to Table 42, the crucial Central Russian system around Moscow lacks virtually any emergency reserve capacity (the ratio of maximum load to total capacity being almost .98) and must rely on borrowed power at times of outages. The Northwest, Caucasus and Middle Volga systems (Fig. 9) with high shares of waterpower, serve important reserve functions for the whole of the unified European Grid, but irregular stream flow may also play a part in the great deal of idle capacity shown.

Very instructive are the data on low availability in Central Asia, the South (Ukraine) and much of Kazakhstan despite the high load factors, which testify to small fluctuations in demand. The relationship between load and capacity is unfavorable even in the Urals and Central Siberia, despite the regular consumption pattern. Table 42 thus supports qualitative findings about the low efficiency of much generating capacity in the Moslem areas of Soviet Asia, the western half of the Ukraine and, to a lesser extent, even in the Urals. Nearly one-third of all Ukrainian power stations for example, are said to be obsolete[30] and apparently operate unreliably. The problem is more serious in Central Asia where, in addition, peaking units are in particularly short supply, requiring much condensing capacity to cover only semi-peak demand and large water turbines

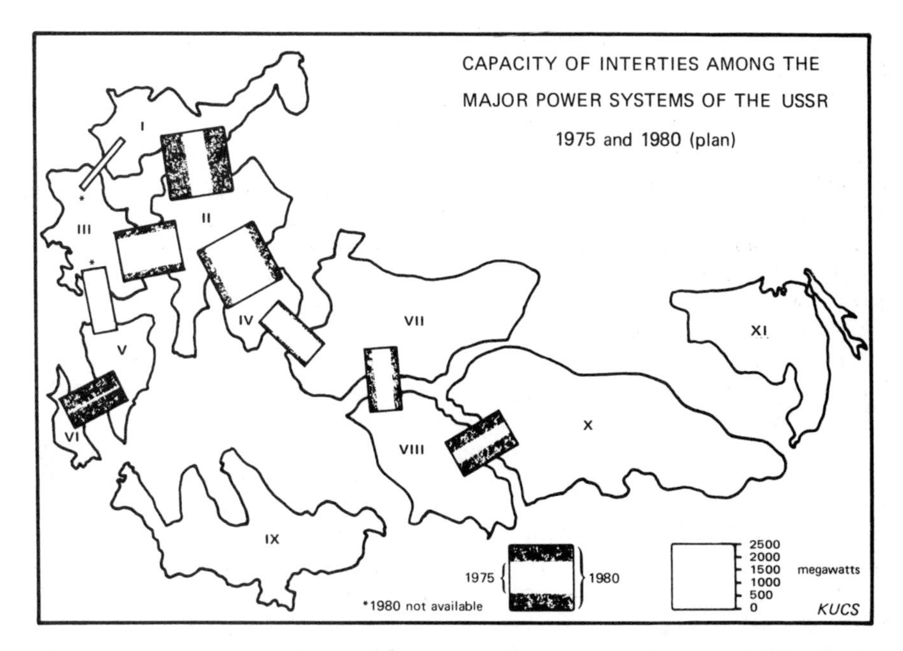

Fig. 10. Capacity of interties among Soviet power systems (1975 and 1980 plan).

I–Northwest	V–North Caucasus	IX–Central Asia–South Kazakhstan
II–Central Russia	VI–Transcaucasia	X–Central Siberia
III–South (Ukraine)	VII–Urals	XI–Far East
IV–Middle Volga	VIII–North Kazakhstan	

standing by for the peaks.[31] Water projects in these Moslem republics also serve multipurpose functions, and irrigation tends to take precedence over power needs, lowering stream flow and increasing idle capacity over much of the year. A good example was the idleness of installed generating capacity in the Toktogul' and Nurek plants in the middle 1970s as reservoir capacity had to be drawn down to meet irrigation needs (see pp. 144-146).

In most parts of Siberia covered by the unified Siberian grid, the importance of industries with continuous processes means high load factors in power consumption. Coupled with more regular water flow at major hydro stations, this should result in higher capacity factors in the Siberian system compared with those of the unified European grid. Yet the two unified systems operate with virtually the same capacity factor, and as late as the early 1970s, availability in the Siberian grid was actually lower.[32] As in the past, the Russians seem to continue to build ahead of local needs and suffer delays in the construction of heavy power-consuming industries. In addition, the high degree of railway electrification, a sector with a low load factor, resulted in a significant increase of idle capacity.[33] The unfavorable indices for the Far Eastern grid are a reflection of many things: the immature development of the system (it is only a few years old), its isolation, the seasonal nature of much economic activity in the region and irregular water flow.

Intersystem transfers today are effected through 330, 500 and 750 KV links, with their 1975 and expected 1980 throughput capacities shown in Fig. 10. The USSR has 1,370 miles of 750-800 KV extra high voltage lines and 11,700 miles of 500 KV lines, and their combined length is planned to grow to 19,000 by 1980. The 1976-80 plan also called for the full coordination of the East European countries with the European grid of USSR via a 750 KV line from Vinnitsa (Ukraine) and Albertirsa (Hungary), which went into operation in late 1978. With the construction of the first 1150 KV AC line (Central Siberia–North Kazakhstan) and the long delayed 1500 KV DC line (Ekibastuz–Tambov), the power transfer potential between the Siberian and European systems would ultimately rise above the relatively limited capacity achieved by completion of a 500 KV link in 1978.

The amount of power shunted through high voltage interties among the major systems is large and will continue to increase in the future as further electrification of the economy, particularly of formerly low-priority sectors, will inevitably worsen the load factor. In 1965, 20 billion KWH and, in 1970, 40 billion KWH (5.4% of total output) was shunted among the economic regions and this is expected to reach 150 to 180 billion KWH by the next decade.[34] Since regional consumption figures are available only for republics, and can be derived for grids only for 1970, our knowledge of net intersystem and interregional flows is scant.[35] The Central Russian system, which occupies the middle of the European USSR and contains the Soviet capital, imports the largest amount of power from neighboring systems, about 14-15 billion KWH in 1975.[36] The Southern (Ukraine) and Middle Volga systems were the largest exporters in the early 1970s, but the surplus of the Volga system has now been drastically reduced to perhaps 2 billion KWH, most of which goes to Central Russia.[37] Vigorous power plant construction combined with only moderate growth of power consumption in the Ukraine,[38] multiplied the surplus of the Southern grid over the last 10 years five to six fold to some 20 billion KWH. Over half of this surplus is transmitted abroad, the rest is fed into the Central Russian system.[39] Despite the surplus and the presence of large fuel resources (coal and gas), only the rapid development of nuclear energy can stave off problems of power supply in the future.[40] Relative to production, the Baltic region, chiefly Estonia, is the most important provider of power for other areas, although this surplus disappears within the Northwest system as a whole.[41] It is via electricity that the nontransportable oil shale of the Estonian SSR makes an important contribution to the energy balance of the northwestern part of the Soviet Union, traditionally poor in energy resources (see Chapter 4).

Large systems, well integrated by extra high-voltage (EHV) ties, not only provide substantial scale economies and savings on reserve capacity, but also broaden the range of choice of generating sources. Significantly they make nontransportable hydropower and, as in the case of Estonian shale, low calorific fuels usable for distant centers of consumption. A Western observer computes that already in 1970 "over 44% of the generating capacity for EHV networks and over 57% of EHV transmission distances were based on low transferability sources of generation and another 22% and 14%, respectively, on a mixture of high and low transferability types. The scale and growth rate of development on the Volga and Angara-Yenisey, for example, were contingent on large-scale EHV networks as are the planned developments on the Ekibastuz and Itat (Kansk-Achinsk) coalfields."[42]

The USSR comes closer to the United States in the total length of transmission lines than in generating capacity or output and has a substantial lead in the length of high-voltage lines (over 220 as well as 330 KV).[43] This is partly a result of a conscious, long-standing drive to forge a national grid, partly a reflection of larger size and less favorable resource distribution relative to population centers. The significant benefits of the unified systems described are abated, to some extent, by substantial line losses. Each year through the first half of the 1970s these amounted to about 9% of all electricity produced by the Ministry of Electric Power, with a growing absolute volume reaching 80.8 billion KWH in 1975 (station consumption added 5.1%, or 5.3 billion KWH to the loss,[44] but this share is primarily a function of the type of fuel and cooling system used and not the function of the lengths of power lines). Soviet planners project only a slight decline in the share of line losses for the rest of this decade, leading to an almost 50% absolute growth in transmission losses by 1980.[45] (Fig. 11 shows the general pattern of power plants and transmission lines of the USSR.)

THERMAL POWER PLANTS: TRENDS AND PROBLEMS

Conventional thermal plants are destined to shoulder the biggest share of total power production at least till the end of the century. Through the quantity and type of fuel burned, they also have a direct and immediate impact on the entire fuel economy. Comprising less than 80% of total capacity installed, they generated almost 86% of power in the mid-1970s. Apart from the increasingly insignificant diesel and other mobile units, thermal plants are more intensively used than hydro, and up to now even nuclear, stations (compare average hours of use, Table 44). Between 1970 and 1975, the utilization of thermal capacity intensified, while that of hydro declined by almost one-fifth, largely because of a series of dry years in the mid-1970s. In the 1976–80 plan, the small decrease projected in thermal station use reflected the greater peaking needs, while the increase for hydro plants was predicated on more abundant stream flow, particularly in the European USSR. The contribution of conventional thermal stations appears even more important relative to the value of fixed assets. Not counting outlays on transmission and distribution, needed for all plants but more important for hydroelectric stations, thermal generating plants accounted for only 63% of all fixed assets in the electric power industry in both 1970 and 1975[46] and of new fixed investment also in much of the 1960s.

As Table 44 shows, conventional steam turbine stations in the USSR are divided between heat and power stations and condenser stations in rather stable

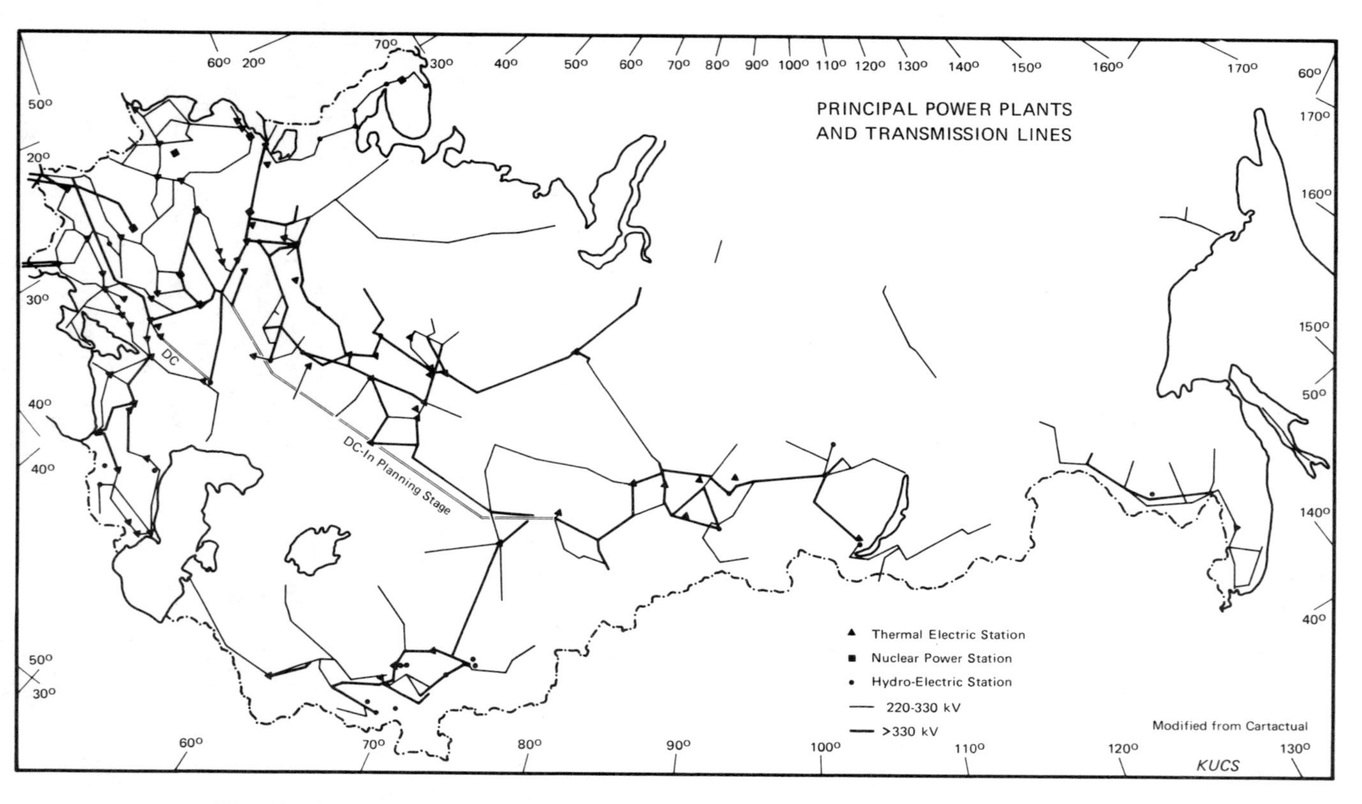

Fig. 11. General pattern of electric power plants and transmission lines in the USSR.

proportions. The former supply both power and heat to industry and/or urban neighborhoods via special bleeding turbines and the capture of their waste heat. They thus attain an overall conversion efficiency of up to 70% compared with 40% for modern steam plants supplying only electricity. Some condenser stations also produce heat (some boiler capacity being connected directly to utility networks). As a rule, condenser stations are larger, require more cooling water and, increasingly, cooling towers and ponds. Consequently, they are located outside major population centers, in contrast to heat and power stations, which, apart from some purely industrial stations, must be placed in large and medium-sized cities to distribute their heat. Accounting for less than a third of all capacity, heat and power stations comprise over half of all power stations found in the interconnected systems.[47]

Central heat is thus an important output of Soviet generating plants and over a third of their fuel consumption must be charged to heat rather than power.[48] In 1975, 73% of the 2,426 million gigacalories of heat produced in the USSR was delivered to consumers in a centralized way from heat and power stations and from industrial and municipal boilers of over 20 gigacalories/hour, the rest from decentralized heaters, mostly individual stoves, whose share is declining.[49] Heat and power stations produce well over half of this centralized heat (918 million gigacalories in 1975 out of a total of 1,767 million) and, according to 1970 data, deliver over three-quarters of it to industry and the rest to residential and municipal consumers.[50] These joint-purpose power plants thus play a significant and increasing role in total heat supply, particularly in urban areas. At the beginning of 1975, they furnished 42% of all the heat (for both industrial and residential-municipal uses) for Soviet cities and settlements (Table 4).

The energy efficiency of Soviet power plants has improved rapidly since World War II. Heat rates (i.e., the amount of fuel used per KWH generated) have dropped more rapidly than in the United States. Today the Soviet average is said to be lower than the mean heat rate for United States utilities, though, because of Soviet emphasis on joint heat and power production, such a comparison is somewhat misleading.[51] Heat rates declined greatly for both condenser stations and for heat and power stations, but more dramatically for the latter. As late as 1961, they used more fuel per KWH than condenser stations, but by the mid-1970s almost one-quarter less, an efficiency all the more remarkable since heat and power stations are, as a rule, significantly smaller and work at lower steam pressure (Table 45).

Technological improvements toward higher steam pressure and temperature and larger turbine size were evidently a factor in the improvement in the heat rate, but they affected both types of stations. The dramatic progress for heat and power stations must be sought primarily in the change in their fuel supply, that is, in the switch from coal to oil and gas, leading to more efficient heat capture and conversion. Except in Siberia and northern Kazakhstan, these stations today run overwhelmingly on liquid and gaseous fuels. As mentioned, they are mostly located inside urban areas, where the concern for clean air has become a real planning issue hastening conversion. Even in the Kuznetsk Basin, the intended conversion of some stations to gas (which, after long delays, has finally arrived)

Table 44. Power Production and Capacity by Generating Types (1970, 1975 and 1980 plan)

Year	Type of generating capacity						
	Thermal (turbine)			Thermal (diesel and others)	Hydro-electric	Nuclear[a]	Total
	Condenser (KES)	Heat and power (TETs)	Total				
	Capacity at year end (thou. MW)						
1970	76.3	45.0	121.3	12.5	31.4	0.9	166.1
1975	103.9[b]	59.0	162.9	9.4	40.5	4.7	217.5
1980 plan	128.0[c]	75.2[c]	203.2	8.4	54.0	18.4	284.0
	(% of total)						
1970	45.9	27.1	73.0	7.5	18.9	0.6	100.0
1975	47.8	27.1	74.9	4.3	18.6	2.2	100.0
1980 plan	45.1	26.5	71.5	3.0	19.0	6.5	100.0
	Output (bill. KWH)						
1970	c. 393[d]	c. 214[d]	607.0	6.0	124.4	3.5	740.9
1975	586.8[b]	299.0[d]	885.8	6.6	126.0	20.2	1038.6
1980 plan	n.a.	n.a.	1097.3	5.7	197.0	80.0	1380.0
	(% of total)						
1970	53.0	28.9	81.9	0.8	16.8	0.5	100.0
1975	56.5	28.8	85.3	0.6	12.1	1.9	100.0
1980 plan	n.a.	n.a.	79.5	0.4	14.3	5.8	100.0
	Average hours of use						
1970	c. 5151	c. 4553[e] (5290)	5136[f]	c. 480	4146[f]	n.a.	n.a.
1975	5648	5068[e]	5675	c. 802	3360	5520	n.a.
1980 plan	n.a.	n.a.	5590	c. 775	3820	5910	n.a.

Notes:

[a]Utility nuclear stations only (see Note d, Table 34).

[b]Computed as a residual.

[c]A small discrepancy among sources required adjustment.

[d]1970 output, according to a percentage breakdown given in *Teploenergetika*, 1976, No. 11, pp. 2–3, apparently refers only to pure heat and power stations, i.e., those having only dual-purpose turbines. The data therefore apply to a slightly smaller universe than those for 1975, from *Elektricheskiye Stantsii*, 1977, No. 8, p. 15, which cover all heat and power stations.

Notes: (Table 44. cont'd)

[e]Figures in parentheses refer to hours of utilization for the production of both heat and power (some condenser stations also produce some heat), as given in *Teploenergetika*, 1976, No. 11, pp. 2-3. Figures without parentheses refer to hours for power production only and are computed from capacity and output.

[f]Refers only to utility stations, which accounted for 97% of all power production.

Sources: Major source is *Energetika SSSR v 1976-1980 godakh*, op. cit. (1977), pp. 11 and 61. For the planned 1980 breakdown between condenser stations and heat and power stations, data are given in *Izvestiya VUZ. Energetika*, 1976, No. 9, p. 6, and in *Elektricheskiye Stantsii* as reported in *Soviet Geography*, December 1976, p. 718. Small differences were ignored in favor of the main source, which appears to be the most authoritative analysis of the Soviet electric power industry.

---

has been announced.[52] Coal, lignite and other solids are more significant in the fuel supply of condenser stations which, in general, are placed well outside city limits. The recent policy to economize on hydrocarbons and allocate them primarily for mobile and furnace uses will enhance the role of low-quality coals in condenser thermal plants. Twenty-one new electric stations fired by solid fuels were to be started during the 1976-1980 plan compared with 15 during the previous five years,[53] and most of them will have to be condenser stations.

A detailed comparison of regional heat rates can also be made, since data for 1974 are available for 52 subsystems of the nation's networks, covering all interconnected grids, save the Far East (Table 46). While fuel input per unit volume of heat produced does not differ greatly, fuel used per KWH of power shows wide variations. Invariably, systems that depend on low-quality solid fuels suffer from unfavorable heat rates even when their stations are large and modern (e.g., the systems of Estonia, Tula, Karaganda and most of the South, burning low calorific shale, lignite and poorly sorted coal fines for the most part). By contrast, the Moscow, Latvian and most of the Volga systems and Bashkiria (part of the Urals grid), which rely almost exclusively on gas and oil, show admirable

Table 45. Heat Rates at Soviet Utility Stations[a]
(in grams of SF per KWH)

Year	All utility stations	Condenser stations	Heat and power stations
1961	468	465	475
1971	366	388	324
1975	340	369	282

[a]Utility stations account for 92% of all thermal electricity.

Source: V. P. Korytnikov (1976), p. 11.

Table 46. Regional Heat Rates and Fuel Consumption of Soviet Utility Stations by Power Systems (1974)[a]

Systems	Fuel input		Total fuel used (mill. tons of SF)
	(grams of SF per KWH of power)	(kg of SF per Gcal of heat)	
Northwest	334	177	34.02
Belorussian	320	174	11.56
Estonian	437	180	6.89
Lithuanian	341	167	4.18
Latvian	284	178	0.89
Central Russia	332	175	58.03
Moscow	286	168	19.87
Tula	372	174	9.20
Kalinin	340	180	5.46
Kostroma	327	176	4.85
Gor'kiy	340	181	4.19
Yaroslavl'	328	178	2.79
Volga	330	177	24.17
Kuybyshev	295	177	9.10
Volgograd	291	172	4.44
Tatar	338	173	5.34
Saratov	253	176	3.20
South	373	173	71.61
Donets Basin	380	179	19.65
Dnieper	355	175	14.07
L'vov	357	171	6.89
Khar'kov	349	169	7.72
Kiev	322	167	7.52
Rostov	387	181	5.79
North Caucasus	365	170	7.83
Stavropol'	340	164	3.06
Groznyy	302	166	2.38
Dagestan	426	180	0.17
Krasnodar	380	171	2.38
Transcaucasia	390	180	12.48
Georgian	377	184	2.93
Azerbaijanian	396	173	6.31
Armenian	361	175	3.24
Urals	336	173	58.04
Sverdlovsk	360	179	19.32
Chelyabinsk	364	173	10.24
Bashkir	278	170	13.50
Perm	340	172	7.33
Orenburg	325	170	5.15
Kirov	380	188	1.87
Udmurt	300	177	0.63

Table 46. (cont'd)

Systems	Fuel input		Total fuel used (mill. tons of SF)
	(grams of SF per KWH of power)	(kg of SF per Gcal of heat)	
North Kazakhstan	386	181	14.92
Pavlodar	336	177	5.34
Karaganda	437	199	4.71
Altay	282	181	1.39
Tselinograd	407	185	1.75
Central Asia–South Kazakhstan	375	177	18.83
Uzbek	362	174	11.22
Tadzhik	380	173	0.46
Kirghiz	342	177	1.56
Turkmen	492	254	1.82
South Kazakhstan	367	174	2.30
Alma-Ata	383	178	1.57
Central Siberia	331	176	40.42
Krasnoyarsk	342	173	7.66
Irkutsk	310	175	8.64
Kuznetsk Basin	358	175	11.48
Novosibirsk	319	177	3.87
Omsk	309	175	4.47
Barnaul	300	179	2.85
Tomsk	354	179	0.83
Buryat	323	175	0.62

[a]Utility stations (stantsii obshchego pol'zovaniya) account for 92% of all thermal electricity.

Source: V. A. Shelest, *Regional'nyye energoekonomicheskiye problemy SSSR* (M. Nauka, 1975), pp. 278–302.

heat rates. One may also note that Belorussia, where gas and oil have largely ousted peat from electric stations,[54] today enjoys high fuel efficiency in power plants. However, Table 46 also shows several grids where heat rates are still unfavorable despite the extensive use of natural gas and/or oil. This is most noticeable in Central Asia–South Kazakhstan, the Caucasus, with Turkmenia and Dagestan showing particularly low rates, and Azerbaijan not too far behind. Obviously, obsolescence, a great number of small stations (many of which may even use agricultural waste), poorly maintained equipment–problems already noted with respect to the low capacity factor–depress fuel efficiency despite the availability of gas and fuel oil. These same problems also play a part in North

Kazakhstan and, to an extent, in the Southern grid, where the burning of solid fuels already results in unfavorable heat rates.

In general, the quality of coal utilized by Soviet power plants is low and continues to deteriorate with the growing emphasis on lignites. In 1974–1975, the average calorific content of coal furnished for stations under the Ministry of Electric Power, which then generated 93% of all electricity in the USSR, reached only 4,137 kilocalories per kilogram (7,443 BTUs per pound), while two-fifths of all solid fuels consumed by Soviet power plants had a heat content of less then 3,500 kilocalories per kilogram (6,297 BTUs per pound). By contrast, less than a tenth of United States electric stations worked on solid fuels with heat content of under 9,895 BTUs per pound.[55] This huge discrepancy casts doubt on Soviet assertions that fuel expenditure per KWH in the USSR today is appreciably below that in the United States.

The problem of low heat content is aggravated by insufficient control on the consistency of fuel—an increasing problem with growing plant size and the standardization of plant design and equipment. In older coal fields, such as the Donets Basin, where several mines may supply a large station, inadequate coordination results in significant variations in fuel quality, particularly since the most important power plant fuel consists of coal fines. In Kazakhstan and Siberia, the rapid development of strip mining is leading to both a rise and much greater variation in the share of ash content. In a mere four years, for example (from 1967 to 1971), the mean ash content of Ekibastuz coal rose from 37% to over 42%, with wide oscillation around the average both at the mine and at the thermal plants.[56]

ENVIRONMENTAL PROBLEMS OF THERMAL POWER STATIONS

As elsewhere, thermal power stations are a major source of environmental deterioration through air pollution, slag disposal and waste-heat removal. Soviet electric stations, for example, are responsible for about one-fourth of all air pollutants originating from stationary sources in the country.[57] As in the Western nations, former negligence has given way to heightened environmental concern and stricter protection measures. This is reflected in growing investment cost per unit capacity and greater restriction on plant location, already severe because of ever increasing plant size. In 1975, 70% of particulate matter and gaseous pollutants were removed, with the 10th five-year plan aiming at 81% by 1980. To raise this share to over 90% for sulfur oxides alone is estimated to increase capital cost per KWH by 40 to 60%, depending on the scrubbing methods employed, none of which as yet are considered very reliable.[58] With increasing station size, ever taller stacks have had to be employed to disperse pollutants. Most of the 20 thermal plants of 2 million KW capacity and over (in 1977) require 250-meter smokestacks to meet sanitary standards (though it is not known how many are actually equipped with such stacks), while the largest new stations exceeding 3 million KW and burning Donets and similar high-sulfur coals, will need stacks of 330 meters.[59]

In the European USSR, sulfur is the primary problem. No more than a fifth of all coal reserves contain less than 1.5% sulfur, with much of the petroleum also being sulfurous. More than half of all the coal produced has a sulfur content above 2.5%,[60] and in 1973 the European USSR and the Urals were responsible for over four-fifths of the 10.7 million tons of sulfur released from coal (the Donbas alone for 25%).[61] Because about 60% of all noncoking coal in the USSR is consumed in power stations,[62] and the domestic sector and small industrial plants receive primarily low-sulfur fuel,[63] electricity generation will continue to release huge quantities of sulfur oxides to the environment for many years to come. Low sulfur western coals in the United States can be and are now transported to the Middle West and South in growing quantities to replace local coals high in this mineral. In the USSR the transport problem is more intractable, and, as will be shown, cheap, low-sulfur Siberian lignites cannot soon play a similar role in the European part of the Soviet Union.

Ash and slag removal and disposal are a growing problem everywhere, imposing increasing restrictions on the location of power plants. Ash lagoons occupy large expanses of land that are irretrievably lost to the economy and are sources of noxious dust formation. Total yearly accumulation of ash and slag at Soviet thermal stations reaches 70 million tons, with single large plants ranging from 1.5 to 2.5 million m^3 annually. Capital and operating costs of ash removal come to 100 million rubles per year, i.e., up to 1.5 rubles per ton.[64] In populous western regions, finding new land for ash lagoons has become an acute problem. Though land is more readily available east of the Urals, disposal is no less difficult. As a rule, eastern coals have higher ash content and/or chemical compositions making them likely to cake onto boiler surfaces and also render the removal of residue troublesome. A high ash content (over 40%) with abrasive characteristics is a particular bane of the Ekibastuz Basin, now the third largest coal producer. The Ekibastuz-based 2.4 million KW Yermak power plant leaves 6.8 million tons of residue per year, currently dumped into the Irtysh valley; the new Ekibastuz No. 1 station, planned for 1979, will leave almost as much.[65] Caking is especially troublesome with Kansk-Achinsk coal, and significant technical problems will have to be solved before these lignites can be used in the proposed mammoth thermal plants of 6.4 million KW installed capacity.[66]

Thermal plants need vast quantities of water. In the USSR, their water demand roughly equals that of all other industries combined, though still only a fraction of that consumed for irrigation.[67] With an intake and discharge of 130 to 150 m^3 an hour for every 1,000 KW capacity,[68] large power stations significantly raise the temperature of adjacent water bodies unless expensive cooling towers are used. Industrial regions in the water-short steppe zone, such as the southeast Ukraine, the southern Urals and Karaganda Oblast may already have serious thermal waste problems.[69] A Soviet source says that in the Ukraine, for example, cooling water resources are virtually exhausted. The share of once-through cooling (i.e., immediate discharge into waterbodies) in 1970 was already down to 50% and was to decline to 25% by 1980.[70] Water needs and land availability for cooling towers and ponds will thus place serious restrictions on the location of power plants in the future. Even in some regions of Siberia,

Table 47. Sectoral Distribution of Electricity Consumption in the USSR (1965–1980)[a]

Sectors of economy	1965		1970		1975		1980	
	Bill. KWH	Total net consumption (%)	Bill. KWH	Total net consumption (%)	Bill. KWH	Total net consumption (%)	Bill. KWH	Total net consumption (%)
Industry	314.2	72.2	437.9	69.8	588.0	63.5	738.5	63.5
Construction	11.9	2.7	15.0	2.4	21.0	2.4	25.0	2.1
Transport	37.1	8.5	54.4	8.7	74.0	8.4	100.0	8.6
Agriculture (including farm residential)	21.1	4.9	38.5	6.1	74.0	8.4	130.0	11.2
Urban residential and commercial	50.6	11.6	82.1	13.1	119.0	13.6	170.0	14.6
Total net consumption	434.9	100.0	626.9	100.0	876.0	100.0	1163.0	100.0
Operating losses	70.3		108.8		151.0		ca. 196.0[b]	
Export	1.5		5.2		11.6		ca. 21.0[c]	
Total production	506.7		740.9		1038.6		1380.0	

Notes and sources:

[a]Operating losses are derived as the difference between gross and net consumption. Line losses, but not station consumption, are also given separately as percent of output. *Energetika* . . . , op. cit. (1977), pp. 13 and 51. For line losses pp. 181–82. Export from corresponding issues of *Vneshnyaya torgovlya SSSR*. Annual. Production from *Nar. khoz. SSSR v 1975 g.*, p. 235.

[b]Line losses are planned to comprise 8.6% of output (i.e., 119 billion KWH) by 1980. Some reduction in the share of station consumption was assumed.

[c]The 750 KV Vinnitsa-Albertirsa intertie, completed in 1978, will be able to transmit 17 billion KWH to East European neighbors at a 75% capacity factor. Bulgaria is assumed to continue importing about 4 billion KWH by another line through Rumanian territory.

such as the Kuzbas, the supply of water has become strained. And the proposed giant power plants in the western half of the Kansk-Achinsk Basin would require the regulation of several streams in the area.[71]

ELECTRICITY CONSUMPTION

Gross power consumption in the Soviet Union equals that in Japan, West Germany and Britain combined and is slightly below half that in the United States. On a per capita basis, however, all these countries exceed the Soviet Union easily, as do several others in West Europe.[72] Lenin's dictum notwithstanding, full-scale electrification, like theoretical communism, is still some way off. Electricity, being a highly refined and the most flexible form of energy, has a high opportunity cost value for most classes of consumers in all regions. This value may take the form of convenience and savings of consumer time, as in the household sector, or of enhanced productivity by substitution of electricity for other forms of energy in production processes. In the first case, this value cannot easily be captured by central planners and devoted to ends they are bent on maximizing. In the second case, the economic benefits from electrification can be made to coincide closely with planners' objectives.

The Soviet pattern of electricity consumption differs greatly from the American. In the USSR, industry consistently received the greater share of power, in contrast to the United States where the residential-commercial economy has long surpassed manufacturing as the largest power user. While industry's portion in Soviet net consumption is gradually declining, it still received 67% of all power in the mid-1970s compared with 40% in the United States.[73] Soviet industry is electrified to near four-fifths of the level of United States industry,[74] even though per capita consumption in the USSR is less than half of the American level. The residential-commercial sector received only 13% in 1975, with the domestic economy itself taking less than 5%.[75] Demand in that sector grew faster than the national average during the 1960s, but only marginally faster between 1970 and 1975, and the 10th five-year plan, too, provided for only moderately accelerated growth (Table 47). In constrast to the high level of electrification of industry in the USSR, per capita consumption of power in Soviet urban households reaches but a third of that in United States homes, and in rural households less then 20%.[76]

The significance of other economic sectors in power consumption also differs greatly from that in the United States. Relative to the American experience, transport in the USSR is heavily electrified. It accounts for a rather stable 8.5% of net power consumption compared with a fraction of 1% in the United States. Over two-thirds of consumption in that sector is by railroads and city mass transit, but the fastest growth in power use is shown by pipelines, with a particularly large jump planned during the second half of the 1970s.[77] The most rapid rise in electricity demand, however, is taking place in agriculture and the increasingly generous supply of electric power is one of the cornerstones of agricultural intensification. Today, 99% of the farms are connected to a grid,

Table 48. Electric Power Consumption and Hours of Use in Soviet Industry (1970–1980)

Branch of industry	Year	Consumption of power					House of use per year[a]	
		All uses		Motors (bill. KWH)	Non-motors			
		Bill. KWH	(%)		Electro-technical processes (bill. KWH)	Lighting (bill. KWH)	Motors	Non-motors
All industry	1970	437.9	100.0	275.0	125.0	37.9	1570	2920
	1975	588.0	100.0	367.0	170.6	50.4	1560	2920
	1980 plan	738.5	100.0					
Fuel industries	1970	46.0	10.5	42.5	0.9	2.6	1750	1000
	1975	60.8	10.3	56.0	1.7	3.1	1740	1700
	1980 plan	79.8	10.8					
Ferrous metallurgy	1970	70.7	16.1	45.9	18.7	6.1	1405	3670
	1975	94.5	16.1	61.6	25.4	7.5	1350	3930
	1980 plan	131.0	17.7					
Nonferrous metallurgy	1970	67.3	15.4	19.3	43.4	4.6	1680	5870
	1975	92.4	15.7	26.0	61.5	4.9	1730	5750
	1980 plan	113.7	15.4					
Chemical and petro-chemical	1970	64.2	14.7	44.8	15.7	3.7	2350	3740
	1975	91.6	15.6	65.1	21.1	5.4	2470	4140
	1980 plan	128.0	17.3					
Machine-building and metal working (engineering)	1970	65.6	15.0	33.3	25.1	7.2	787	1306
	1975	90.3	15.4	45.7	34.8	9.9	810	1280
	1980 plan	116.0	15.7					

[a]Maximum possible = 8,760. *Source: Energetika* . . . , op. cit. (1977), pp. 27, 37, and 51.

receiving some electricity. Milking and watering operations are now overwhelmingly electrified, the cleansing of stalls and feedlots from one half to two-thirds, though the bulk of feeding must still be done manually.[78] In 1975, nearly three-fourths of the power consumed in agriculture was expended on farm operation and irrigation, the remainder going to household uses of the farm population. The absolute volume of electric power supplied to farm work in the USSR and the United States is roughly comparable, but Soviet agriculture claims a greater share of total power supply than American agriculture, at least within the farm gate.[79]

Because of its overwhelming importance for labor productivity, the geographic variations in demand and differences in load intensity, electricity consumption in industry warrants a few additional comments. From 1928 to 1970, the growth rate of labor productivity in the USSR and that of industrial power per worker have shown a close relationship. Relative to a moving base to minimize index number distortion, the relationship is near one to one, though, of course, not a simple cause and effect.[80] It should remain fairly close for some years until manual labor, which is still responsible for a fifth of basic industrial and four-fifths of auxiliary operations,[81] is mostly eliminated. (After that, when machines replace not human hands but other machines, the relationship between power consumption and labor productivity becomes less tractable.) M. A. Vilenskiy argues forcefully that a rising share of power consumption in electrotechnical processes is one of the hallmarks of scientific-technological advance. As the first phase of electrification is gradually accomplished and electric motors provide most of the mechanical force in industry, further electrification is achieved by the spread and deepening of electrotechnical processes, characteristic of much of advanced manufacturing. He is deeply disturbed by the fact that, as industrial work is becoming more fully electrified, the share of electrotechnical processes has failed to grow since the mid-1960s, stabilizing at 28 to 29%. This, combined with the slow spread of electrification to auxiliary work, has led to a sharp deceleration in the growth rate of power consumption (Table 1), which augurs ill for improving the technical level of Soviet industry.[82]

Vilenskiy's concern is well founded. The 10th five-year plan called for an average 5.8% yearly rise in net power consumption and only a shade more in output (Table 47) at a time when the working age population will expand by only 1.7% a year. During the 1980s the latter will increase by a mere half percent a year.[83] Industrial growth in the forthcoming years is, therefore, predicated on a rapid improvement of labor productivity and the drastic reduction of hand labor, still dominant in auxiliary processes. The share of electrotechnical processes should rise a little, from 29 to 31%,[84] provided the construction of such new power-using factories, most of which are slated for Siberia, will not suffer their accustomed delay. Meanwhile, electricity consumption in the populous European USSR is planned to grow by only 5.5% per year,[85] with 35% of the increment originating from nuclear stations,[86] for which such ambitious construction plans will be difficult to achieve.

As Table 48 shows, five major industrial groups use three-quarters of all power consumed by industry. Ferrous metallurgy, nonferrous metallurgy, the chemical and engineering branches each received over 90 billion KWH in 1975, between 15 and 16% of the industry total. The fuel group received over one-tenth. The proportion of these power intensive branches in total consumption increased substantially through the 1960s but stabilized in the early 1970s. It was planned to rise again by 1980 because of the rapid growth in the shares of ferrous metallurgy and the chemical industries.[87] Given the relatively slow growth of total electricity consumption and the rising share of nonindustrial sectors, particularly agriculture, this would, of course, result in a slow expansion of power consumption in the light, food and other consumer goods industries. Any detrimental consequence to labor productivity increases from the declining growth of power consumption would most likely affect these latter industries first.

It is also revealing to look at consumption distribution between motors and electrotechnical processes and average hours of power use. (Lighting accounts for less than 10% everywhere except in the engineering group.) Demand by electric motors dominates in all industrial branches except nonferrous metallurgy, where electrotechnical processes are more significant (Table 48). However, the increasing importance of precision engineering, demanding ultra-vacuum, electric induction and laser technologies, the electrochemical fashioning of metals, etc., also resulted in a high share of these processes in the machine-building and metal-working branches. As expected, capacity is more continuously utilized by electrotechnical apparatus than by motors and is highest in ferrous and nonferrous metallurgy and the chemical industry. Hence the high load factor in regions, such as the Urals, the eastern Ukraine, Siberia and Central Asia–Kazakhstan, where these branches dominate (Table 42). Prime movers in all these branches tend to be concentrated in the mining stage, where seasonal variation is also a problem. This depresses the load factor in such regions as the Far East, where the mining stage of these industries is more important than the smelting and electrochemical stages.

Influenced by the dominant position of industry in power demand, the geographic variation in per capita consumption closely follows the distribution of industry, particularly heavy industry. And since well over 80% of all electricity is generated from conventional fossil fuels, it also follows per capita variation in fuel consumption as treated in Chapter 2. The Spearman rank correlation coefficient between the per capita value of industrial fixed assets and per capita power demand is almost identical to that between industrial fixed assets and fuel demand over the major economic regions on a per capita basis (Rs = 0.90 vs. Rs = 0.93 for 1965). Unfortunately, power consumption figures in more recent years are available only for union republics, making detailed regional comparisons within the vast Russian Republic impossible. As shown below, however, only a modest power exchange and virtually no net transfer of electricity as yet takes place between Siberia and the European RSFSR or between the Far East and the rest of Siberia. For these three regions of the Russian Republic, therefore, production minus international export may be used as a surrogate for gross consumption (i.e., inclusive of operating losses).

Variations in per capita consumption among regions and republics remain high, with little evidence of convergence. The coefficient of variation between 1965 and 1975 changed from 0.53 to 0.51 and between 1970 and 1975 (based on a somewhat different set of areal units) from 0.52 to 0.51. Significantly, all non-Russian ethnic regions and republics, with the exception of the eastern Ukraine (Donets-Dnieper), Estonia and Kazakhstan (the latter with a Russian majority) rank far below the Soviet average (Table 49). Except for the Baltic republics, they are still strongly rural, with barely half or less than half of their population in cities, and tend to lean heavily on a rather backward, undercapitalized agriculture and the technologically weak food-processing industries. From the mid-1960s to the mid-1970s, the relative position of the most backward republics, i.e., Moldavia and those of Central Asia, improved slightly, but the relative standing of Transcaucasia and of the Ukraine in per capita power consumption declined. Within the vast Russian republic, the European provinces, taken as a whole, are converging toward the Soviet average, becoming relatively less power-intensive in their consumption pattern. By contrast, Siberia is widening its lead over all other provinces. This is, of course, consistent with the planned restriction on the location of certain power-intensive industries in the European USSR and their expansion east of the Urals.

The development of power grids and the great flexibility of electric energy have resulted in a nearly complete divorce between generating and consuming sites. Yet, beyond moderate distances, electricity remains costlier to transport than most fuels. Consequently, the broad spatial patterns of consumption and production (in contrast to a fine regional mesh) coincide more closely for electricity than for fossil fuels. Energy-short provinces receive large quantities of fuel over thousands of miles. Despite increasing transmission distances, the same provinces essentially can bring in power only from beighboring regions and must generate most of their electric power within their boundaries. This situation may change in the next 20 years, but today it is still much in evidence. As mentioned earlier in this chapter, the aggregate amount of electric power shunted among economic regions spanned by the unified networks reached about 40 billion KWH at the beginning of the 1970s, or 5.4% of the total generated. While this exceeded the output of Kazakhstan (or Rumania, Switzerland and Belgium), it is a small share when compared to the percentage of total fuel output moving across regional boundaries. Moreover, about half of this consisted not of net transfer but of exchanges[88] (peak sharings and exchanges between thermal and hydro plants and water basins with different hydrological regimes), while most of fuel transfers were and are net transfers.

The European USSR imports over two-fifths of its aggregate fuel requirement,[89] mostly from Soviet Asia, but to a degree also from abroad (chiefly gas from Iran and coal from Poland). There is no evidence that it brings in any electric power; virtually all Soviet electricity exports originate from the western provinces, overwhelmingly from the Ukraine and Moldavia. Transcaucasia, which participates heavily in seasonal and daily interregional exchanges, is neither an exporter nor an importer on an annual basis. The Moslem republics of Kazakhstan and Central Asia, which send about half of their combined fossil fuel output to the European USSR are, when taken together, self-contained in electric

Table 49. Index of Industrial Fixed Assets and Per Capita Electricity Consumption by Economic Region and Republic

Region and republic	Value index of industrial fixed assets per capita	Index of per capita electricity consumption		
	1965 100 = 646.5 rubles in 1955 prices	1965 100 = 2179 KWH	1970 100 = 3016 KWH	1975 100 = 4021 KWH
USSR	100	100	100	100
RSFSR	n.a.	123	122	116
European RSFSR	n.a.	110	109	104
Northwest	148	118		
Central Russia	95	93		
Central Chernozem	65	64		
Volga-Vyatka	83	78		
Volga	132	112		
North Caucasus	83	64		
Urals	155	203		
West Siberia	112	141	} 200	} 206
East Siberia	143	266(311)[a]		
Far East	103	77(63)[a]	79	83
Ukraine		94(92)[a]	96(94)[a]	92
Donets-Dnieper	165	164(159)	170[b]	
Southwest	46	39	35[b]	
South	74	52	44[b]	
Baltic	102	57	77	74
Estonia			114	
Latvia			66	
Lithuania			55	
Belorussia	58	43	52(52)	60
Transcaucasia	94	85	72	65
Georgia			63	
Azerbaijan			72	
Armenia			89	
Kazakhstan	90	96	92	100
Central Asia	46	43	43	45
Uzbekistan			45	
Kirghizia			37	
Tadzhikistan			37	
Turkmenia			35	
Moldavia	39	26(41)[a]	35	42

Notes:

[a]Data in parenthesis (from Shelest) are apparently in error or refer to pre-1965 regional boundaries. For the Far East, for example, lack of any interconnection, export, or import

Notes (Table 49. cont'd)
mean that consumption had to equal production in all the years cited. For Moldavia, the figure from *Energetika*. . (1977), and for East Siberia, Ryl'skiy's figure was used.

[b]Per capita consumption for the economic regions of the Ukraine distributed according to 1967 data. From P. Voloboy and V. Popovkin, "On indicators of the economic level of regions and oblasts," *Ekonomika Sovetskoy Ukrainy*, 1968; No. 10, p. 59.

Sources: Index of industrial fixed assets from Akademiya nauk SSSR and Gosplan SSSR, *Ekonomicheskiye problemy raxmeshcheniya proizvoditel'nykh sil SSSR* (Moscow: Nauka, 1969), p. 63. Average for the USSR from *Nar. khoz. SSSR v 1965 godu*, pp. 12 and 64. Electricity consumption from *Energetika*. . . , op. cit. (1977), p. 13; V. A. Shelest, *Ekonomika razmeshcheniya elektroenergetiki SSSR* (Moscow: Nauka, 1965), pp. 66–67 and V. A. Ryl'skiy, *Ekonomika mezhrayonnykh elektroenergeticheskikh svyazey v SSSR* (Moscow: Nauka, 1972), p. 88.

---

power consumption and output, though some electricity is exchanged with neighboring provinces.[90] Siberia, which ships vast quantities of oil and gas plus large amounts of coal westward (and some to Pacific ports) does not yet transmit any appreciable volume of power either westward or to the Far East, though exchange with the Urals and Kazakhstan is taking place.

Large net flows of electric power over distances beyond, at most, a thousand miles are predicated on the success of extra high voltage transmission lines. The technology is undergoing intensive research and development. The start in the construction of the Ekibastuz-Tambov 1,500 DC line has been promulgated as a decree of the party's Central Committee. With a 6 million KW capacity, it could annually transmit up to 40 billion KWH of power,[91] close to the consumption of Belorussia and the Baltic Republics combined and nearly equal the electricity generated by all hydro stations of the European USSR north of the Caucasus.[92] Still more grandiose and distant plans envisage 2,200 KV DC lines, which could transmit 80 billion KWH per year.[93]

It these projects come to fruition, the bulk transmission of electric power on a continentwide scale would become a reality. As with fuels, the broad regional pattern of consumption and production will diverge substantially. However, another thrust of technological development in the power industry, namely nuclear electricity generation, is having a diametrically opposite impact. Large nuclear plants certainly increase the need for interconnection and system exchanges, since they must have a large market area and must operate at high availability. But they are built and will continue to be built near major load centers, within regions of demand, and thus are in direct competition with EHV lines of transcontinental dimensions. As Chapter 6 showed, Soviet atomic power technology is well advanced and the leadership is fully committed to an accelerated construction of nuclear plants through the cis-Volga provinces. The Soviet nuclear program has experienced considerable delays. However, the main weakness seems to be not technology but manufacturing capacity and organization, the speed of delivery and construction, and probably funds.

Whether in the years ahead Soviet planners will push both of these competing technologies equally hard, thus risking shortages of capital and more delays for both, seems a moot point. A long period of testing and evaluation will probably follow the construction of the first EHV line from Soviet Asia, if it indeed is built. Final decision on the function and importance of this bulk transport medium in the energy system is unlikely to be made before the middle 1980s when the impact of the nuclear program and the relationship of the two technologies, too, can be more fully assessed.

NOTES

[1]Joel Darmstadter, "Energy consumption: Trends and patterns," in Sam H. Schurr, ed., *Energy, Economic Growth and the Environment* (Baltimore: Johns Hopkins, 1972), pp. 155–223.

[2]*Nar. khoz. SSSR v 1975 godu*, p. 223.

[3]*Nar. khoz. SSSR v 1974 godu*, pp. 198–199 and *Nar. khoz. SSSR. 1922–1972*, pp. 152–153.

[4]V. G. Treml, et al., "Interindustry structure of the Soviet economy: 1959 and 1966," in U.S. Congress, Joint Economic Committee, *Soviet Economic Prospects for the Seventies* (Washington: Government Printing Office, 1973), pp. 246–269.

[5]V. I. Weitz and R. E. Mints, *Electric Power Development in the USSR* (Moscow: I.N.R.A. Publishing Society, 1936), p. 16 and 136–137.

[6]Leon Smolinski, "The Soviet economy," *Survey*, No. 59 (April 1966), p. 91.

[7]*Nar. khoz. RSFSR v 1962 godu*, p. 77, and G. B. Yakusha and A. M. Tverskoy, "Fuel supply problems of electric stations in the Ukrainian SSR," in Gosplan Ukrainskoy SSR and SOPS, *Voprosy ratsional'nogo ispol'zovaniya toplivno-energeticheskikh resursov* (Kiev: Naukova Dumka, 1964), p. 14.

[8]According to M. A. Vilenskiy, *Po leninskomu puti sploshnoy elektrifikatsii* (Moscow: Nauka, 1969), pp. 130–131, small stations below 500 KW capacity operated at an availability only 41% of the national average in 1965. Availability of all Soviet power stations in that year, figured as the number of hours of utilization divided by 8760, i.e., the number of hours in a year, was 0.54. Hence $0.54 \times 0.41 = 0.21$. The corresponding availability for utility stations was 0.60. Data from *Nar. khoz. SSSR v 1976 godu*, p. 203.

[9]M. A. Vilenskiy (1969), p. 135.

[10]Ibid., p. 132 and P. I. Bogdashkin, *Elektrifikatsiya sel'skogo khozyaystva SSSR* (Moscow: Sel'khozgiz, 1960), p. 243.

[11]*Ekonomicheskaya Gazeta*, No. 4 (January 1967), p. 18.

[12]A. A. Stepankov, *Ekonomicheskaya effektivnost' proizvodstva kapital'nykh vlozhenii* (Moscow: Akademiya Nauk SSSR, 1963), pp. 413–414.

[13]*Nar. khoz. SSSR v 1976 godu*, p. 160 and *Nar. khoz. RSFSR v 1971 godu*, p. 60 and *1975 godu*, p. 42.

[14]*Nar. khoz. SSSR v 1969 godu*, p. 393 and *Nar. khoz. SSSR v 1976 godu*, p. 160.

[15]*Nar. khoz. SSSR v 1976 godu*, p. 160.

[16]P. K. Savchenko and A. R. Khodzhayev, *Toplivno-energeticheskiy kompleks Sredneaziatskogo ekonomicheskogo rayona* (Tashkent: Uzbekistan, 1974), pp. 120 and 125; A. P. Tret'yakova, "Characteristics of development of the production infrastructure in development of gas fields in West Siberia," *Ekonomika gazovoy promyshlennosti*, 1976, No. 1, pp. 12–14, and *Sotsialisticheskaya Industriya*, March 20, 1977, p. 2.

[17]Table 42 and *Nar. khoz. SSSR v 1975 godu*, p. 235.

[18]*Energetika SSSR v 1976–1980 godakh* (Moscow: Energiya, 1977), p. 164.

[19]A. Troitskiy, "New advances in Soviet electric power," *Planovoye khozyaystvo*, 1976, No. 12, p. 20 and *Elektricheskiye Stantsii*, 1977, No. 1, p. 2.

[20]*Elektricheskiye Stantsii*, 1977, No. 1, p. 2 and *Energetika* . . . , op. cit. (1977), pp. 214–215.

[21]A. M. Nekrasov and V. Kh. Khokhlov, "Electric power in the 10th five-year plan," *Izvestiya VUZ, Energetika*, 1976, No. 9, p. 7.

[22]M. Vilenskiy, "Some problems of full electrification," *Voprosy Ekonomiki*, 1960, No. 8, p. 55 and *Sotsialisticheskaya Industriya*, June 21, 1970, p. 2.

[23]*Energetika* . . . , op. cit. (1977), p. 166.

[24]Computed from capacity and output of all power stations given in *Nar. khoz. SSSR v 1976 godu*, p. 201.

[25]B. L. Baburin and I. I. Fain, *Ekonomicheskoye obosnovaniye gidroenergostroitel'stva* (Moscow: Energiya, 1975), pp. 90–117; L. A. Karol', *Gidravlicheskoye akkumulirovaniye energii* (Moscow: Energiya, 1975), pp. 17–19 and R. W. Campbell, "Technological levels in the Soviet energy sector," in *East-West Technological Cooperation.* NATO Directorate of Economic Affairs, Brussels, 1976, pp. 250–251. Only one pumped storage plant is completed today in the USSR near Kiev on the Dnieper, although two more were to be started by 1980. By contrast, the United States and Japan each had 20 in the early 1970s and West Germany 28. Japan had 4.5% and West Germany 5.2% of their total generating capacities in the form of pumped storage stations; such capacity in the United States and Japan was to grow sharply by 1980. B. L. Baburin and I. I. Fain (1975), pp. 114–117 and *Energetika* . . . , op. cit. (1977), p. 142.

[26]B. L. Baburin and I. I. Fain, op. cit. (1975), pp. 88–89.

[27]*Gidrotekhnicheskoye stroitel'stvo*, 1970, No. 4, pp. 7–8.

[28]*Statistical Abstract of the United States.* 1975, p. 553.

[29]R. W. Campbell, op. cit. (1976), p. 249, reports a lower level of utilization for Soviet plants in 1970 once the load factor and reserve capacities for the two countries are equalized.

[30]M. A. Zurabov and M. Ya. Ayzenberg, "Five-year plan problems in fuel economics," *Organizatsiya i planirovaniye otrasley narodnogo khozyaystva*, Vol. 26–27 (Kiev University, 1972), pp. 79–80.

[31]Savchenko and Khodzhayev, op. cit. (1974), pp. 101–102, 120 and 135.

[32]Data on capacity and output of the two unified grids for the early 1970s is given in *Ekonomicheskaya Gazeta*, No. 4, January 1973, p. 2 and *Elektricheskiye Stantsii*, No. 12, 1972, p. 4.

[33]A Western study in the late 1960s estimated that the power demand of electric railways of West and East Siberia, little more than 12% of the regions' total in the mid-1960s, required an installed capacity well in excess of that needed to satisfy demand by the aluminum, iron and steel, pulp and paper, synthetic rubber, cement and nitrogen fertilizer industries combined. In contrast, actual consumption by the railways amounted to only half of that consumed by these industries. Brenton Barr and James Bater, "The electricity industry of Central Siberia," *Economic Geography*, 1969, No. 4, pp. 352–355.

[34]V. A. Ryl'skiy, *Ekonomika mexhrayonnykh elektroenergeticheskikh svyazey v SSSR* (Moscow: Nauka, 1972), p. 23.

[35]Interregional flows for 1965 and intersystem flows for 1970 were derived by the author in "Soviet Electric Power: Problems and Trends in Resource Use," paper presented at the Conference on Soviet Resource Management and the Environment, University of Washington, June 6–7, 1974, pp. 26–30.

[36]Table 42 gives 1975 output as 156.1 billion KWH. Consumption is said to have been 170 billion KWH. V. A. Shelest, *Regional'nyye energoekonomicheskiye problemy SSSR* (Moscow: Nauka, 1975), p. 282.

[37]A. A. Adamesku, ed., *Problemy razvitiya i razmeshcheniya proizvoditel'nykh sil Povolzh'ya* (Moscow: Mysl', 1973), p. 114.

[38]Per capita electricity consumption in the Ukraine during 1965-1970 ranged from 93 to 96% of the Soviet average and fell to 92% of the mean by 1975. By contrast, per capita production between 1965 and 1975 grew appreciably faster in the Ukraine than in the country as a whole. Computed from production, consumption and population data, *Nar. khoz. SSSR*, various issues and *Energetika* . . . , op. cit. (1977), p. 13.

[39]The Southern grid covers Moldavia and the Ukraine and part of adjoining Rostov Oblast in the Russian Republic. Production and consumption figures are available for the first two but not for Rostov Oblast. However, it cannot greatly influence the total net surplus. Data from *Nar. khoz. SSSR v 1975 godu*, p. 236 and *Energetika* . . . , op. cit. (1977), p. 13 and *Vneshnyaya torgovlya SSSR v 1975 godu*, p. 25. Virtually all net export of electricity originates in the Ukraine and Moldavia.

[40]In the Ukraine, which accounts for over nine-tenths of the capacity of the Southern grid, half of the increment in power station capacity during the 1976-1980 plan is to be nuclear. M. M. Seredenko, *Puti povysheniya effektivnosti osnovnykh fondov i kapital'nykh vlozheniy v narodnoye khozyaystvo* (Kiev, 1976), p. 16 and *Sotsialisticheskaya Industriya*, March 23, 1977, p. 2.

[41]*Energetika* . . . , op. cit. (1977), pp. 13 and 265 and Gosplan SSSR, SOPS, *Problemy formirovaniya i razvitiya territorial'no-proizvoditel'nykh kompleksov* (Moscow, 1976), p. 209.

[42]Francis M. Leversedge, "The implication of extra-high voltage transmission lines on power system management in the USSR," paper presented at a conference on *Energy in the USSR and Eastern Europe* by the University of Alberta, Division of East European Studies, May 27-28, 1977, pp. 11-12.

[43]*Energetika* . . . , op. cit. (1977), p. 159, and Federal Power Commission, *The National Power Survey* (Washington: Government Printing Office, 1971), Part I, Chapter 13, p. 4.

[44]*Energetika* . . . , op. cit. (1977), pp. 52 and 181-182.

[45]Ibid., p. 182.

[46]Ibid., p. 237; R. Ya. Bril' and I. M. Kheyster, *Ekonomika sotsialisticheskoy energii* (Moscow: Vysshaya Shkola, 1966), p. 342 and A. A. Stepankov, op. cit. (1963), p. 432.

[47]V. P. Korytnikov, ed., *Rabota TETs v ob"yedinennykh energosistemakh* (Moscow: Energiya, 1976), p. 7.

[48]The time series on heat inputs per KWH and per gigacalorie of output are given for utility stations in *Nar. khoz. SSSR v 1975 godu*, p. 238 and in earlier volumes. Utility stations have accounted for over nine-tenths of all power production during the last 10 years. See also P. S. Neporozhnyy, ed., *Elektrifikatsiya SSSR* (Moscow: Energiya, 1972), p. 537.

[49]*Energetika* . . . , op. cit. (1977), pp. 63 and 65.

[50]Neporozhnyy et al., eds., *Elektrifikatsiya SSSR 1967-1977* (Moscow: Energiya, 1977), p. 45. The distribution of heat supply by heat and power stations to industry, household and municipal consumers is computed from data in *Energetika SSSR v 1971-1975 godakh* (Moscow: Energiya, 1972), pp. 80-82.

[51]Soviet utility stations in 1976 were said to require, on the average, 337 grams of standard fuel or 2,359 kilocalories per KWH generated. The Edison Electric Institute reports 10,449 BTUs, i.e., 2,664 kilocalories, per KWH as the average for United State utilities in 1977. *Nar. khoz. SSSR za 60 let*, p. 203 and Edison Electric Institute. Private communication. For earlier years in the two countries see Ibid., p. 203 and Edison Electric Institute, *Statistical Yearbook* (New York), various issues.

[52] *Ekonomicheskaya Gazeta*, No. 30 (July 1977), p. 24.

[53] *Soviet Geography*, December 1976, p. 717.

[54] Between 1960 and 1970 the share of peat in the fuel balance of Belorussian power stations dropped from 68% to 31% (and, according to another source, to 24%), with further decline expected. A. G. Lis, *Problemy razvitiya i razmeshcheniya proizvoditel'nykh sil Belorussii* (Moscow: Mysl', 1972), p. 79 and A. G. Lis, *Razmeshcheniye proizvoditel'nykh sil ekonomicheskikh rayonov* (Moscow: Mysl', 1977), p. 118, and *Energetika...*, op. cit. (1972), pp. 172-173. The smaller share probably refers to stations under the Ministry of Electric Power only.

[55] Neporozhnyy et al., *Elektrifikatsiya SSSR 1967-1977,* p. 157 and *Teploenergetika*, 1977, No. 11, p. 10.

[56] *Sotsialisticheskaya Industriya*, March 23, 1977, p. 2; V. A. Zaydel', *Osnovnyye napravleniya i perspektivy razvitiya teplovykh elektrostantsiy* (Moscow: Energiya, 1976), p. 32, and N. Zalogin, et al., *Energetika i zashchita atmosfery* (Moscow, 1976), p. 7.

[57] *Energetika...*, op. cit. (1977), p. 272.

[58] Data refers only to stations under the Ministry of Electric Power, which generate 93% of all electricity. Ibid., pp. 272 and 275-276 and *Elektricheskiye Stantsii*, 1975, No. 1.

[59] Standards quoted in F. Leversedge, op. cit. (1977), p. 27. Number of plants over 2 million KW from *Elektricheskiye Stantsii*, 1977, No. 12, p. 5.

[60] V. S. Al'tshuler, *Novyye protsessy gazifikatsii tverdogo topliva* (Moscow: Nedra, 1976), pp. 191-192.

[61] I. P. Krapchin, *Effektivnost' ispol'zovaniya ugley* (Moscow: Nedra, 1976). p. 68.

[62] Estimated from scattered data and distribution of all coal and other fuels in various uses for the late 1960s in M. S. L'vov, *Resursy prirodnogo gaza SSSR* (Moscow: Nedra, 1969), p. 13.

[63] Al'tshuler, op. cit. (1976), p. 193.

[64] N. Fey'tel'man, "Timely problems in the development of the fuels and energy complex," *Ekonomicheskiye Nauki*, 1976, No. 4, pp. 32-43.

[65] *Trud*, November 4, 1976, p. 2; Shelest, op. cit. (1975), pp. 246-247; Zaydel', op. cit. (1976), pp. 31-32 and Zalogin, et al., op. cit. (1976), p. 7. In addition, the removal of particulate matter appears more troublesome at stations burning Ekibastuz coal than at those using other solid fuels. The electrofilters currently in use do not work effectively for Ekibastuz coal. L. I. Kropp and I. N. Shinugol', "Enhancement of the efficiency of operation of electrofilters in collecting ashes from Ekibastuz coal," *Teploenergetika*, 1977, No. 6, p. 14.

[66] Zaydel', op. cit. (1976), p. 31; S. K. Karyakin, "An investigation of the mineral composition of Kansk-Achinsk coals in connection with power-station use," dissertation (Tomsk Polytechnical Institute, 1975), pp. 105 and 129-135.

[67] V. S. Zamakhayev, *Ispol'zovaniye vody v narodnom khozyaystve* (Moscow: Energiya, 1973), pp. 16-37 and 62; S. Oziranskiy, "Payment for water resources," *Planovoye Khozyaystvo*, 1968, No. 9, p. 67.

[68] A modern fuel-burning generating plant produces with, each KWH of electricity, 3900 BTUs of heat energy that needs to be carried away. A gallon of water can remove about 120 BTUs if its temperature is allowed to increase 15°F. Thus, cooling for a 1 million KW, 40% efficient plant (near the present efficiency limit) requires 9,000 gallons of water per second. In cubic meters, such a plant requires 40.9 per second. John M. Fowler, *Energy and the Environment* (New York: McGraw-Hill, 1975), pp. 174-175.

[69] Craig ZumBrunnen, "An estimate of the impact of recent Soviet industrial and urban growth upon surface-water quality," paper presented for the Conference on Soviet Resource Management and the Environment, University of Washington, June 6-7, 1974, pp. 28-38.

[70]T. S. Vishnevskiy, "Contemporary problems of construction of large fossil-fueled and nuclear electric stations," *Sovremennyye problemy energetiki* (Kiev, 1976), p. 64.

[71]Shelest, op. cit. (1976), p. 227.

[72]Computed as production minus export. *Nar. khoz. SSSR v 1975 godu*, pp. 142–145 and *Vneshnyaya torgovlya SSSR v 1975 godu*, p. 25.

[73]Table 47 and *Statistical Abstract of the United States*, 1975, p. 559.

[74]*Nar. khoz. SSSR v 1976 godu*, pp. 108–109.

[75]*Energetika . . .* , op. cit. (1977), p. 50.

[76]Ibid., p. 50 and *Statistical Abstract of the United States*, 1975, p. 559. If it is assumed that United States rural households are electrified to the same degree as urban ones, the level of electrification in Soviet rural households is only 17–18% of their American counterparts.

[77]*Energetika . . .* , op. cit. (1977), p. 43.

[78]Ibid., pp. 43–45.

[79]Ibid., pp. 44 and 46 and A. M. Nekrasov and Khokhlov, op. cit. (1976), p. 11. Irrigation alone took 13 to 14% according to these two sources. Federal Power Commission, *The 1970 National Power Survey*, Part I (Washington: Government Printing Office, 1971), Ch. 3, p. 18.

[80]M. A. Vilenskiy, op. cit. (1975), p. 36–39.

[81]Ibid., p. 40.

[82]Ibid., pp. 138–145.

[83]U.S., C.I.A., *Soviet Economic Problems and Prospects.* ER 77-10436U (Washington, July 1977), pp. 3–5 and Murray Feshbach and Stephen Rapawy, "Soviet population and manpower trends and policies," in U.S. Congress, Joint Economic Committee, *Soviet Economy in a New Perspective* (Washington: Government Printing Office, 1976), pp. 130–134.

[84]Table 48 and *Energetika . . .* , op. cit. (1977), p. 31; *Soviet Geography*, December 1976, p. 719.

[85]Production in the European USSR is to grow from 740 billion KWH in 1975 to 975 billion in 1980. All the 21 billion KWH of export will originate in the European USSR. No appreciable transmission of power is expected from the Asian to the European USSR before the early 1980s.

[86]*Energetika . . .* , op. cit. (1977), p. 115.

[87]M. A. Vilenskiy, op. cit. (1975), p. 144 and Table 48.

[88]Estimated from Table 47.

[89]According to *Energetika . . .* , op. cit. (1977), p. 149. Total fuel output amounts to 860 million tons of standard fuel equivalent. Assuming that about 78% of all fuel was consumed in the European USSR, 1975 fuel consumption amounted to 1108 million tons of standard fuel, and 1978 consumption to 1120 million. From *Nar. khoz. SSSR v 1975 godu*, p. 112.

[90]*Energetika . . .* , op. cit. (1977), pp. 13 and 265.

[91]*Kazakhstanskaya Pravda*, April 22, 1977; Shelest, op. cit. (1975), pp. 248–249; and Ryl'skiy et al., op. cit. (1974), pp. 64–65.

[92]*Energetika . . .* , op. cit. (1977), pp. 13 and 125. The first half of the 1970s, however, was characterized by unusually low water regimes.

[93]Ryl'skiy et al., op. cit. (1975), pp. 67–71.

Chapter 8

ALLOCATION PROBLEMS ON THE DOMESTIC AND EXPORT MARKETS

In the energy sector the problems of allocation and optimization confront resource planners particularly sharply. Energy is indispensable in all stages of the production process, it may be applied in the form of heat, electrical power, process steam and mechanical work and derived from several primary sources and geographic areas. Both primary sources and secondary forms are in large measure substitutable for one another, though at different levels of cost and efficiency. Economic decision-making involves sectoral and regional allocation, combination and degree of substitution throughout the energy system so as to maximize overall advantage from the use of these resources. However, with high economic and political stakes in the Comecon bloc as a whole and the rapid growth of Soviet foreign trade with the non-Communist world, Soviet planners today cannot attempt solutions to these problems as if their economy were entirely a closed system. A least-cost allocation of energy sources and forms on the domestic market may maximize hard-currency earnings or the economic and political benefits accruing from increased cooperation within the bloc.

The evidence indicates that step by step the USSR is "allowing itself to become more intertwined into the world economy," although its commitment to irrevocable, full-scale participation in world trade is still uncertain.[1] Because of the crucial role of energy in Soviet exports, increased trade orientation should have a significant impact on the perception and aspirations of planners concerned with energy allocation. The degree to which the domestic consumption pattern can be modified in favor of export is controversial and requires detailed study. Nor is the foreign market for Soviet fuels homogeneous and exports to East Europe, at least, should be clearly distinguished from those beyond the bloc. It is these problems of allocations this chapter investigates. The analysis is based on energy trade data through 1976. Starting in 1977, the Soviet Union imposed secrecy on exports and imports of energy goods in physical units, publishing only ruble value.

THE DEMANDS OF FOREIGN TRADE ON SOVIET ENERGY RESOURCES

The world's second largest economy, the USSR, is still a relatively minor factor in world trade: in 1977, it ranked 8th, just ahead of Canada and behind the Netherlands, in total turnover, of which half was with the Comecon countries.[2] By all signs, Soviet leaders now give great importance to trade. They regard commerce with the West as valuable and, within Comecon, they are keenly aware of the influence of economic relations on political stability. In this increasingly important sector of the Soviet economy, the energy industries play a dominant part. Even before the 1973 price explosion, mineral fuels comprised close to 30% of all Soviet exports to the industrial West. Thanks to the huge price rise, which also spurred an increase in the volume of oil and gas exports, this share reached over half (53%) in 1976, nine-tenths of it being crude oil and refined products.[3] Energy exports to the socialist countries of East Europe comprised 27% of all Soviet exports to these states; they were still below world price levels, though the price difference was expected to vanish by the end of the 1970s.[4]

For the East Europeans, Soviet energy supplies are crucial. With Rumania excepted, Soviet petroleum in the 10th five-year plan was still slated to meet 86% of the East European countries' oil imports and all of their natural gas imports.[5] Despite the announced slowdown in the rate of growth and probable stabilization of Soviet petroleum supplies to East Europe, Soviet oil was still expected to account for at least 60% of these countries' import through much of the 1980s (again excluding Rumania).[6] A significant absolute reduction in the volume of Soviet oil to East Europe (which could be replaced only for hard currency on top of the already huge indebtedness, estimated to be some $25 billion in 1976) would probably lead to economic and political difficulties.[7] Even in the face of a slow growth or possible peaking in Soviet oil production, this is not a course Soviet leaders can undertake lightly. With a stabilization of Soviet oil exports to the Comecon members, the importance of Soviet gas should grow appreciably, and large increases in its export were scheduled for the 1980s.

The foregoing illustrates the significance of energy, particularly hydrocarbon, exports to Soviet planners. With vigorously rising petroleum output, Soviet leaders have so far been spared the really difficult decisions about priority allocation between exports and domestic needs. The dampening of growth in domestic oil consumption in the middle 1970s in favor of exports involved only small quantities.[8] It could be effected by drawing down on stocks, by some substitution of natural gas and by relatively easy conservation measures. Moreover, Soviet oil production was growing annually by 25 to 29 million tons, increments never reached before 1973, while natural gas output was again accelerating after several years of declining increments.[9] Soviet leaders, therefore, could augment oil and gas exports to both the Communist and hard-currency markets. In 1975 and 1976, oil exports to non-Communist countries surged more than to Communist states, and thanks to the new world prices, pushed hard-currency earnings to a record high. However, oil sales to Comecon as a whole also grew by more than 10% both years and increased to every country of the bloc.[10]

No one can confidently forecast Soviet priorities between exports and internal demand in the event of a slowdown in the expansion, let alone peaking or decline, of petroleum output. Three Western experts recently voiced somewhat contrary opinions[11] and Soviet planners themselves may not have an answer to this question. The regular and quite rapid decline in the increment of petroleum production after 1975 and the absolute drop in output in all but two or three of the Soviet petroleum provinces[12] suggest that this choice may soon have to be faced. If Soviet petroleum output indeed peaks or manages to grow slowly, only a part of oil exports could be preserved. Most, though perhaps not all, hard-currency sales would have to be eliminated. In the more controversial case of an actual drop in oil production, predicted by the C.I.A., the economic and possibly political consequences throughout the bloc are likely to be so profound that speculations become pointless.

East European leaders seem to be equally in the dark. The original Soviet commitments to other Comecon states for the 1976–80 period amounted to 364 million tons of oil, about 90 billion cubic meters of natural gas and 67 billion KWH of electricity, all to East Europe except about 10% of the oil destined for Cuba and Mongolia.[13] These commitments may have been too ambitious or based on faulty planning. After semi-official warnings, word of a slowdown in the growth of Soviet oil supply was reportedly passed to member states at the June 1977 meeting of the Council of Economic Mutual Assistance, and Soviet-bloc nations began stepped-up negotiations with American companies.[14] On the other hand, a late 1977 Hungarian source reported over 400 million tons of crude oil and refined products and about 100 billion cubic meters of Soviet gas consignments to the Comecon partners during the 1976–1980 period.[15] Given such confusing and contradictory reports, one wonders about the firmness and consistency of planning within the bloc. Long-range planning is even more tentative. Among the sketchy announcements concerning Czechoslovakia's long term economic program, accepted in 1977 by that country's Politburo, is the unequivocal statement that the energy sources necessary for planned growth were as yet available only until 1980.[16]

Although East Europe is still spared the burden of paying in hard currency for Soviet oil received under the five-year contract, the price concessions are coming to an end. By the end of the 1970s, Soviet prices to the Comecon will fully include the drastic price increases since 1973 and should approximate world levels. Nor can the hard-currency burden be wholly avoided, since Comecon obligations include the delivery of pipe and equipment for Soviet oil, products these countries must obtain from the West. Soviet-bloc countries also enjoy no price advantage on their purchases of natural gas. East European states in 1976 paid as much for Soviet gas as Austria and more than West Germany, Italy and France, though again the Communist countries used hard-currency only in part, in the form of their equipment supplies.[17] Substantial East European funds, and even some labor, are also being harnessed to help Soviet resource development and build facilities that will also aid Soviet hard-currency exports. Still, the USSR must unquestionably forgo significant economic benefits from every ton of oil and cubic meter of gas it sells within Comecon rather than the West.

The Value of Oil Exports to the West vs. Soviet Domestic Uses

Even during the 1960s, with the prevailing low world prices for hydrocarbons, Soviet petroleum exports beyond the Comecon area evidently made good economic sense. Not only did they become an important source of foreign exchange, but declining production costs (at least until the mid-1960s) and the expansion of the pipeline network enabled the Soviet Union to sell well above costs, Western accusations of dumping notwithstanding. In addition, the cheapness and increasing abundance of natural gas resulted in low replacement costs for petroleum at home, wherever gas became available.

The fourfold jump in the world price of oil and mounting Soviet hard-currency needs multiplied the value of petroleum on the export market. Today the Russians can easily justify the replacement of oil for export even by expensive deep-mined coal and would also find it economically rational to displace more gas from at least some part of the domestic market. Based on opportunity-cost criteria, the ability of petroleum to earn foreign exchange has made it too expensive to be consumed at home, at least in most stationary uses, whatever the substitute fuel.

As shown later (Chaper 9), in under-boiler uses the marginal cost of run-of-the-mine coal per ton of standard fuel (7 million kilocalories) through much of the 1980s is expected to approximate 20–22 rubles. Reduced efficiency of operations when compared to plants burning hydrocarbons raises this cost by some 15%, to which an additional but uncalculable environmental penalty must be added.[18] For much of the 1980s, therefore, the marginal cost of coal is likely to range between 24 and 26 rubles per ton of standard fuel for run-of-the-mine coal, though more after washing and sorting. With the current world market price for oil of $64 per 7 million kilocalories or 93 rubles at the official exchange rate, and an expected price of around $100 per 7 million kilocalories by the early or middle 1980s, the economic rationality of petroleum export appears strong. The full benefits are greater still, since the value of hard-currency earnings far exceeds the official ruble rate.

A greater use of coal to free oil and gas for export, however, is a difficult matter, encountering both technological and geographic obstacles. Even petroleum products and natural gas are not freely interchangeable in many uses. Speculation about substitution policies, therefore, is meaningless without some knowledge of the functional breakdown of consumption, i.e., the technological uses to which energy is applied. In recent years enough information has been released to permit such an analysis and to project it foreward on the basis of

Soviet plans and various assumed rates for aggregate energy demand.

Fuel Substitution and the Functional Distribution of Demand

A functional distribution of aggregate fuel input into the Soviet economy, with projections for the future is shown in Table 50. The table follows Soviet usage and is also consonant with the useful breakdown given for 1973 in the 1975 C.I.A. report, *Soviet Long-Range Energy Forecasts*. The forecasts for 1990, assembled and fleshed out in that report from highly tentative and optimistic Soviet figures published in the early 1970s, were not followed here. Soviet expectations since that time appear to have been scaled down and, at any rate, 1990 was judged too far ahead for projections of categories of technological use. The 1980 figures for total fuel demand as well as for fuel consumption by electric stations are also somewhat smaller than originally anticipated by the 10th five-year plan (1976–80). The performance of the fuel industries in the first three years of the plan period indicated that, even with the inclusion of self-produced firewood and peat, aggregate fuel output would fall short of expectation, and consumption, too, must be correspondingly reduced. Total consumption here matches that given in Table 53 but incorporates an additional 20 million standard tons of self-produced fuel, the amount that was furnished in 1975 and

Table 50. Functional Breakdown of Aggregate Fuel Consumption in the USSR (in million tons of standard fuel; 1 ton = 7 million kilocalories)

Functional uses	1975	1980	1985		
			A		B
Electric stations	472	565	640	660	670
All other boilers	146	220	310	290	310
Industrial furnaces and technological equipment	285	325	360		380
Small heating devices	220	230	245		250
Mobile units and power tools	240	275	325		335
Nonenergy uses	80	105	140		145
Total	1440	1720	2020		2090

Notes and sources: Electric stations in 1975 and 1980 from *Energetika . . .* , op. cit., pp. 149 and 151. Fuel for all other boilers in 1975 appears as a difference between fuel use for the production of electric and thermal energy (*Nar. khoz. SSSR v 1975 godu*, p. 112) and fuel demand by electric stations. Other uses and projections are worked out on the basis of percentage distribution in Neporozhnyy et al. (1974), p. 156 and the rationale is explained in the text. Losses given separately in Neporozhnyy's breakdown are allocated proportionately to the different uses. The 1975 total incorporates self-produced fuel by the population. The 1975 distribution is also consonant with that made by the C.I.A. in *Soviet Long-Range Energy Forecasts*, p. 19 for the year 1973.

is not likely to be drastically diminished. The planned share of fuels burned by power stations in 1980 was retained but applied to the somewhat reduced total.

Industrial furnaces and technological equipment. In 1975, industrial furnaces and related equipment (kilns, ovens) consumed 285 million standard tons of fuel, slightly less than one-fifth of the Soviet total. Such users also accounted for two-thirds of industrial fuel demand when industry is defined to exclude thermal power stations. Since 1960, most of the increment in fuel consumption by furnaces has been satisfied by hydrocarbons. Aside from coking coal, the demand for solid fuels in these uses increased only slightly, if at all. This has been particularly true in the European USSR, where, according to one source, solid fuels have become virtually eliminated from industrial furnaces and technological equipment, though they still cover one-third of boiler needs.[19] They are still important in the Urals and continue to dominate in Siberia and Kazakhstan, but the latter two account for less than one-fifth of all stationary fuel demand (Table 5). Whatever Soviet perceptions may be about the adequacy of hydrocarbon resources or about oil and gas as hard-currency earners, Soviet writers show no indication that a significant relative expansion of coal use in the technological and furnace sector is viewed as a rational course. And in blast furnaces, at least, the further penetration of oil and gas to save coke, always in short supply and produced at great cost and large energy expenditure, is unavoidable.

Unlike Japanese, West European and also increasingly United States blast furnaces, where coke is saved by using vaporized oil spray, Soviet furnaces in the 1960s used mainly natural gas for that purpose. Since 1971, however, the consumption of fuel oil has risen more rapidly than that of natural gas, and by 1980, petroleum should comprise one-eighth of all hydrocarbons injected into blast furnaces. Heavy winter demand and consequent strain on gas supplies is one reason, but beyond a certain point Soviet researchers also report diminishing cost-effectiveness when only gas is used.[20] In open-hearth furnaces and in those for the production of rolled steel, pipe and general metal goods, gas is expected to be a substitute for fuel oil, particularly since no equipment changes are required.[21]

Following Soviet experts, we can project a slower growth of fuel use in industrial furnaces and related equipment than in the economy as a whole both in the 10th five-year plan and beyond 1980 (Table 50). There is likely to be further noticeable penetration of hydrocarbons, above all natural gas, until 1980 (especially in the Kuzbas and the Urals), but little further growth at the expense of coal afterward. On the other hand, reverse substitution back to coal is also ruled out over the next decade. With these in mind, of all furnace and related uses in industry shown in Table 50, the share of solid fuels is projected to decline from 44 to 46% in the first half of the 1970s to roughly 35% through the first half of the 1980s.

Small heating devices include single home and detached apartment house furnaces in the domestic-municipal sector, furnishing virtually all heat to rural communities and small towns, but also equipment in agriculture, construction and even industry in isolated locations. Because of inefficient energy use and enormous manpower requirement (estimated to reach 400 million labor-days in

1976),[22] Soviet planners intend to reduce the share and eventually the magnitude of fuel burned this way. But the substitution of centrally supplied heat, mostly from heat and power stations and large-scale boilers, is a slow process and, at any rate, is unlikely to affect rural areas. A slow and declining rate of absolute growth is, therefore, projected until the mid-1980s.

Aside from pipeline gas and LPG, the dispersed small heating devices depend almost entirely on solid fuels, i.e., coal, peat, and firewood, about half of the combined supplies of the latter two being gathered by the population at huge expense of labor time. An estimate by the Academy of Communal Economy places fuel consumption by detached units for space and water heating in the household-municipal economy at 135 million tons of standard fuel (70–75% of all energy use in the household and municipal economy) or about four-fifths of the total energy consumed by small heating devices.[23] Cooking takes a few more million. A growth in gas consumption here must be expected and, indeed, it is the announced program. It is also safe to assume that coal and hydrocarbons will gradually supplant at least part of the large volume of firewood and self-produced fuel, reportedly amounting to 45 million standard tons at the beginning of the 1970s, virtually all of which is used by decentralized heating devices and small boilers grouped with the above.[24] Despite their declining share and small increment in total fuel demand, such equipment, too, will contribute slightly to the increase of both hydrocarbon and coal consumption.

Mobile demand, power tools and nonenergy uses. Because of the low level of motorization, direct fuel demand by mobile machines in the USSR is relatively modest. Significantly, less than half of this consumption was accounted for by the transport sector, with agriculture and construction taking the larger share.[25] The future development of this demand is subject to opposing influences. On the one hand, the expansion of Soviet automobile production has slowed and the 10th five-year plan envisaged an annual growth of only 3.2%. Similarly, the cultivation and harvesting of field crops is now fairly highly mechanized and the emphasis with respect to field machinery has shifted from rapid growth of quantity to improvement of quality, assortment and performance.[26] At the same time, the accelerated development of the resource-rich but remote Siberian regions is raising hydrocarbon consumption by transport and construction machinery while also increasing waste. It also enhances demand by stationary engines and power tools, such as pipeline compressors and pumps and diesel units for drilling in remote locations. Except for some 10 million tons of coal still used in the transport sector,[27] mobile demand for fuels is tied to petroleum, primarily to light products but also to middle distillates and fleet fuel oil. Petroleum and natural gas also are used as fuels for stationary power tools not run by electricity.

Most of the *nonenergy uses* comprise raw materials for the chemical industry, though bitumen, lubricants, petroleum coke, etc., are also included. It is difficult to separate the fuel components within nonenergy uses. Again, most of the fuel consists of refinery products, LPG and pipeline gas, with petrochemical synthesis taking the greater part. It is also certain that hydrocarbons will account for most of the increment in these uses over the coming years. It is reasonable to project

Table 51. Functional Breakdown of Oil and Gas Use (1976 and 1980) (million tons of standard fuel; 1 ton = 7 million kilocalories)

Use categories	1976		1980
	Natural gas	Oil	Natural gas and oil
Mobile uses	–	195[a]	230*
Liquefied gases	5.3[b]	7[b]	16[c]
Field use and losses	3.1	20[d]	32*
Gas pipelines–fuel	18.0[c]	–	
Gas pipelines–losses	6.6[c]	–	43[c]
Gas processing plants–fuel and losses	3.6	–	6*
Petroleum refining–fuel and losses	2.5[e]	29[e]	38[c]
Blast furnaces	10.1	0.5[f]	15[g]
Open-hearth furnaces	6.8	3.5[f]	11[g]
Chemical raw materials, industrial lubricants, bitumen (hydrocarbons other than liquid gases)	22.2	45[h]	85*
Household furnaces and stoves	15.8	negl.	24*
Subtotal (A)	94.0	300	500
Steel pipes	1.8		
Rolled steel	6.4		
Furnaces in nonferrous metallurgy	2.8		
Cement kilns	12.4		
Furnaces and ovens for glass, ceramics, other construction and refractory materials	12.2		
Forges and related devices for shaping metals	8.3		
Other applications	11[i]		
Subtotal (B)	55.0	60*	170*
Subtotal A + B	149.0	360	670
Electric Power Ministry and industrial power stations	111.3	136[j]	315[j]
Industrial boilers	49.1		
Municipal boilers	33.0		
Small boilers and isolated power stations	10.1[i]	45*	110*
Total power station and boiler use (subtotal C)	203.5	181	425
Net addition to stock and storage and to fill new pipelines	15.5[k]	1[l]*	45*
Apparent consumption (production minus net export)	368.0	542	1140*

Notes and sources: Except where otherwise noted, all 1976 figures for natural gas are from an authoritative study in *Gazovaya promyshlennost'*, 1978, No. 6, pp. 12 and 29. Figures noted with letters are a mixture of Soviet and Western data and the author's calculations derived from a variety of sources as explained on opposite page.

Notes and sources (Table 51. cont'd)

*Denotes arbitrary estimates by the author that appear reasonable but cannot be supported by solid evidence. They are also designed to help the columns add up to apparent total consumption.

Apparently consumption for 1976 is factual, converted to standard ton equivalents from physical tonnage given in *Nar. khoz. SSSR za 60 let*, p. 205 and *Vneshnyaya torgovlya SSSR v 1976 q.*, pp. 26 and 38. 1980 data correspond closely to projections in Table 53.

[a]Estimated by assigning all gasoline and most kerosene to mobile uses. Fuel oil and diesel fuel were assigned according to their percentages in the transport sector and, for diesel fuel, also in the agricultural, construction and other sectors. The output of refinery products for 1975 is available from R. W. Campbell, *Soviet Fuel and Energy Balances* (Santa Monica CA.· RAND Corporation, Research Report R-2257, 1978), Appendix. The consumption breakdown by sectors is given in *Vestnik Statistiki*, 1978, No. 1, p. 9.

[b]Total for 1975 given in *Energtika SSSR v 1976–1980 godakh* (Moscow: Energiya, 1977), p. 149, was adjusted upward and broken down between oil and gas according to Campbell, op. cit. (1978). Appendix.

[c]For 1976, *Ekonomika gazovoy promyshlennosti*, 1977, No. 11, p. 27; for 1980, *Energetika . . .* , op. cit. (1977), pp. 149 and 151.

[d]Field and transport losses are conservatively estimated to be about 3%. V. V. Arenbrister, *Tekhniko-ekonomicheskiy analiz poter' nefti i nefteproduktov* (Moscow: Khimiya, 1975), pp. 18–19, gives 5% for total losses (in addition to internal fuel use in the various operations) from fields through supply depots of refined products. Of this 5%, less than one-third, or 1.6%, was lost during the refinery operation.

[e]Total for 1975 is given by *Energetika . . .* , op. cit. (1977). pp. 149 and 151. It was adjusted upward slightly and the consumption of natural gas separated out according to Campbell, op. cit. (1978). Appendix.

[f]Consumption of oil in blast and open-hearth furnaces projected from 1972 and 1974 data as given in *Ekonomika chernoy metallurgii*, No. 5 (Moscow: Metallurgiya, 1976), pp. 91–93, and N. I. Perlov et al., *Tekhnicheskiy progress i toplivo-energopotrebleniye v chernoy metallurgii* (Moscow: Metallurgiya, 1975), p. 131.

[g]Natural gas consumption projected from 1970, 1975 and 1977 volumes as given in *Gazovaya promyshlennost'*, 1978, No. 6, p. 12. Growth in oil consumption is given in *Energetika* (Prague), 1977, No. 9, p. 455.

[h]Natural gas is said to provide 44% of all raw material for chemical synthesis. Most of the rest originates from petroleum. *Ekonomika gazovoy promyshlennosti*, 1977, No. 11, pp. 30–31 and G. F. Borisovich and M. G. Vasilev, *Nauchno-tekknicheskiy progress i ekonomika khimicheskoy promyshlennosti* (Moscow: Khimiya, 1977), p. 29. Heavy, nonfuel refinery products from Campbell, op. cit. (1978). Appendix.

[i]The categories "other branches of industry" and "other branches of the economy" given in *Gazovaya promyshlennost'*, 1978, No. 6, p. 29, were divided between furnace-type applications and boiler use as in the table. This conforms to data about gas consumption by all types of electric stations everywhere. *Energetika . . .* , op. cit. (1977), p. 111.

[j]*Energetika . . .* , op. cit. (1977), p. 151.

[k]Residual for the columns. Data unavailable, but this figure is reasonable, since net additions to storage have been increasing from 1.7 million tons of SF equivalent in 1973 to over 6.8 million tons of SF equivalent in 1975. The 1976–80 plan called for rapid growth of storage. Data from S. A. Orudzhev, *Gazovaya promyshlennost' po puti progressa* (Moscow: Nedra, 1976), p. 71.

[l]According to *Nar. khoz. SSSR za 60 let*, p. 83, stocks of all fuels declined during 1976, evidently in response to the export drive. Since the latter concerned only hydrocarbons, and primarily petroleum, it is reasonable to assume that petroleum stocks were drawn down somewhat. On the other hand, 2,000 km of new oil pipelines had to be filled up in 1976.

an increase for mobile uses somewhat below the rate for aggregate fuel demand, and for nonenergy uses, an increase appreciably faster than the rate for fuel requirement as a whole.

As the foregoing discussion has shown, a large part of aggregate fuel demand in these functional categories is covered and must continue to be covered by hydrocarbons. Not only are mobile and nonenergy uses (particularly chemical synthesis) dependent technologically on petroleum and, in the latter case, also on natural gas, but hydrocarbons are also essential and indispensable in a wide range of stationary uses. Table 51 provides a more detailed breakdown of the consumption of natural gas and oil in 1975, with plans and projections for the two hydrocarbons combined in 1980. Data limitations preclude the precision and detail for oil that are now possible for gas. The first subtotal and its breakdown represent technological areas of demand where hydrocarbons are irreplaceable. The second subtotal includes industrial furnaces and related equipment where replacement of oil and gas, though theoretically possible, could be accomplished only with great difficulty and at enormous cost, even in the long run. As Table 51 shows, the 10th five-year plan anticipated hydrocarbon demand in these nonsubstitutable categories to grow from about 510 million standard tons equivalent in 1975 to 670 million standard tons by 1980 (subtotals A + B). On the basis of projected increases in total fuel demand by major functional categories (see Table 50 and accompanying discussion), we can estimate that by the mid-1980s, hydrocarbon requirements apart from electric stations, industrial and municipal boilers will rise to between 800 and 850 million tons of standard fuel equivalent, depending on the rate of growth of aggregate energy consumption.

It is clear that such nonsubstitutable demand (from mobile, furnace, and chemical uses, pipeline compressors and the like) shown in Table 51 puts no real strain on hydrocarbon supplies. Nor does it compromise the continued export of petroleum at present or even higher levels or stand in the way of a huge surge of gas export. While substantial, these categories of demand amount to a little less than half of the expected 1980 oil and gas output, i.e., total consumption in Table 51 plus net export. The 1985 requirements by these consumers again come only to 50–55% of total hydrocarbon output even if Soviet oil production peaks in the early 1980s (Table 53). A serious strain is building up, however, in the category of boiler demand (power generation, process steam and hot water), where petroleum and gas show the smallest economic benefit and where their replacement by alternative and, from a technological point of view, perfectly feasible solid-fuel sources has a strong economic argument. It is here where export opportunities come into clear conflict with a domestically suboptimal use.

Thermal Power Stations and Industrial Boilers: Coal versus Hydrocarbons

In the USSR today, both petroleum and natural gas provide much larger shares of the fuel supply to electric stations than in the United States. The 10th five-year plan (1976–80) called for a stabilization in the relative proportions of both hydrocarbons in the generation of electric power, with a corresponding marginal increase in the share of coal (Table 10). These constant proportions, however, imply large absolute increments–more than 20 million natural tons of oil (about 30 million tons standard) and some 24 billion cubic meters of natural gas (28 million tons standard).[28] Even with diminishing shares in the contribution of these quality fuels to power generation after 1980, their absolute growth in power station use cannot be arrested before mid-decade. Altogether, 31 new oil and gas-fired plants were to be started in the 10th five-year plan, most to be completed only in the 1980s,[29] though some of these may, of course, be scrapped if oil supplies fail to expand as planned.

Oil and gas requirements by power stations will develop differently in the European and Asian parts of the country. Consequently, the expected Soviet demand by thermal electric plants must be aggregated from regional projections reflecting the dissimilar needs of these areas. A close examination shows that, in East Siberia, the Far East and most of Kazakhstan, hydrocarbons will have to cover no more than 4–5% of total power station demand throughout the 1980s. In both West Siberia and the Urals, natural gas and, to a smaller degree, oil will have to comprise 30–33% of all fuel consumption for the generation of electricity.[30] When these percentage requirements are combined with a little over 5% yearly growth rate of power demand in the Asian USSR and allowance is made for the output of hydroelectric plants under construction and preparation, the conclusion emerges that the Urals and the Asian regions must retain over 40 million standard tons equivalent of their gas and oil resources for power stations each year by the early and middle 1980s.[31] Given the historical relationship of fuel demand between electric stations and other boilers, the generation of electricity, steam and hot water for all purposes is likely to require 55 million tons of standard fuel in the form of gas and oil products even in the east.

West of the Urals, the demand for gas and oil in boiler uses will be vastly greater. By 1975, boiler fuel demand exceeded 400 million standard tons, with Soviet plans projecting close to 500 million tons by 1980.[32] Beyond the early 1980s, the installation of new condenser stations (which nationwide account for two-thirds of thermal electricity production) is planned to cease throughout the cis-Volga provinces, with nuclear reactors and heat and power plants assuming all the growth in nonpeak generating capacity. Allowing for the consequent substantial reduction in incremental fuel demand in electric stations and assuming a mere 4.5% annual growth in power consumption in the European regions and the Caucasus, total fuel demand for the generation of steam, electricity and hot water west of the Urals should reach 500 million standard tons in 1980 and 570 million standard tons five years later.[33]

In contrast to these demands, the provinces west of the Urals have available no more than 190 million standard tons of solid fuels annually for noncoking purposes, even including centrally furnished firewood and the shipment of coal from the Asian USSR. And some of these fuels, perhaps 50–60 million standard tons, have to be assigned to the household-commercial economy and for nonboiler, noncoking uses in industry, agriculture and transport. A huge gap, therefore, exists in the European USSR between power-station and boiler needs and the quantities of solid fuels available for such use. This gap, which reached 250–260 million tons of standard fuel in 1975, will widen by 100 million in a few years. It may stabilize after the early 1980s if ambitious plans for coal transport from the Asian USSR and the tremendous "coal by wire" EHV project from Ekibastuz materialize in time.[34]

It is gas and oil that must close this gap west of the Urals between boiler-fuel requirements and available solid fuels. As shown above, the hydrocarbons must also provide modest quantities (50–55 million tons of SF per year through much of the 1980s) for electricity, steam and hot water even in the Asian Soviet Union and the Urals. This adds up to at least 400 million standard tons of hydrocarbons for the boiler fuel market, which compares closely to the Soviet plan for 1980 (425 million in Table 51).

There is little that Soviet planners can do about the situation until and unless Siberian lignites or the electricity generated from them can reach the Volga River in really massive quantities, and nuclear plants can take over most of the base load in the European USSR and not just the increment. The ambitious nuclear program cannot be accelerated greatly. It is also a moot question whether an economically acceptable solution can be found for the transport problem of Siberian lignites and whether the Soviet leadership is willing to make the huge effort needed for the venture. On the other hand, recent research is providing solid, quantitative data about the low economic gains from the use of hydrocarbons as compared to coal for the production of steam and electricity.[35] Soviet fuel policy people may be spurred to greater efforts than in the past to increase the output of solid fuels and the transport of eastern coals to the European USSR. It is certain that the requirement for oil and gas in boiler uses will continue to be determined by the size of the shortage of solid fuels west of the Urals and by the rate at which Siberian and Kazakhstan coal fields can expand production and supply consumers in the European provinces.

As the foregoing analysis intended to prove, the conflict between domestic and export needs and the consequent impending strain on oil and gas resources will thus center on fuel supplies to power plants, industrial and municipal boilers plus centrally furnished household fuels. Fuel allocation to power plants, industrial users and households has already been tightened and, in 1977, supplies to industry as a whole, including electricity generation, were reportedly cut by 7%.[36] A direct causal connection cannot be proved, but the 3% growth in power production in 1977, as opposed to the 6.95% annual average from 1970 to 1976, and the poor performance of a number of energy-intensive heavy industrial branches must have been partly related to reduced fuel allocation and to the virtual stagnation of coal output in the late 1970s.[37] Soviet electric-energy economists are disturbed by the fact that the growth rate of electricity consumption in the USSR has fallen much faster than that of total energy demand (a situation not found either in developed or in rapidly industrializing countries, at least until now) and the progress of electrification is too sluggish for the desired rate of scientific and technological change.[38] The concern over the slow rate of electrification appears well founded when it is realized that in Soviet industry alone roughly 12 million persons (34% of the labor force) are still engaged in manual work, that in animal husbandry, stationary nonmechanized processes require half of all manpower,[39] and that manual labor abounds also in other sectors at a time when the population of working age is expanding by only 1.5% per annum and will grow by a mere half percent per year during the 1980s.[40]

Substitution: Gas versus Oil

In stationary processes, substitution between the two hydrocarbons encounters no technological problems. In both boiler and furnace uses the required changes in equipment are simple and cheap. Many, perhaps even the majority, of plants are also designed to burn both fuels since, until recently, inadequate storage capacity for natural gas resulted in significant seasonal oscillation of supply. However, the impending exhaustion of most gas fields west of the Urals, making them available for seasonal storage, will largely eliminate this difficulty in the future. The long-distance transmission of crude oil costs less than one-fifth that of natural gas.[41] In addition, as of now, on an even calorific basis, crude oil exports to the West earn twice the hard currency of gas exports at precent prices. If such price relationship were to continue, economic logic would work against a massive surge of gas deliveries for the foreign market beyond the existing long-term contracts (mostly barter deals for large-diameter pipe) and for the growing use of gas to replace oil at home.[42]

Soviet energy planners and officials are aware of the opportunities offered by the wider use of natural gas. Recent research by scientists of this industry, however, indicates a growing concern about the present pattern of allocation and particularly about the vast quantities of gas burned under boilers.[43] They may thus resist the further expansion of this fuel in the boiler market even when it substitutes for oil, demanding more coal for that purpose, and rather push to replace petroleum for export in smelting, heat treatment and petrochemicals, i.e., in furnace-type equipment and as raw material.[44]

Beyond a general tightening of fuel allocation and norms and a cutback in the growth of aggregate energy consumption, the upsurge of Soviet oil exports was made possible by the already noticeable substitution of natural gas for residual fuel oil; this has been evident particularly in the Volga-Ural regions,[45] whose locational advantage with respect to the Central Asian, West Siberian and Orenburg gas fields is greater than that of more westerly provinces. Judging from the size, number and direction of gas pipelines under construction, such substitution will continue most strongly in the Volga-Urals, but will be increasingly observable also in the northern half of the Russian Plain.

The present wide price difference between oil and gas on the export market, however, is unlikely to last much longer. In addition the increasingly apparent difficulties in the Soviet petroleum industry and rising West European interest in Soviet gas both point to a rapid expansion in the importance of that commodity in foreign commerce. Most experts now believe that gas will become the leading hard-currency earner by the mid-1980s, replacing oil. According to one analyst, Soviet gas imports in West Europe under contracts signed before 1977 should exceed 50% of Soviet oil imports in calorific terms by 1980.[46] East Europe, too, is evidently accepting increases in gas deliveries in place of most or all of the increments in petroleum supply beyond 1980.

Because transport and distribution account for most of the cost of gas and because the final distributor network is still skeletal, significant oil-by-gas displacement inside the USSR should be expected only in concentrated bulk uses. Planners will find such a switch for scattered, smaller consumers more costly and in many cases physically impossible. Many industrial and smaller municipal boilers may find it difficult to change to gas from fuel oil, while in power stations, cement plants and other industrial furnaces the substitution of gas presents less of a problem. As a corollary, this also suggests that hard coal could be a more economical substitute for oil in dispersed uses than in concentrated ones. The market comprised by industrial boilers (many of which are small and scattered) and smaller power stations will be shared by all three of the major fuels, though one of them may dominate in particular geographic regions. In big thermal power plants a determined effort should indeed reduce the proportion and, eventually, the absolute amount of petroleum products. It is doubtful, however, that lignites and coal fines will be available in sufficient quantities in most Soviet regions to substitute for the oil and provide the increment, at least for the next 12–15 years. The relief will have to come as much from natural gas as from solid fuels. Despite the low relative productivity of gas in electric stations, a significant growth of gas consumption in thermal plants must be expected.

To close this analysis about the functional distribution of consumption and the future of fuel substitution, it may be useful to return for a moment to the pre-1973 situation. In the late 1960s, Robert Campbell, alluding to the relatively slow growth of Soviet motor fuel need compared to the expansion of crude output, noted that petroleum in furnace and boiler fuel uses "will compete with exports rather than with motor fuel use. This may well mean a much bigger role for oil as boiler fuel and also implies that the future behavior of Soviet oil exports will be much influenced by how rapidly, and at what level of cost, they [the Russians] are able to expand natural gas production."[47] The author also showed that at the then prevailing market prices the Russians could not really afford to export petroleum if its opportunity cost was to be computed on the basis of replacement cost by coal, but they could not affort not to export if its cost was figured on the basis of replacement cost by natural gas.[48] In the 15 years before 1975, petroleum had indeed acquired a bigger role as boiler and furnace fuel, its share rising from 12% in 1960 to 19% in 1975. In power stations alone, the percentage increase was from 7.5 to 28.8,[49] despite a long-standing philosophical dislike for burning oil in such a way. Stationary demands for oil were indeed competing with export needs rather than with motor fuel uses even before the rise in world prices and they compete much more so today.

The fourfold jump in the world price of oil and the growing Soviet hard-currency needs have shattered the assumptions behind the second part of Campbell's cogent analysis. Today the Russians can easily justify the replacement of petroleum for export not only by gas but by expensive deep-mined coal as well, and would in fact find it economically rational to displace even gas from a large part of the boiler market. And yet, in this writer's opinion, opportunities to increase, even to maintain, the level of petroleum export will continue largely to depend on the rate at which gas output can be expanded and, specifically, on the speed with which big supplies from northwest Siberia will reach market areas. Neither coal nor nuclear power can step into the breach, at least in the next 10 years. In contrast to the 1960s, the substitution of coal for hydrocarbons is eminently rational today. On the export market, hydrocarbons have greater value than in most stationary domestic uses even when valued at the cost of the most expensive coals. The constraints are no longer economic but physical and technological, at least in the short and medium term. They involve the feasibilities of transport, particularly for Siberian lignites, the speed of mine construction and expansion of the labor force. As in the United States, geographic constraints, long lead times and the increasing lumpiness of the factors of production, combined with problems in the institutional-political environment, hugely complicate the implementation of policies dictated by theoretical economic rationality.

REGIONAL ALLOCATION PROBLEMS: LOCATIONAL RENT AND REGIONAL PRICING

It has already been shown that the sectoral and functional allocation of Soviet energy resources is vastly complicated by the geographic or regional variable that confronts decision-makers at every step. The European USSR and the Urals, with three-fourths of the energy market, are struggling with a rapidly increasing fuel deficit, which must be closed by a massive and growing volume of energy flow from the Asian provinces. In addition, the substantial spread among the marginal costs of various fuels (and of electric power produced from them), the huge differences among production and transport costs of the same fuel from the various coal basins, petroleum fields and petroliferous areas have generated a large, though regionally sharply varying, set of economic rents. This applies particularly to hydrocarbons, with their cheapness, utility and convenience relative to possible substitutes. In the last two decades, virtually every major geographic province could have reaped large savings by using more gas and oil than available, which imparted a high opportunity cost value to these fuels through most of the USSR. For gas, which is expensive to transport, economic rent has been mostly, though not exclusively, an expression of the locational advantage of given fields. For oil, which of all fuel is the cheapest to transport, it has chiefly indicated extraction cost differentials both among oil fields and relative to other energy sources.

In the 1967 price reform, field prices for oil and gas were raised substantially to include such rent payments, as well as finding costs and capital charges. The regional prices for hydrocarbons in stationary uses were made to approximate their opportunity cost value, i.e., the cost of replacing them with inferior fuels.

Table 52. Production Costs and Prices of Different Fuels in Soviet Regions (1973-74)
(in rubles/ton of standard fuel)

Cost and price	Central Russia	Baltic	Ukraine	Siberia	Central Asia	Kazakh-stan	Urals
			Coal				
Average cost[a]	18.20		11–13	8.6–9.5	15.0	14.0	15.4
Long-run marginal cost	21–23	23–25	19–23	9–15	14–16	13–20	15–18
Wholesale price	18.5–20.0		9.1–29.5	7.0–12.7	10–29	8.5–23.0	13–31
			Fuel oil				
Average cost[a]	4.20[b]			10.20[c]			
Long-run marginal cost	22–25	23–26	21–25	15–17	14–18	15–18	18–20
Wholesale price	18.20	18.20	18.20	20.40	18.20	18.20	18.20
			Natural gas				
Average cost[ad]	11.30	9.80	9.70	9.90	8.40	9.20	5.60
Long-run marginal cost[d]	23–26	24–27	22–26	15–18	14–19	18–20	19–21
Wholesale price	18.00	18.80	16.20	14.50	9.40	12.80	14.50

Notes:
[a]Including 15% capital charge.
[b]From Volga crude.
[c]Tyumen' Oblast.
[d]Costs for gas include cost of transport.

Source: Ye. N. Il'yina and D. Utkina, "Methodological problems of creating an information system on the cost-effectiveness of gas use," *Ekonomika gazovoy promyshlennosti*, 1975, No. 8, p. 23.

Although the new prices still did not fully reflect the technological advantage, cleanliness and convenience of hydrocarbons, the spread between delivered cost and price was very large. Prices of natural gas and fuel oil ranged from two to five times their delivered costs per calorific unit through the European USSR at the beginning of the present decade. The spread was greatest in the North Caucasus, the Ukraine and the Volga Region, i.e., in regions of significant or heavy consumption that were also the major producers of hydrocarbons five to ten years ago. For over a decade, therefore, Soviet prices of hydrocarbons and of boiler and furnace fuels in general have been sufficiently sound to come close to maximizing revenue and the utility of oil and gas through the spatial system as a whole.[50]

In recent years, supply and cost relationships in the Soviet energy system have been changing rapidly, creating pressures for another price reform. As wasting, indeed rapidly exhaustible, resources, oil and gas have upward sloping production-cost curves. After the plateau phase is reached, costs rise in each reservoir, field and petroleum province, sometimes swiftly, and the marginal cost of the branch as a whole increases as production moves to geographically less accessible locations and to deeper strata. In the USSR, since the late 1960s, extraction costs from mature fields have risen sharply for both oil and gas. Also, with declining production in most provinces west of the Urals, expansion in output is coming entirely from remote eastern and Arctic regions, which must also compensate for the exhaustion of old reservoirs. The marginal delivered costs of gas and oil today on a calorific basis approaches that of coal in many European regions and even their average costs are not so far below those of solid fuels than in the past. As the hydrocarbon resources of the Volga Region, Caucasus and Ukraine dwindle and their costs rise, the rent payments generated by their favorable location are also fading in importance.

With sharply rising marginal costs, the existing prices, which were originally designed to spur the further conversion of equipment to hydrocarbons, particularly gas, appear far too low today, hindering the effectiveness of administrative measures to stimulate greater economy of use. On the average, the wholesale price of gas at least is actually below that for steam coal on an even calorific basis and appears so in several regions.[51] Coal prices, however, are also lower than marginal costs in a number of provinces and seem to approximate average costs in the various producing regions during the early 1970s, with cost rises likely since then (Table 52). A new system of fuel prices is in preparation, and it is said that the new prices will also reflect the cost of environmental protection, land reclamation, and at long last, the full cost of geological prospecting.[52]

NOTES

[1] Marshall I. Goldman, "Autarchy or integration–the USSR and the world economy," in U.S. Congress, Joint Economic Committee, *Soviet Economy in a New Perspective* (Washington: Government Printing Office, 1976), pp. 81–96. Quotation from page 95.

[2]U.S. CIA., *Handbook of Economic Statistics* (ER 78-10365, 1978), pp. 53–54, 63.

[3]A. J. Lenz and H. Kravalis, *Soviet/EE Hard Currency Export Capabilities*, Office of East-West Policy and Planning, Bureau of East-West Trade, U.S. Department of Commerce, Oct. 11, 1977.

[4]*Vneshnyaya torgovlya SSSR v 1976 godu* (Moscow: Statistika, 1977), p. 18.

[5]John R. Haberstroh, "Eastern Europe: Growing energy problems," in U.S. Congress, Joint Economic Committee, *East European Economics Post-Helsinki* (Washington: Government Printing Office, 1977), p. 395. In 1976, energy imports from the USSR comprised 30% of East Europe's import bill from the Soviet Union, a share that has grown significantly since. U.S. C.I.A., *Handbook of Economic Statistics*, 1977, p. 62.

[6]Haberstroh, op. cit., p. 395.

[7]Without Yugoslavia. Ivan V. Matusek, "Eastern Europe: Political context," In U.S. Congress, Joint Economic Committee, *East European Economies Post-Helsinki*, p. 7.

[8]It may be computed from the change in apparent consumption, i.e., production minus net export, that out of the 18.2 million tons growth in petroleum export during 1976, a little over 10 million tons was diverted from domestic consumption.

[9]*Narodnoye khozyaystvo SSSR za 60 let* (Moscow: Statistika, 1977), p. 205 and *Ekonomicheskaya Gazeta*, 1978, No. 6, p. 5.

[10]*Vneshnyaya torgovlya SSSR v 1976 godu*, pp. 63–64 and *Vneshnyaya torgovlya v 1975 godu*, pp. 68–69.

[11]Robert E. Ebel writes, "What must give, then, must be the volume of oil available for exports to non-Communist buyers, which in the past have been defined as a residual, and will continue to be in the future." *The Oil and Gas Journal*, Special Supplement: Petroleum/2000, August 1977, p. 505.

John P. Hardt avers, "Soviet needs and planned commitments require it to be a modest exporter of oil and natural gas to hard-currency Western nations and to East Europe throughout the period of the seventies and eighties," in "Soviet oil and gas in the global perspective," in U.S. Congress *Project Interdependence: U.S. and World Energy Outlook Through 1990*. Report printed at the request of John D. Dingell, Henry M. Jackson and Ernest F. Hollings by the Congressional Research Service, Library of Congress (Washington: Government Printing Office, 1977), p. 787.

Marshall Goldman also expects the USSR to persist under foreign trade pressure to export oil for hard currency by continuing to restrict domestic oil consumption. *Wall Street Journal*, Feb. 27, 1978, p. 6. Neither of the last two scholars projects a plateau, let alone a drop in Soviet petroleum output.

[12]*Narodnoye khozyaystvo SSSR za 60 let*, p. 205; *Ekonomicheskaya Gazeta*, 1978, No. 6, p. 5 and *The Oil and Gas Journal*, Dec. 26, 1977, p. 70.

[13]In F. Kormnov, *Razvitiye sotsialisticheskoy ekonomicheskoy integratsii* (Moscow: Ekonomika, 1976), p. 31.

[14]*Petroleum Economist*, October 1976, p. 393, and December 1976, p. 454, and *The New York Times*, Nov. 21, 1977.

[15]*Figyelö*, No. 50, December 1977, p. 7.

[16]Quoted by *Figyelö*, No. 38, September 1977, p. 7.

[17]*The Oil and Gas Journal*, Aug. 15, 1977, p. 31.

[18]A. A. Makarov and L. A. Melent'yev, *Metody issledovaniya i optimizatsii energeticheskogo khozyaystva* (Novosibirsk: Nauka, 1973), p. 107.

[19]P. S. Neporozhnyy et al., "Fuel and power economy of the Soviet Union," *Ninth World Energy Conference. Transactions*, 1974, Vol. 2, p. 160.

[20]*Ekonomika chernoy metallurgii*, No. 5 (Moscow: Metallurgiya, 1976), pp. 91–93. A Czechoslovak source says that natural gas used in Soviet blast furnaces is slated to grow from

8 billion cubic meters in 1975 to 11.7 billion in 1980, i.e., a 46% increase, while the consumption of petroleum in these furnaces is planned to grow by 96%. *Energetika* (Prague), 1977, No. 9, p. 455.

[21] *Gazovaya promyshlennost'*, 1978, No. 6, p. 12.

[22] *Ekonomicheskaya Gazeta*, No. 48, November 1977, p. 13.

[23] Ibid., p. 13. This is consistent with another source, which gives 1040 million gigacalories, i.e., 148.5 million tons of standard fuel, as the plan for heat consumption in the household-municipal economy as a whole in 1975. A. S. Pavlenko and A. M. Nekrasov, *Energetika SSSR v 1971-1975 godakh* (Moscow: Energiya, 1972), p. 177.

[24] *Elektricheskiye Stantsii*, 1970, No. 12, p. 29, gives the total for firewood and self-produced fuel as 45 million tons standard. *Narodnoye khozyaystvo SSSR za 60 let*, p. 204, gives 26.6 million tons standard for centrally supplied firewood in 1970 and 24.6 million in 1976.

[25] *Pravda*, Oct. 28, 1976, p. 2.

[26] E. M. Rubenking, "The Soviet tractor industry: Progress and problems," in U.S. Congress, Joint Economic Committee, *Soviet Economy in a New Perspective* (Washington: Government Printing Office, 1976), pp. 600–619.

[27] V. Voropayeva and S. Litvak, "On the fuels and energy balance of the USSR," *Vestnik Statistiki*, 1978, No. 1, p. 9 and *Nar. khoz. SSSR za 60 let*, pp. 204 and 206.

[28] Computed from *Energetika SSSR v 1976-1980 godakh* (Moscow: Energiya, 1977), pp. 149 and 151.

[29] *Soviet Geography*, December 1976, p. 717.

[30] In East Siberia, the Far East and most of Kazakhstan, demand for hydrocarbons will come mostly from peaking needs and isolated diesel units in remote areas. On the other hand, the accelerated and increasingly energy-intensive development of northern oil and gas fields, where coal is unavailable, is reducing the share of solid fuels in power stations of West Siberia (Table 11) and will continue to do so at least through part of the 1980s. The industrial complex of the Urals is rather easily accessible to both eastern oil and gas and to coal from the Kuznetsk Basin and Ekibastuz. The latter, in particular, is playing an increasing role as an electric station fuel in the region. However, much of this coal is needed to replace output from local deposits nearing exhaustion and will contribute only partly to growth. *Elektricheskiye Stantsii*, 1977, No. 2, p. 6.

[31] No nuclear stations are operating or are planned for these regions, excepting a small one in the remote northeast and a secret military plant identified by the Russians as "Siberian," but believed to be in the Urals near Troitsk. Therefore, the difference between total power production and hydroelectric output must come from conventional thermal plants both today and in the foreseeable future. The 1975, 1976 and planned 1980 production figures of electric power for all republics, Siberia and the European USSR are given in *Soviet Geography*, April 1977, pp. 271–277. Hydroelectric capacity and output potential through 1980 in the different regions is given by *Energetika* . . , op. cit., p. 129.

[32] Estimated from data in previous footnote and in *Energeticheskoye stroitel'stvo*, 1976, Nos. 11/12, pp. 19–20, and *Razrabotka neftyanykh i gazovykh mestorozhdeniy*, Vol. 4 (Moscow: VINITI, 1972), pp. 44–45.

[33] *Energetika* . . . , op. cit., pp. 87–89. Past and present distribution of capacity and output among the different types of plants, including heat and power stations and condenser stations, and the 1980 plan for some are given in A. M. Nekrasov and V. Kh. Khokhlov, "Electric power in the 10th five-year plan," *Izvestiya VUZ, Energetika*, No. 9, 1976, p. 6 and *Energetika* . . . , op. cit., p. 11 and 61. The distribution of output among condenser stations and heat and power stations can be computed according to data in *Teploenergetika*, 1976, No. 11, pp. 2–3.

[34]If these plans are realized on time, by the early 1980s about 250 million standard tons of solid fuels annually will be available for regions west of the Urals for noncoking purposes in place of the present 190 million. These amounts include both local production and coal shipped and transmitted indirectly via electricity from Soviet Asia. *Soviet Geography*, April 1977, p. 264, and April 1978, p. 282. Physical tons converted to calorific equivalents.

Rail loading of eastern coal destined for regions west of the Urals is expected to rise from 31.3 million natural tons in 1975 to 45.7 million in 1980 (or from about 27 million tons standard to 39 million). *Soviet Geography*, November 1977, p. 701. Much of this, however, will have to be coking coal, since the expansion of an integrated iron and steel industry in European Russia is proceeding largely on the basis of Kuzbas coal.

The construction of the Ekibastuz-Tambov 1,500 KV DC line, 1,500 was supposed to start in 1978. It is quite possible that, even if the line is completed by 1985, the 6 million KW power capacity to supply it will not be. *Kazakhstanskaya Pravda*, April 22, 1977. At any rate, the 40 billion KWH of power to be transmitted from Ekibastuz is equivalent to the shipment of 13 million standard tons of fuel, according to the planned heat rates of large thermal stations.

[35]At current prices, the use of gas vs. coal is computed to yield only 3.6 rubles of saving per 1,000 m^3 in the stations of the Electric Power Ministry. In industrial boilers raising process steam, the comparable gain is 6.4 rubles per 1,000 cubic meters. By contrast, in cement kilns, gas saves almost 24 rubles and in glass-making more than 64 rubles. In 1976, 161 billion m^3 of gas burned under power station industrial and municipal boilers produced an aggregate economic saving of 0.8 billion rubles. By contrast, the 110 billion m^3 used as technological fuel (in furnaces, kilns, etc.) and as raw material saved 2.5 billion rubles. *Gazovaya Promyshlennost'*, 1978, No. 6, pp. 28–30.

[36]*Newsweek*, Jan. 2, 1978, p. 10.

[37]*Ekonomicheskaya Gazeta*, No. 6, February 1978, p. 5, and *Narodnoye khozyaystvo SSSR za 60 let*, pp. 201 and 206.

[38]M. A. Vilenskiy, *Ekonomicheskiye problemy elektrifikatsii SSSR* (Moscow: Nauka, 1975), pp. 141–145; M. A. Vilenskiy, "Electrification and scientic-technical progress," *Voprosy Ekonomiki*, 1977, No. 4, pp. 5–13. The general relationship between electrification and technological progress is discussed in some detail in Beschinskiy and Kogan, op. cit., pp. 103–130, 187–406, *passim*.

[39]*Sotsialisticheskaya Industriya*, Jan. 27, 1977; *Narodnoye khozyaystvo SSSR v 1975 godu*, p. 532; Beschinskiy and Kogan, op. cit., p. 293.

[40]Murray Feshbach and Stephen Rapawy, "Soviet population and manpower trends and policies," in U.S. Congress, Joint Economic Committee, *Soviet Economy in a New Perspective* (Washington: Government Printing Office, 1976), pp. 113–154, particularly p. 152.

[41]A crude oil pipeline can transport more than five times as much calorie content as a gas pipeline of the same diameter. In addition, pumping stations for oil are simpler and cheaper than compressor stations for gas and need less fuel. Finally, steel pipe for gas lines is subject to more stringent quality requirements than for crude oil lines.

[42]Robert E. Ebel, "Soviet oil looks to the West," *The Oil and Gas Journal*, Special Supplement: Petroleum/2000, August 1977, p. 508.

[43]*Gazovaya Promyshlennost'*, 1978, No. 6, pp. 27–30.

[44]In several such fields and equipment of nonferrous metallurgy, metalworking, steel and chemical industries, the use of gas is calculated to yield large economies even when compared to using oil products or electricity. These economies are several times greater than the ones that gas can achieve in power stations compared with coal. Ibid., pp. 11–13, 17–18, 21–22 and particularly pp. 28–29.

[45]S. A. Orudzhev, *Gazovaya promyshlennost' po puti progressa* (Moscow: Nedra, 1976), p. 59 and V. I. Manayev, "Bashkir ASSR: Its complex today and in the future," *Ekonomika i Organizatsiya Promyshlennogo Proizvodstva*, 1977, No. 2, pp. 46–67.

[46]U.S., C.I.A., *USSR: Development of the Gas Industry* (ER 78-10393, July, 1978), pp. 22–23.

[47]Robert W. Campbell, *The Economics of Soviet Oil and Gas* (Baltimore: Johns Hopkins Press, 1968), p. 180. The quotation reads," A much bigger role for oil as boiler fuel relative to gas than in the United States." This, too, was true until 1970, when the peaking of gas production and tightened supply to utilities in the United States began to reduce the share of natural gas in electric power generation. There was a corresponding surge of petroleum, largely imported, in electricity production until the oil embargo, at which time a decline in the share of oil also began. Nevertheless, even in 1973, petroleum products in the United States accounted for less than 21% of the fuel supply of thermal power stations compared with 28% in the USSR (28.8% in 1975). *Statistical Abstract of the United States*, 1976 (U.S. Dept. of Commerce, Washington), p. 553 and *Energetika* . . . , op. cit., p. 151.

[48]Ibid., pp. 236–240.

[49]N. V. Mel'nikov, *Mineral'noye toplivo* (Moscow: Nedra, 1971), p. 199 and *Energetika* . . . , op. cit., p. 149. Mel'nikov's percentage was corrected upward to eliminate the share of manufactured gases, which he gives separately.

[50]For a detailed treatement of these issues, see Robert W. Campbell, "Price, rent and decisionmaking: the economic reform in Soviet oil and gas production," *Jahrbuch der Wirtschaft Osteuropas*, Vol. 2, 1971, pp. 291–314 and L. Dienes, "Geographical problems of allocation in the Soviet fuel supply," *Energy Policy*, Vol. 1, June 1973, pp. 3–20, especially pp. 6–16.

[51]*Gazovaya Promyshlennost'*, 1977, No. 2, pp. 14–17 and *Ekonomika gazovoy promyshlennosti*, 1975, No. 8, p. 23.

[52]*Gazovaya Promyshlennost'*, 1977, No. 2, pp. 14–17.

Chapter 9

ENERGY MODELING AND THE EXPANSION OF SUPPLIES

As early as the middle 1960s, Prime Minister Aleksey N. Kosygin noted that the successful steering and management of the Soviet economy required greater emphasis on long-range planning. Contemporary developments in science and technology, the increasing size and complexity of investment projects were said to demand a long-term view. Following this lead and subsequent proposals that "the five-year plans be worked out within the framework of a system of long-range plans," a party-government decree instructed the appropriate government agencies "to work out 10-year to 15-year forecasts of scientific and technical developments to be used in planning." Gosplan "set up special task forces to do the preparatory work for drafting the major sections of the long-range plans" (1976-90). Put on a proper "scientific" basis with the use of modern mathematical techniques, such planning was to improve the performance and upgrade the technological level of the Soviet economy.[1]

Energy modeling, forecasting and scenario-building have occupied a central place in the burst of modeling and forecasting work that followed these official strictures. As the subsequent pages show, much is revealed through this modeling work about Soviet thinking, the perception of alternatives and the future course of development as seen by planners despite the schematic nature of the information. And, however tentative the inputs and constraints to the forecasts may be, within the framework of the assumptions made, they are likely to be more accurate than those available to Western researchers.

SOVIET ENERGY MODELS AND FORECASTS

The comparatively few and locationally restricted nature of primary energy sources and the relatively narrow output mix (reducible to common calorific equivalents) make energy planning quite suitable for such techniques. In addition, since the Soviet leadership exercises greater control over demand than is possible in a market-type economy, the difficulties stemming from consumer response, from income and price elasticities with respect to various energy forms barely arise. Soviet energy modeling consists of various scenarios matching specified, discrete levels of aggregate demand with different combinations of supplies in a linear programming framework. The objective function is to minimize the total cost of meeting the desired gross demand, which in more sophisticated, recent models may include the requisite capacity expansion by the supplying industries. Such models are applied both to the main energy-producing branches individually, even to separate coal fields and petroleum provinces, and to the energy sector as a whole.

In Soviet attempts to model the energy system as a whole, some efforts are apparently made to distinguish energy requirements by type of use, such as the functional categories treated in Chapter 8, but the extent of these refinements and the feasibilities of substitution are not clear. Nor is the regional detail evident in these forecasts, though, in Campbell's words, the shadow prices (i.e., long-run marginal costs) differentiated by regions and energy source may depend heavily on such details and refinements.[2] It is likely that while the type of use will continue to influence the solution and objective functions yielded by such models, the controlling influence will be assumed increasingly by the geographic or spatial variable—an issue that will be taken up in more detail later.

In all scenarios described in Soviet works, the least-cost path calls for a rapid expansion of Siberian oil and gas output but also of surface coal and lignite east of the Urals and nuclear power west of the Volga. Direct investment required for incremental output in standard tons (7 million kilocalories) is said to be lowest for lignite from the Kansk-Achinsk Basin (and by implication also for Ekibastuz coal). Energy increments from Siberian oil allegedly need somewhat greater investments, and increments from Siberian gas about twice as much. In addition to direct investments, such increases require large capacity expansion in the supplying industries, with the volume depending on specific conditions and varying with the speed of exploitation. Again, for nearly all industrial inputs, such indirect effects will be greater for natural gas than for oil or for surface-mined coals east of the Urals. The major inter-industry inputs for the hydrocarbons are and will continue to be pipe and other metal products; those for eastern lignite and coal, equipment and machinery for production, transport and enrichment (the last not yet solved for Kansk-Achinsk lignite) as well as electric power.[3]

The huge inter-industry demands put serious constraints on the optimal plans, particularly with respect to gas. Even with relatively minor drops in production at older gas fields, such plans call for an increment of 100 billion m^3 of West Siberian gas every two to three years, and its delivery to market. This, in turn, would require the following increases in annual production capacity by 1990 to service the West Siberian gas industry alone: 32 million tons of rolled steel (almost one-third of present capacity), 300,000 tons of nonferrous metals and close to 25 million tons of cement (one-fifth of present capacity).[4] As in the past, the Russians are understandably counting on imports to ease those constraints, and some of their optimization efforts attempt to model the import effect. One study, for example, suggests about half a billion rubles of total saving for every two million tons of imported, large-diameter pipe.[5] (Imports of compressor stations may be even more crucial, but this writer has seen no work trying to quantify their impact on alternative energy paths.) At any rate, Soviet planners are well aware that the attainment of the optimum levels of hydrocarbon output called for by their models may be problematic. Thus they also incorporate more pessimistic output and cost constraints for oil and gas (as well as nuclear power) in subsequent iterations to arrive at second and third-best solutions. The latter, as a rule, result in a greater role for coal, particularly Siberian lignite. Such second-best solutions, assume that investment resources between the hydrocarbon and coal variants are reasonably transferable and that an economically acceptable resolution to the difficult enrichment and transport problem for self-igniting Siberian lignite is imminent. These assumptions are far from certain, and indeed, very unlikely, at least within the required time frame and going from an optimal to a second-best alternative probably will not be feasible within the next decade.

As mentioned, the controlling variable in Soviet energy scenarios, the variable with the greatest impact on feasible choices and total costs, is likely to be increasingly the geographic factor. This variable now plays its role in a more straightforward and blunt fashion than was the case during the 1960s. The salient fact of this spatial variable now is the relentless growth of the energy deficit throughout the whole of the European USSR. In this core area, which contains most of the Soviet population and economic output, it is no longer possible to speak of energy-rich and energy-poor regions, since even the Ukraine has become a massive net importer of fuel from the eastern provinces. West of the Urals, the gap between demand and supply is actually widening faster in regions that used to have a surplus (Ukraine, Caucasus and, to a lesser extent, parts of the Volga basin) than in those that have always had to struggle with a shortage of energy.[6] The output of solid fuels in the European USSR is virtually stationary and that of hydrocarbons dropping, with the decline likely to accelerate from 1980 on.[7] The production cost of fuels, especially oil and gas, in these parts of the country is also rising sharply, eliminating most of the large economic rents earned by well-located western fields during the 1960s.[8] Even in the southern half of the European USSR, the cost of gas from the Ukraine and the North Caucasus is not too much less than that of gas from Central Asia and Tyumen' Oblast.[9] West Siberian oil, whose incremental cost also began to increase rapidly, is actually cheaper west of the Urals than oil from European fields, and likely to remain so.[10]

Under these circumstances, breaking down forecast demand regionally within the three-fifths of the Soviet energy market found west of the Urals has lost most of its importance. It has little impact on the feasible choices, costs and inputs of various energy scenarios. Incremental supplies of fuel and increasingly also Soviet exports must originate from remote provinces of the Asian USSR, whose environmental extremes, distances and lack of infrastructure make projections of output and delivery costs uncertain and tentative. Aggregate future costs in the Soviet energy economy as a whole will be influenced by the speed and expense of Siberian development and fuel transport to such an extent that the minor consequences on total energy costs resulting from demand variations within the European regions will be almost entirely submerged. By contrast, the growth of fuel demand in the cis-Ural territories as a whole compared with that in the eastern regions will become even more decisive for Soviet energy policy.

A TENTATIVE REGIONAL MODEL FOR FUEL ALLOCATION UNTIL 1985-90

To gauge the evolution and main regional configuration of optimal supply mixes and marginal fuel costs, a simplified programming solution was formulated for the spatial allocation of boiler and furnace fuels (which account for virtually all stationary uses) until 1985 and 1990. The work follows two earlier attempts by the writer (in 1971 and 1973) to model the task of regional fuel allocation for the beginning of the 1980s but with some modifications.[11] Incremental costs for hydrocarbons and their projections for the future have clearly risen more radically and constaints on supplies have proved greater than expected in earlier Soviet work. Expansion of supplies, especially west of the Urals, therefore had to be scaled down and the incremental cost of energy, especially hydrocarbons, increased substantially.

Unfortunately, Soviet estimates on projected volumes and costs of output from the different deposits and provinces, producing and yet to be developed, are extremely scanty. Some sources of supplies, therefore, had to be combined and assigned a uniform cost, particularly with respect to oil. In place of the more than 30 supply variables used in the 1973 study, the present model deals with little more than 20. One measure of improvement in the cost matrix, however, could be effected: in addition to the variation in calorific content, on the basis of new data, efficiency coefficients for use were also incorporated in the cost matrix for coals from different deposits. The present model also distinguishes between the differential opportunity costs of gas and fuel oil in the blast furnaces of iron and steel regions (where hydrocarbons save expensive coke) and other uses.[12]

Similarly to the supply variable, the number of consuming regions was reduced by aggregation. Since the expansion of fuel supplies must originate east of the Urals, 1,000 to 3,000 miles from consuming centers, the relatively compact area of the European USSR south of Lat. 60°N was broken down to only three geographic units in demand projections, a decision also encouraged by data limitations. These large regions each have aggregate demand on the order of West Germany or Japan. Yet the transport cost differential for hydrocarbons from trans-Ural fields varies less than 2 rubles per ton of standard fuel (7 million kilocalories) among them. For Siberian coals, transport cost differentials among these regions would be greater, but only Kuzbas coal is expected to be delivered west of the Volga in significant quantities even in the late 1980s. By contrast, the Asian USSR, with less than 30% of aggregate fuel consumption, was broken down into five regions. This is dictated by larger geographic size and by the fact that cheap, abundant but barely transportable lignite resources are able to fuel the rapid expansion of demand locally but not some distance away.

Aggregate fuel consumption was projected to grow at annual rates varying between about 2.5 and 4%. However, given the strained supply situation in the European USSR and Soviet intentions to restrict the expansion of energy-intensive industries west of the Urals, fuel demand in the European regions was projected to increase at an appreciably slower rate, ranging from 2 to 3.2% a year. Correspondingly, demand in the Asian USSR was assumed to rise much faster, at annual rates ranging from 3.2 to 5.6% a year until 1985 and 1990.

These projections also indicate an important spatial consequence of a tightening in Soviet energy supplies, a consequence apparent from the data even before the program is attempted. It is hardly possible to accept less than a minimum 2–3% annual growth rate in fuel consumption for the European USSR, the core of the Soviet economy. This means that a possible supply crunch will affect the higher rates set for the Asian USSR and, therefore, the ambitious development plans envisaged for Siberia and Kazakhstan, because most of the projects in these provinces are highly energy-intensive. However, since resource exploitation and primary processing are the main economic specialties east of the Urals and, in particular, these regions are critically important for the expansion of the energy supply, the circle is closed. A slowdown in Soviet fuel production in the short and medium term will endanger the expansion of energy and raw material resources in the longer run by retarding the development of Siberia and Kazakhstan.

Cost-minimizing solutions were derived for 15 scenarios differing mainly in the volume of oil and gas output from the principal producing provinces, the levels of regional demand and oil-export levels. Extreme cases were avoided both concerning volumes of output and the directions of possible energy flows. Exports were treated first as a residual, then as a priority up to the calorific total prevailing in the late 1970s, and they were also differentiated between hydrocarbons and coal. Finally, to examine the impact of export priorities combined with possible production shortfalls, the lower limit for consumption in each region was treated in several scenarios as an unbounded variable.

Even aside from the data problem, it must be emphasized that the optimum patterns yielded by these models do not represent a global optimum. The exercise rests on the simplified assumption that the same increment in stationary energy demand has roughly equal return throughout the spatial system. Insofar as optimum allocation is sought only among large economic areas with a wide range of consumers, the inability to account for differential values arising from alternative technological uses is probably not prohibitive. And, as mentioned, blast-furnace use of oil and gas could be treated separately. Because of differences in skill levels, age and composition of equipment and past investment in fixed facilities, labor and capital productivity also varies areally. Return or incremental energy supplies thus will not be uniform geographically even in the same uses. But this valid objection, too, is mitigated by the long-term secular trend indicating a lessening of regional productivity differences in maturing economies.

INTERPRETATION AND ANALYSIS OF MODEL

As expected, the programming solution assigns coal to remain the primary fuel source for stationary use in every major region of the Asian USSR except Central Asia. It should continue its overwhelming dominance in East Siberia, increase substantially its already high present share in Kazakhstan (63% in 1975, but as high as 71% if gaseous by-products of coal processing are included[13]) and stay dominant, if less overwhelmingly, in the Far East as well. As today, coal should remain the primary energy source for West Siberia also, but the optimum mix of this region will be sensitive to the growth rate of hydrocarbon supplies relative to that of aggregate energy demand and the degree of emphasis of oil exports. With more abundant oil and gas output, petroleum should retain an important role in West Siberia's boiler and furnace fuel balance, with as much as a quarter of total fuel consumption, roughly the same as today.[14] With less optimistic projections for hydrocarbons and/or higher priority for oil export, petroleum products in the province's fuel mix appear decidedly suboptimal. Among all economic regions, it is in Kazakhstan and West Siberia that the clearest case for the substitution of coal for petroleum can be made. Subject to constraints concerning the reconstruction of existing equipment, it is here that oil can be replaced by coal most economically. Beyond the use of some oil-well gas, natural gas should play only a negligible role in West Siberia, since the region's vast reserves yield greater effect farther west. In no scenario did gas emerge as an optimum fuel in Kazakhstan, except in its southernmost parts bordering Central Asia. However, the economic penalty to the Soviet energy system as a whole from a significant role for this fuel in Kazakhstan appears slight. Given the existence of already or partly amortized pipelines through parts of Kazakhstan, the present contribution of natural gas to its energy balance (23% in 1975[15]) may well be preserved.

In Central Asia (together with the adjoining oases of southern Kazakhstan), natural gas already satisfies most (roughly three-fifths) of all boiler and furnace-fuel requirements,[16] but according to the model this share ought to go much higher still. Further substitution of gas for solids and heavy petroleum products seems to be called for, but, more importantly, gas should satisfy all growth in stationary fuel requirements. The spread between the marginal delivered cost of gas and those of alternative fuels is and will remain greater in Central Asia than in the regions to which this gas is transmitted. This means that fuel costs for the whole Soviet regional system are not lowered by the export of Central Asian gas to the Russian Republic at the expense of local needs. This has been true in the past for the amount used in Urals blast furnaces, where gas saves scarce and expensive coking coal. However, with increasing quantities of West Siberian gas available in the RSFSR, only surplus gas should leave Central Asia. And it should be directed to the Ukraine and adjoining provinces of the Russian Plain (as indeed it is beginning to happen) rather than to the Moscow area and even Leningrad, which have been major destinations of Central Asia gas. However, the economic penalty of using Central Asian gas so far north is slight and, because of the existing facilities, deliveries will probably continue to take place. Similarly, the dwindling gas and still extensive but increasingly costly coal resources of the Ukraine and North Caucasus should be kept entirely within those regions, a development that is already observable. Local demand ought to have priority also on possible future discoveries.

The basis of this realignment is the massive flow of Siberian gas from around the Ob' estuary through the world's largest pipeline system now in formation. Both Soviet optimization attempts and the present writer's own studies indicate that this resource yields the greatest economic effect in the Urals and in the northern half of the European USSR, and all present and future gas pipelines from northwest Siberia are heading toward these provinces. Despite the accelerated development of nuclear power, the European provinces will become dependent on hydrocarbons from the Asian USSR to a greater extent than today. The importance of petroleum in the European USSR will be preserved (though its relative contribution will decline compared to gas), but the model indicates some change in the regional pattern for optimality to be attained. In the northern half of the cis-Volga plain (the western provinces of the RSFSR, Belorussia and the Baltic Republic) increments in fuel supply should come from gas rather than oil products; indeed, optimality would demand some substitution even in existing facilities.

By contrast, in the Ukraine and adjoining provinces of southern European Russia, incremental growth in boiler and furnace fuel demand ought to be shouldered primarily by refinery products originating from West Siberian crude. (In the data matrix, the output of local hydrocarbons is assumed to be declining or at least to remain stationary.) In all 15 scenarios, most of the heavy products from West Siberian oil are assigned to the European South. Clearly, the easy transportability of crude oil compared with natural gas and coal from Soviet Asia and the existence of large refinery complexes in the southern European USSR mean that oil products from West Siberian crude produce the greatest economic effect, other things being equal, in regions most distant from the source of petroleum. It must be emphasized, however, that because of the existence of a fairly extensive and partly amortized gas pipeline network west of the Urals, the economic penalty suffered if the relative proportions between oil and gas deviate from the theoretical optimum is slight in most regions of the European USSR (half a ruble per ton of standard fuel in most scenarios and one ruble in a few). This gives Soviet planners considerable flexibility to maximize the use of existing equipment designed for either oil or gas, to give due weight to storage needs, size of consumers, etc. without greatly violating spatial optimality. Smaller consumers, for example, may have to be allocated fuel oil rather than gas if the distribution pipeline network fails to develop sufficiently.

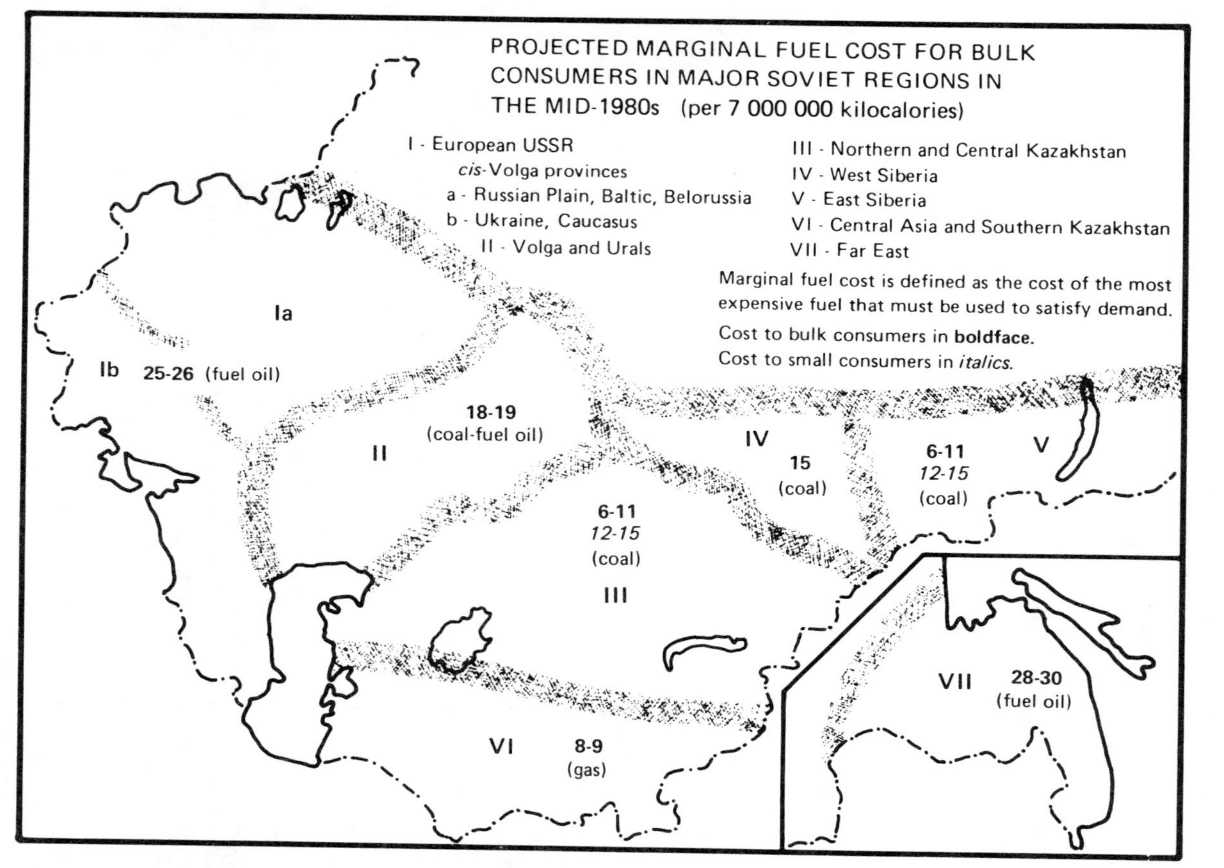

Fig. 12. Projected marginal fuel cost for bulk consumers in major Soviet regions in the mid-1980s.

Conditions of spatial optimality, which in most regions of the European USSR can be closely approached with a range of proportions between gas and oil from various sources, are more restrictive in the case of coal. As the gap between coal supply and boiler demand west of the Urals widens, solid fuels are increasingly restricted in their geographic range. They can effectively contribute only to the fuel mix of the economic areas where they are mined, but their share is destined to decrease in each major region. At the same time, the cost of closing or even diminishing that gap by the shipment of eastern coals is high or very high everywhere west of the Volga River. In all scenarios, the model restricts the transport of eastern steam coal, exclusively from the Kuzbas and to a much lesser degree from Ekibastuz, to the Volga-Urals area. And the economic penalty from a suboptimal pattern of regional shipments for these energy-hungry western provinces is greater than in the case of oil and gas.

In the Volga-Urals area, the model calls for a substantial increase in coal consumption, but the share of solid fuels in the energy mix should stay roughly the same as today. Therefore, it is incorrect to expect substitution of coal for hydrocarbons even in these provinces, which are more favorably located for the transport of eastern coals than the rest of the European USSR. Under conditions of optimality, coal will merely hold its own in the fuel mix of these important industrial regions, its consumption expanding at about the same rate as the demand for hydrocarbons. The model calls for large-scale substitution of gas for petroleum products, especially if oil supplies tighten, a move that is already under way.

If one takes into account variations in heat capture and combustion efficiency of the different fuels per standard calorific unit, marginal costs for stationary uses except coke are expected to range geographically by the late 1980s (perhaps by mid-decade already) in the following fashion. In the European USSR, fuel costs from 18 to 25-26 rubles per ton of SF, i.e., 7 million kilocalories, are expected, and it is conceivable that the upper end of the range will be higher. Unlike today, oil products will no longer be able to earn any sizable economic rent compared with solid fuels anywhere in the European USSR. Judging from the rate of cost escalation, petroleum may well become more expensive on a calorific basis even considering the extra cost to users from the lower efficiency of coal. The expansion of coal output, however, will continue to be constrained by geological and labor difficulties to rates slower than the growth of aggregate fuel demand. Therefore, throughout the cis-Volga provinces, heavy products from marginal crudes (of the Volga region, Caucasus, Ukraine, the European North), together with equally or almost equally expensive local coals, will define marginal fuel costs for the second half of the 1980s and this cost should approximate 25-26 rubles per 7 million kilocalories. This also means that we should expect only negligible regional variations in these costs for the entire European area west of the Volga, which still contains more than half of the Soviet population, industry and aggregate energy market (Fig. 12). The Volga-Urals region, more accessible to cheaper Kuzbas and Ekibastuz coals, and enjoying lower-cost fuel oil now largely made from cheaper West Siberian crudes, will certainly show lower marginal cost, closer to 18 rubles per 7 million kilocalories (one ton of standard fuel).

In the northern half of Kazakhstan and adjoining zone of southern Siberia (eastward to Chita), surface coals will define marginal fuel costs for bulk users. These resources are cheap or very cheap to mine, costing from 3 to 10 rubles per 7 million kilocalories for run-of-the-mine coal. Their low heat and high ash content and troublesome physical properties, however, make them difficult or impossible to transport and complicated, often inefficient, to burn even near the place of extraction. These factors raise users' cost, hence the marginal costs in that zone, into the range from 6 to 11 rubles per ton of standard fuel. Small consumers will have to depand on more expensive quality coals, generally from underground, and even on fuel oil to some extent. Marginal cost of fuel to them will thus reach 12-15 rubles per ton of SF (but a few rubles higher in isolated pockets needed to be supplied by liquid products), still much lower than in the European USSR.

In the Far East, even in its southern zone, marginal fuel costs are destined to remain high, rivaling and exceeding those west of the Volga. Coal is expensive, though less so than in the European parts, and its production cannot keep up with the growth of energy demand. The vast territory and the skeletal transport network also restrict coal distribution and will do so even after the completion of the Baikal-Amur Mainline. The future of the planned internationally financed gas pipeline from Yakutia is still uncertain and, at any rate, it will not be built before the late 1980s. This gas would also be more expensive than the calorific equivalent from coal, even with credits for its higher efficiency, and its distribution would be highly concentrated. Therefore, marginal costs will have to be defined by fuel oil, which will be more costly than west of the Urals whether it is to be derived from West Siberian or offshore Far Eastern crudes.

For the Moslem Republics of Central Asia, all scenarios assign natural gas to be the marginal fuel for bulk users and even smaller consumers in large and middle-size urban centers, at least until 1990. Incremental costs of Central Asian gas have also risen, but through most of the 1980s should certainly remain below 9-10 rubles per 1,000 m^3, i.e., 8-9 rubles per ton of standard fuel. Marginal fuel costs, therefore, will continue to rival those in northern Kazakhstan and southern Siberia. Beyond 1990 the reserve situation and technical problems cloud the horizon, unless significant advances in deep drilling, in dealing with complex reservoirs and high sulfur content are made before that time. In several Soviet works, it is recommended to calculate the marginal cost of Central Asian gas in those republics at the cost of replacing it with Kuzbas coal in Moscow minus the transport charge,[17] but this suggestion is clearly discriminatory for local consumers and was not followed here. Undoubtedly, however, users in small towns and rural areas will have to bear higher marginal costs. The gas distribution network is expected to remain inadequate and these consumers must continue to rely on coal and fuel oil (in addition to noncommercial sources) at roughly double the cost of natural gas.

Important insights were derived also from the treatment of the export variable. It is clear that as long as the delivered cost of gas throughout the European USSR remains appreciably below that of heavy petroleum products per calorific unit, oil exports yield a higher return for the economy than gas exports, even when world market prices for the two hydrocarbons are equalized. The rising share of gas at the expense of petroleum, therefore, must have basically noneconomic motives, or at any rate motives impossible to quantify, i.e., anxiety about the adequacy of oil reserves, West Europe's desire for diversification, perhaps Soviet perception of strategic dividends from the greater economic interdependence that long-term gas export entails. As expected from the minimization of transport costs, oil export ought to originate from fields of the European USSR and, if prospecting is successful, from those of the Far East, while West Siberian crude, originating near the geographic center of the country, should be consumed entirely at home.

During the 1980s, if petroleum and gas exports receive priority at the expense of domestic requirements, the major Soviet regions will be differentially affected. In the European USSR, the overwhelming portion of both oil and gas will be burned in stationary uses. Within this area, the provinces west of the Volga must bear the brunt of any restriction, since even if significant equipment changes toward coal use are made, substitution possibilities will be minimal. Eastern coals can reach the area only in small quantities and at high economic penalty to the energy system as a whole. When the lower limit of regional demand was treated as an unbounded variable (because of export priority and/or supply restrictions), consumption was cut back by the largest quantities in the northern half of the European plain west of the Volga valley. However, with further diversions from the domestic market for export in the model, demand restrictions in the southern European USSR began to approach those in the northern, central and western parts of the European USSR. Clearly, the energy situation in these formerly fuel-rich and exporting provinces is almost as serious as farther north, a problem analyzed in detail elsewhere.[18] As the compact area west of the Volga and south of the 60° parallel still concentrates half of all Soviet industry and contains a disproportionately high share of the skilled labor, it is a moot point how far Soviet planners could reduce consumption to give exports priority.

The third major region to be affected by such cutbacks (assuming a spatial optimum is pursued) would be the Far East. Although the absolute volumes that could be diverted are small, the percentage relative to original demand is higher even than in the northern European USSR. The explanation lies not only in the high marginal cost of local coals, but also in the isolation and distance of the Far East from other centers of energy production. This makes the substitution of fuels (transported from other regions) for those exported extremely costly to the economy as a whole. BAM will improve matters, but will not invalidate this conclusion.

PROSPECTS FOR EXPANSION OF DOMESTIC FUEL SUPPLIES

In Chapter 2 it was shown that for the last two decades there was no evidence of any decline in the true energy intensiveness of the Soviet economy. Net energy consumption rose slightly faster than national income as officially reported and much faster when national income is estimated according to

Western concepts. The rate of improvement in heat capture and conversion efficiency has slowed drastically with only marginal gains expected over the next 15 years or so.[19] At the same time, worsening resource conditions and the colossal efforts to develop remote Siberia are increasing the nation's energy requirements. It is extremely unlikely, therefore, that Soviet energy consumption can grow appreciably slower than GNP for any sustained period and it may even grow faster. The rise in energy consumption will basically depend on projections for GNP and on the rate of economic growth Soviet leaders push for and will find politically acceptable. In turn, however, the growth rate will in no small measure be influenced by the amount and quality of energy that will be delivered to key sectors of the economy, in short, by energy supplies. A precise and detailed description of primary energy resources, their geography and development, has already been provided in earlier chapters. This section focuses on the long-term prospects of supplies and presents what, in the author's opinion, will be the most likely scenario for the future.

The difficulties of substituting solid fuels for hydrocarbons have been amply demonstrated. Notwithstanding pronouncements about the need to economize on hydrocarbons, the 10th five-year plan (1976–80) assigned oil and gas the same share of total increment in fuel output that they shouldered in the previous five years and not a much smaller share than these fuels achieved during the 1960s. West Siberian oil and gas alone were planned to account for four-fifths of all the growth in aggregate fuel output on a calorific basis.[20] Obviously, an enhanced role for coal, if it happens, can be envisaged only over the much longer term and Soviet planners are not unaware of this salient fact. We will take up the long-term prospect of the expansion in coal production first, then turn to hydrocarbons, whose prospects have a more profound and far-reaching impact.

Prospects for Coal. Soviet planners will not be able to accelerate coal production appreciably before the late 1980s. Until then, underground mines can expand production only very slowly. Between the late 1970s and the late 1980s, a fifth of current capacity must be replaced just to sustain recent output levels (in 1972, 28% of the deep mines, with 22% of all coal mined, had been in operation for over 23 years).[21] Labor constraints are also serious and only a quarter of the miners are in the 20 to 35 year age group.[22] In the European USSR, the production of noncoking coal is planned to remain essentially stagnant on a calorific basis until 1980 and only modest growth may be expected afterward because of difficult and worsening geological conditions and rising sulfur content. The famed Donets Basin stagnated in the late 1970s and the Pechora Basin, most of which lies beyond the Arctic Circle, continued slow growth in output. We can anticipate modest expansion (in the case of the Donbas barely over 1% per annum) through most of the 1980s, such increments being mainly coking coal. Most other fields in the European USSR and the Urals show declining production.[23] Old seams, operating for two to three decades, are being worked out faster than they can be replaced under rapidly deteriorating geological conditions. In the Donbas, over one-third of the mines have essentially exhausted their reserves.[24]

Nearly all growth in Soviet coal output will have to come from mines, primarily strip mines, beyond the Urals, and most of the coal and lignite would have to be transported, or transmitted in the form of electricity, westward over distances of 1,000 to 2,000 miles. The present writer has shown elsewhere that Soviet coal production by 1990 cannot hope to exceed, probably even reach, 1 billion metric tons without a crash program to develop methods for processing and enriching vast quantities of the self-igniting, troublesome lignites of Kansk-Achinsk deep in Siberia. A program for the processing of 250 million tons of lignite was estimated by Soviet experts to require 15 billion rubles without social overhead cost and associated transport expenses.[25] Yet even that vague minimum target of 1 billion tons for Soviet coal output as a whole would be insufficient to arrest the decline in the relative contribution of coal to the national fuel balance.[26]

The latest reports suggest that, despite good laboratory results, research and development work on Kansk-Achinsk lignite is proceeding slowly. By Soviet admission, insufficient attention has been paid to the complex and sharply varying composition of these coals even for local use. Power stations operating on them (some since the mid-1960s) still encounter serious problems with wide-ranging ash-melting points and suffer from reduced efficiency and wasted capacity. The technical difficulties of burning the lignite even on the spot in the new power complexes with 800 MW units are far from solved.[27] In principle, Gosplan reportedly approved the eventual construction of nine giant strip mines with a combined yearly capacity of 300 million tons of lignite for mine-mouth power stations more than double the size of the largest coal-fired plants today.[28] Similar grandiose plans on the High Plains of the United States were, on closer examination, found to have unacceptable environmental consequences.[29] Indeed, the anticipated impact gives pause to even strongly coal-oriented and pro-Siberian Soviet specialists.[30] At any rate, such a colossal project is purely chimerical until the very long distance transmission of electricity by 2200 KV links to the Urals and beyond becomes possible. Such a technological feat is not even proposed before the end of the century and may be totally unfeasible. Technological developments in competing lines of energy research may abort such plans as well.

Since the mid-1970s, coal output as a whole is increasing by a mere 2% per annum, well below planned rates, and labor productivity is improving at a snail's pace.[31] The attainment of the 1980 target is unlikely, reinforcing doubts about hopes for 1 billion tons at the close of that decade. Additional coal-mining capacity completed in the late 1970s has been less than the yearly average in each year of the previous five-year plan and less even than the average since 1960.[32] While some improvement should be expected, it is significant that a Soviet source forecasts an annual growth of fixed assets in the coal industry between 1975 and 1985 at a rate significantly below that experienced during the 1960s and early 1970s.[33] It is clear that, just as the United States, the USSR finds the rapid acceleration of coal production a formidable and difficult task. This writer expects 1990 output to fall somewhat short of 1 billion natural tons, and to be equivalent to some 620 million standard tons, or to 430 million tons of oil in calorific units (Table 53).

Table 53. Projections of the Soviet Fuel Balance with Stationary and with Slowly Expanding[a] Petroleum Output (1980, 1985 and 1990)

Fuel balance	Petroleum				Natural gas		Coal		Peat, shale, and firewood (mill. tons of SF)
	Stationary output		Slowly expanding output						
	mill. tons	mill. tons of SF	mill. tons	mill. tons of SF	bill. m^3	mill. tons of SF	mill. tons	mill. tons of SF	
				1980					
Output	605		605		415		790		50
Net export	120		120		40		25		
Consumption	485	694	485	694	375	448	765	513	50
				1985					
Output	605		655		605		880		52
Net export	90		110		55		30		
Consumption	515	736	545	779	550	657	850	555	52
				1990					
Output	605		710		750		980		57
Net export	65		85		80		34		
Consumption	540	772	625	894	670	800	946	596	57

Total fossil fuel consumption (mill. tons of SF)

Year	With stationary petroleum output	With slowly expanding petroleum output
1980	1750	1705
1985	2000	2043
1990	2225	2347

Average yearly growth rate of fossil fuel consumption

Period	With stationary petroleum output	With slowly expanding petroleum output
1980–85	3.2%	3.7%
1985–90	2.15%	2.8%

[a] 1.6% annual increase.

Petroleum and natural gas resources. The Soviet petroleum industry is entering a critical period, with sharply declining increments. The problems besetting it have been well documented. Although the USSR has become the world's largest oil producer, it has done so at prodigious cost in terms of manpower and material. According to the C.I.A.'s estimate, the USSR may expend as much effort on producing oil as all the non-Communist world combined. With twice as many workers and roughly the same number of rigs, the Oil Ministry in recent years has managed to drill only about one-fifth as much meterage as did American companies.[34] More importantly, to maximize current output, the Russians overemphasized development at the expense of prospecting and exploratory work; development drilling increased almost fivefold since 1950, but exploratory drilling has been less in recent years than in the mid-1960s.[35] The capacity required merely to offset depletion has multiplied (from 110 million tons in the second half of the 1960s to 390–400 million tons during the 10th five-year plan);[36] rash, at times even reckless, production methods inflicted serious damage to a number of key oil fields and, by Soviet admission, resulted in a loss of oil ultimately recoverable.[37] During the last 15 years, the average recovery rate for the USSR as a whole declined by 4%. Barring radical improvements in recovery technology, Soviet authorities expect a continuation of this trend.[38]

The Russians have been proud of their technique of early and intense application of water flooding, reporting higher recovery rates from reservoirs than American producers.[39] However, while maximizing yields during the early life of fields, such methods create production problems later on as rapidly growing volumes of water are lifted to recover the oil and fields are redrilled to replace the flooded wells. The water cut has grown particularly rapidly in the important fields of West Siberia, which were brought into production under great urgency.[40] Crash development and insufficient attention to reservoir characteristics of new deposits are continuing. According to an authoritative source, the new fields in Tyumen' Oblast to be put into production in the second half of the 1970s were to go on line with little study of the geological structure.[41]

The size of proved reserves appears insufficient to sustain expansion during the 1980s, though an upturn later cannot be ruled out. Official reserve figures are secret, but fragmentary information suggests that the proved plus indicated reserves, i.e., $(A + B) + C_1$ categories, total perhaps 10 billion to 11 billion tons (73 billion to 80 billion barrels).[42] A Soviet authority states that from early 1959 to the mid-1970s the share of the indicated (C_1) portion grew sharply at the expense of the proved categories, from less than 33 to over 52%.[43] The reserve-production ratio has been declining for over a dozen years[44] and at the current level of output only about 10 years' worth of fully proved, recoverable resources seem to be at the disposal of Moscow planners.[45] The larger half of these reserves today are in West Siberia and in Arctic areas, and most known fields not yet worked to full capacity are much smaller than those that sustained the Soviet oil boom in the last quarter century. Despite the impressive resource potential of the vast sedimentary basins, the near to medium-term outlook for continued expansion is not encouraging.

In early 1978, Gosplan ordered a significant increase in prospecting and exploration to curb the erosion of the reserve-production ratio.[46] Prospecting, however, continues to suffer from distorted incentives, unresponsiveness of supply agencies and technological weaknesses, particularly in geophysical work. Linear meterage of hole drilled persists as the principal index of material incentives and profits, resulting in a strong bias for shallow holes and resistance to deeper drilling, since speed declines exponentially with depth.[47] The North and Siberia need special cold-weather equipment, which producers are reluctant to supply. And, perhaps most importantly, geophysical work remains woefully inadequate and, by Soviet admission, continues to grow slower than prospecting and exploratory activities as a whole. The latter drastically reduces the effectiveness of exploration, forcing the Russians to concentrate almost entirely on anticlinal traps and forgo reservoirs of more complex and elusive types.[48]

For the last three years in a row, 1976 through 1978, the first time since World War II, the Soviet petroleum industry is experiencing a decrease in the volume of increment, and production plans remained unfulfilled.[49] Still more significant, all but four producing regions (West Siberia, Komi ASSR, Perm' Oblast and Udmurt ASSR) suffered absolute declines, some for a second time.[50] Within five years the reservoirs of the Volga and the Caucasus are expected to run practically dry.[51] Even in West Siberia, some 10 fields have reached or passed their peak, including giant Samotlor, which was responsible for most of Soviet production growth in the 1970s. Future increments must originate from smaller reservoirs scattered in the uncharted swamps. Since the beginning of the 10th five-year plan, paved roads and power lines have been built to only two of the 10 new fields put on line in the region. In 1977, only half of the proposed eight fields were brought into production. To speed up development, thousands of drillers began to be flown in from oil fields in the European USSR, and were made responsible for roughly one-tenth of the planned Siberian meterage in 1978.[52] For the longer term, Soviet crews will have to tap deeper lying strata in the northern half of West Siberia under a thick layer of areally continuous permafrost. They must also discover and prepare reserves in still more remote East Siberia, the offshore Arctic and/or in the super-deep beds of the Caspian lowland. By Soviet estimates, development costs in East Siberia could reach three to four times those in West Siberia[53] and the others are unlikely to be cheaper to tap. None of these can yield large quantities of oil until the late 1980s, by which time much of this capacity will be required to offset depletion in West Siberia.

In the light of the foregoing, the planned oil production target of 620–640 million tons for 1980 probably will be missed by some 10 million tons. Looking further down the road, only very slow growth can be foreseen for the Soviet oil industry during the 1980s, barring some extraordinary luck, and production may indeed peak. The present writer is reluctant to project an actual downturn, as the C.I.A. does, but given the severity of the problems, it is not a development that can be considered inconceivable.

In contrast to petroleum, the problems of the natural gas industry are due almost entirely to geography and the difficulties of transport, since no reserve bottleneck hampers rapid expansion. Explored (proved plus probable, or $A + B + C_1$) reserves in 1976 are said to have reached 28 trillion m^3, i.e., 990 trillion cubic feet.[54] Again, only part of this total is in the fully proved, recoverable category,[55] but Soviet resources can doubtless guarantee a steady rise in output until the end of the century. More than three-fifths of these reserves are concentrated in the permafrost-ridden wilderness of northwest Siberia (mostly in supergiant deposits) and 13 to 14% are located in the deserts of Central Asia, both 1,200 to 1,800 miles from the main urban-industrial concentrations of the European USSR. The reservoirs of the North Caucasus and the Ukraine, which held almost two-thirds of all Soviet gas in the mid-1950s, hold less than 6% today.[56]

The difficulties and delays of transporting West Siberian gas to market areas have forced the Soviet Union to press to the utmost its smaller Caucasian, Ukrainian and Volga fields and later those also in Central Asia, leading to premature pressure drops and exhaustion. Output in the North Caucasus has been declining for a decade, in Uzbekistan and the Ukraine since the middle 1970s. Expansion from two major growth regions west of the Urals (Orenburg and the Komi ASSR) can do little more than offset these declines. Increase in Central Asia is similarly slowing (much of the gain from Turkmenia merely counterbalancing gradual exhaustion in Uzbekistan) and the region's output will peak by 1980. In the future, the older centers of the gas industry will act as a drag on its performance and profitability, tying up capital and requiring large investment to moderate or delay their collapse. During 1971-75, almost one-third of the industry's fixed capital was concentrated in the Ukraine and the North Caucasus[57] and, at least in the Ukraine, the 1976–80 plan had called for several hundred new wells.[58] Moreover, while the vast reserves make the stress on development as against exploratory drilling in the aggregate less serious than in the oil industry, the significant reduction in prospecting in favor of developmental activities in older regions will have its adverse consequences. Similarly, Soviet backwardness in deep drilling and seismic technology is most serious for the mature petroliferous provinces, in many of which the average depth of exploratory wells has reached or surpassed 4,000 meters.[59]

The sole reliable source of major increments in gas production, therefore, is represented by northwest Siberia. Transport and production constraints to deliver this resource to market centers at the requisite volume will continue to be severe. The Russians are installing what amounts to the world's biggest large-diameter trunkline system from a single gas-producing province. Three major subsystems (Northern Lights, Urengoy-Center and Urengoy-Chelyabinsk, all of them being of 48-inch and 56-inch pipes) originate from the supergiant Urengoy and Medvezh'ye fields and will tap smaller neighboring deposits as well. To produce the planned 139 billion m^3 (not including oil-well gas) from Tyumen Oblast by 1980, more than 60 compressor stations, mostly with 25,000 KW units, would have to be installed on these lines alone.[60] Pipeline construction goals for the Soviet Union as a whole for the 1976–80 plan are given as 22,600 miles, almost 30% of it from 48 and 56-inch pipe.[61] Three hundred new compressor stations were called for compared with 148 in the previous five-year period, and Moscow reportedly hopes to purchase 200 of them from the West.[62]

Can these goals be achieved? Not fully. The key Urengoy-Chelyabinsk pipeline was running behind schedule in 1978;[63] the delivery and installation of compressor stations presented an even greater bottleneck. The pipeline to Chelyabinsk is now scheduled to operate at half capacity in 1980, almost certainly resulting in a 15 billion m^3 shortfall for West Siberia.[64] The faster-than-anticipated decline in older regions will deprive the Soviet planners of at least half of that amount. Soviet natural gas output in 1980 is likely to fall near or below the midpoint of the 400–435 billion m^3 range (14.1-15.4 trillion cubic feet) originally approved in the 10th five-year plan guidelines rather than the specific high target at the upper end of the range announced seven months later.

Beyond 1980, increments from Urengoy cannot alone guarantee all of the production increase required from West Siberia, and Medvezh'ye will have reached its peak level. Development must move to deposits farther north and northeast, including the forbidding Yamal Peninsula, an additional 300 miles beyond the Arctic Circle, where permafrost conditions are worse. By that time also, most fields in old producing regions will have run practically dry. Therefore, the Soviet gas industry, which achieved a 91 billion m^3 increment in output during the first half of the 1970s and should achieve a 125-130 billion m^3 increment during the second half, cannot possibly hope for a more than 200 billion m^3 increase under the best of circumstances between 1980 and 1985. Consequently, the mid-1980s should see gas output rise to some 600-620 billion m^3 (21.2-21.9 trillion cubic feet) and at the close of the decade perhaps 750 billion m^3 (26.5 trillion cubic feet), but probably less.

AGGREGATE ENERGY SUPPLIES: IMPLICATIONS

Fossil fuels, accounting for 96% of aggregate energy consumption,[65] will continue to define both the gross volume and the rate of increase of total energy supplies. A dozen years hence their share in primary energy will still be over nine-tenths of the total and barely under four-fifths even at the turn of the century.[66] We may ignore peat, shale and wood, which contribute less than 3% today and will diminish further in relative importance. The previous sections have shown that Soviet planners cannot count on solid fuels to substitute widely for hydrocarbons and cannot effect a rapid acceleration of coal output. Coal production will grow only by somewhat over 2% per year in the 10th five-year plan and will not reach the 1980 target. Growth rates are likely to improve during the 1980s, but a great unsurge is simply unfeasible. The burden on hydrocarbons cannot be relieved. The Soviet oil industry is plainly in trouble. Its future will determine the growth rate for Soviet energy consumption as a whole, prospects for hard-currency earnings and, to a large degree, even the nature of Comecon relations and the primary trade orientation of East Europe. The pressure on natural gas will also intensify, but geographical and technological constraints will prevent the gas industry from making up for the shortfalls in the other fuels.

Table 53 shows that with a 1.6% annual increase in petroleum production through the 1980s, the USSR could continue to achieve a 3.7% yearly rise in domestic fuel consumption from 1980 to 1985 and close to a 3% rise in the latter part of the decade, maintain present levels of oil shipments to East Europe and remain a small petroleum exporter to hard-currency markets throughout the decade. With the peaking of petroleum output, exports beyond the Comecon area will have to be eliminated by the latter part of the decade or shipments to East Europe cut by half. This would appear inevitable, since it would seem inconceivable that the Soviet domestic economy could operate and develop its resources with aggregate fuel demand expanding by less than 2% per annum. If oil production suffers an absolute decline, most of the exports to East Europe might have to cease and the Soviet bloc would have to weather that additional shock on top of the Soviet loss of hard currency.

We have shown that energy-GNP elasticities cannot fall appreciably in the next dozen years. Labor and capital shortages and systemic ills, however, should further retard Soviet economic growth even without an energy pinch, and slower GNP growth also should mean a more sluggish rise in energy demand. The Moscow planners could and will be able to live with an economic expansion of only 3%, even slightly less. As in the past, they can tighten their nation's collective belt. However, the political and psychological consequences of economic growth much below that rate may be profound. A concurrent pinch of labor, energy, some other material and capital resources, together with the slow rise of labor productivity and the perennially unpredictable weather, would herald a new era in Soviet economic planning. In Gregory Grossman's words, "the tension between goulash communism and Gulag communism may again come into sharper focus."[67]

NOTES

[1]Gertrude E. Schroeder, "Recent developments in Soviet planning and incentives," in U.S. Congress, Joint Economic Committee, *Soviet Economic Prospects for the Seventies* (Washington: Government Printing Office, 1973), pp. 13–15.

[2]Robert Campbell, *Soviet Energy R&D: Goals, Planning and Organization* (Santa Monica, CA: RAND, R-2253-DOE, May, 1978), pp. 10–15, especially p. 10.

[3]Some recent summaries of this modeling work on the national level are given in A. Vigdorchik et al., "Methods of optimizing the long-term development of the fuels and energy complex of the USSR," *Planovoye khozyaystvo*, 1975, No. 2, pp. 29–37; L. A. Melent'yev and A. A. Makarov, "Peculiarities of optimizing the development of the fuels and energy complex," *Izvestiya Akademii Nauk SSSR, Energetika i transport*, 1974, No. 3, pp. 12–19; V. Z. Tkachenko, "Methods of investigating the material linkages between the fuels and energy complex and other sectors of the economy," unpublished dissertation (Irkutsk: Energetics Institute, 1975); G. A. Andreyev et al., "Peculiarities of the external linkages of various fuel-producing bases," in *Voprosy vliyaniya razvitiya energetiki na drugiye otrasli narodnogo khozyaystva* (Irkutsk, 1975), pp. 29–45.

[4]Tkachenko, op. cit., p. 128.

[5]G. A. Andreyev et al., "Experimental evaluation of possible changes of demand in

aggregated economic sectors under changing conditions of energy development," in *Voprosy vliyaniya . . .* , op. cit., pp. 68–84, especially p. 79.

[6]See an analysis of the problems of increasing resource exhaustion combined with an industrial structure long biased towards heavy industries. Leslie Dienes, "Basic industries and regional economic growth: the Soviet South," in *Tijdschrift voor Economische en Sociale Geografie*, 1977, No. 1, pp. 2–15, and Chapter 7 (Minerals and Energy) in I. S. Koropeckij, ed., *The Ukraine Within the USSR: an Economic Balance Sheet* (New York: Praeger, 1977).

[7]*Energetika SSSR v 1976–1980 godakh*, op. cit., p. 149; Ya. Mazover, "Location of the fuel-extracting industry," *Planovoye khozyaystvo*, 1977, No. 11, p. 143; *Sovetskaya Tatariya*, Nov. 7 and 27, 1976, and *Soviet Geography*, April 1978, pp. 273–285.

[8]For example, production costs of natural gas in Krasnodar Kray increased from 0.40 ruble per 1000 m^3 in 1964 to 5.47 rubles per 1000 m^3 in 1975. Because these costs do not fully take into account all exploratory expenses, total cost was higher still. Since 1975, the production cost of this gas continued to rise appreciably. During 1971–75, investment made in older fields designed to compensate for pressure drops and production losses exceeded investment made for incremental capacity. A. D. Brents et al., *Ekonomika gazodobyvayushchey promyshlennosti* (Moscow: Nedra, 1975), pp. 49–51 and R. D. Margulov, Ye. K. Selikhova and I. Ya. Furman, *Razvitiye gazovoy promyshlennosti i analiz tekhnikoekonomicheskikh pokazateley* (Moscow: VNIIEGAZPROM, 1976), pp. 23 and 25.

[9]Since the North Caucasus–Center–Leningrad pipeline system, built in the mid-1960s, has long been amortized, the delivered cost of Siberian gas to the south should be close to those in Central Russia and Leningrad.

[10]Crude oil is far cheaper to transport than natural gas. Transport cost differences from West Siberia, therefore, raise the cost of the much cheaper West Siberian crude only slightly.

[11]Leslie Dienes, "Issues in Soviet energy policy and conflicts over fuel costs in regional development," *Soviet Studies*, July 1971, pp. 26–58 and "Geographical problems of allocation in the Soviet fuel supply," *Energy Policy*, June 1973, pp. 3–20.

[12]The program was run by the University of Kansas Computation Center on a Honeywell MPS, Series 60 (Level 66) package. The writer gratefully acknowledges the indispensable and generous help of Jeff Bangert, Charles Eklund and W. Maxwell of the Computation Center and of Professor Arthur Djang of the Business School. The main sources consulted for the data matrix include *Ekonomika gazovoy promyshlennosti*, 1975, No. 2, p. 12; 1975, No. 8, p. 23; *Referativnyy sbornik. Ekonomika promyshlennosti*, 1977, No. 7, V-18; A. A. Makarov and L. A. Melent'yev, *Metody issledovaniya i optimizatsii energeticheskogo khozyaystva* (Novosibirsk: Nauka, 1973), pp. 104–110; *Ekonomika chernoy metallurgii* (Moscow: Metallurgiya, 1976), pp. 91–92; N. D. Lelyukhina, *Ekonomicheskaya effektivnost' razmeshcheniya chernoy metallurgii* (Moscow: Nauka, 1973), pp. 204 and 208. N. V. Mel'nikov, *Mineral'noye toplivo* (Moscow: Nedra, 1971), pp. 175–84; the sources cited in Dienes, op. cit. (1973), Note 21, give costs for the late 1960s and early 1970s for nonspecialized boiler and furnace fuels.

[13]P. Ye. Semenov and V. F. Kosov, *Problemy razivitiya i razmeshcheniya proizvoditel'nykh sil Kazakhstana* (Moscow: Mysl', 1974), p. 67.

[14]Estimated by the author as a residual after the approximate consumption of coal and natural gas is accounted for.

[15]See Note 13 above.

[16]Gas consumption estimated as production minus net export. *Soviet Geography*, April 1978, p. 273; R. D. Margulov et al., op. cit., p. 46 and *Vneshnyaya torgovlya SSSR v 1976 g.*, p. 90. The two Bukhara–Urals pipelines and the Central Asia–Central Russia Nos. 1 and 2 pipelines are known to have worked at or near full capacity for a number of years now. Total boiler and furnace fuel consumption in Central Asia is estimated from its 1970 share as shown in Table 5.

[17]Makarov and Melent'yev, op. cit. (1973), p. 108 and Ya. G. Feygin et al., eds., *Problemy ekonomicheskoy effektivnosti razmeshcheniya sotsialisticheskogo proizvodstva v SSSR* (Moscow: Nauka, 1968), p. 304.

[18]Leslie Dienes, "Basic industries and regional economic growth: the Soviet South," *Tijdschrift voor Economische en Sociale Geografie*, 1977, No. 1, pp. 2–15.

[19]A. A. Beschinskiy and Yu. M. Kogan, *Ekonomicheskiye problemy elektrifikatsii* (Moscow: Energiya, 1976), pp. 23–24 and 200–202.

[20]Ya. Mazover, op. cit., p. 145 and *Narodnoye khozyaystvo SSSR za 60 let* (Moscow, 1977), p. 204.

[21]A. M. Kurnosov et al., "Perfection of the mining stock . . . ," Tsentral'nyy Nauchno-Issledovatelskiy Ekonomicheskiy Institut Uglya, *Nauchnyye trudy* (Moscow, 1975), p. 6.

[22]R. Z. Kosukhin, "Contemporary problems of socio-economic development of a mining enterprise," in *Ekonomicheskiye parametry gornykh predpriyatiy budushchego* (Moscow, 1976), p. 187.

[23]*Energetika* . . . , op. cit. (1977), p. 149; Mazover, op. cit. (1977), pp. 143–144; A. Krichko and L. Semenov, "Prospects of coal mining and utilization," *Planovoye khozyaystvo*, No. 8, pp. 93–94 and *Soviet Geography*, April 1978, pp. 281–284.

[24]Kurnosov, et al., op. cit., p. 6, and Ministerstvo Geologii SSSR, VIEMS, *Ekonomika mineral'nogo syr'ya i geologorazvedochnykh rabot*, No. 8 (Moscow, 1976), pp. 12–13. In addition, the sulfur content of coal in the European USSR is deteriorating. More than half the coal mined west of the Urals already has a sulfur content of over 2.5%. Only a fifth of the coal reserves in the European USSR have a sulfur content of less than 1.5%. V. S. Al'tshuler, *Novyye protsessy gazifikatsii tverdogo topliva* (Moscow: Nedra, 1976, pp. 191–192. The ash content of the coal is also increasing. In the Ukraine, it grew from 24.2% of the total mineral mass in 1965 to 31.7% 10 years later. *Ugol' Ukrainy*, 1976, No. 2, p. 6.

[25]Leslie Dienes, "The USSR: An energy crunch ahead," *Problems of Communism*, September-October, 1977, pp. 41–60, especially pp. 51–57.

[26]Computed from data in *Narodnoye khozyaystvo SSSR za 60 let*, p. 204, and in Table 53; 1.2 billion tons even with a less than 3% growth in aggregate fuel consumption during the 1980s and a stationary level of crude oil output which, by definition, raises the relative share of coal.

[27]*Sotsialisticheskaya Industriya*, November 19, 1977, p. 2.

[28]*Sovetskaya Rossiya*, Jan. 13, 1978.

[29]Alvin M. Josephy, Jr., "Agony of the northern plains," *Audubon*, June 1973.

[30]V. A. Shelest, *Regional'nyye energoekonomicheskiye problemy SSSR* (Moscow: Nauka, 1975), pp. 224–227 and L. S. Popyrin et al., "Electric power in Siberia in the 10th five-year plan," *Izvestiya VUZ. Energetika*, 1976, No. 11, p. 8.

[31]*Pravda*, Jan. 23, 1977, pp. 1–2; *Ekonomicheskaya Gazeta*, No. 6 (February 1978), p. 5 and No. 34 (August 1978), p. 5.

[32]*Narodnoye khozyaystvo SSSR*, various issues and *Soviet Geography*, April 1978, p. 282.

[33]N. I. Nikolayev and A. A. Kosar', *Effektivnost' kapital'nykh vlozhenii v ugol'noy promyshlennosti Kuzbassa* (Kemerovo: Politekhnicheskiy Institut Kuzbassa, 1976), p. 48.

[34]U.S., C.I.A., *Prospects for Soviet Oil Production. A Supplementary Analysis.* ER-77-10425, June 1977, p. 2.

[35]*The Oil and Gas Journal*, Oct. 10, 1977, p. 73 and Feb. 6, 1978, p. 36.

[36]*Neftyanoye khozyaystvo*, No. 7, 1976, p. 4 and P. S. Sapozhnikov and G. D. Sokolov, *Ekonomika i organizatsiya stroitel'stva v neftyanoy i gazovoy promyshlennosti* (Moscow: Nedra, 1976), p. 7.

[37]See *Neftyanoye khozyaystvo*, 1976, No. 10, pp. 27-29; 1976, No. 4, pp. 51-54; 1976, No. 3, pp. 24-25; 1977, No. 7, pp. 5-7; 1977, No. 6, pp. 30-33; 1977, No. 4, p. 9 and *Problemy geologii nefti*, No. 8 (1976), pp. 130-136.

[38]*Geologiya nefti i gaza*, 1977, No. 9, p. 65.

[39]*Ekonomicheskaya Gazeta*, No. 11, March 1974, p. 7.

[40]A. I. Zhechkov and L. B. Kuznetsova, "Analysis of the productivity of old oil wells in West Siberian fields," *Ekonomika neftyanoy promyshlennosti*, 1976, No. 8, pp. 23-25. Soviet scientists have admitted serious difficulties in applying their accustomed techniques of field development to West Siberian reservoirs. V. P. Maksimov, *Exploitation of Petroleum Deposits in Complex Conditions* (Moscow, 1976), as translated by the Joint Publication Research Service, L/7372, September, 1977, pp. 26-27. The difficulties encountered by the petroleum industry in West Siberia are analyzed on pages 5-34.

[41]Maksimov, op. cit., p. 26. A 1977 source also reveals that for the previous two years exploratory work in West Siberia has gone badly and plans to prove up reserves remained unfulfilled. *Geologiya nefti i gaza*, 1977, No. 9, p. 6.

[42]See the exhaustive analysis of Soviet oil (and gas) reserves by David Levin, "Oil and natural gas resources of the Soviet Union and methods of their estimation," in U.S. Congress, *Project Interdependence: U.S. and World Energy Outlook Through 1990*, Congressional Research Service, Library of Congress (Washington: Government Printing Service, 1977), pp. 821-848. For recent Western estimates see also *The Oil and Gas Journal*, Feb. 6, 1978, p. 36; *World Oil*, Aug. 15, 1976, p. 44; *The Christian Science Monitor*, Feb. 25, 1975, p. 4.

[43]G. P. Ovanesov and M. V. Geiguin, "The question of confirming oil reserves in the categories C_1 and C_2," *Geologiya nefti i gaza*, August 1975, pp. 7-13, as quoted by Levin, op. cit., p. 832.

[44]N. Mel'nikov and V. Shelest, "The fuels and energy complex of the USSR," *Planovoye khozyaystvo*, 1975, No. 2, p. 11 and NATO Directorate of Economic Affairs, *Exploitation of Siberia's Natural Resources*, Brussels, 1974, pp. 76 and 110-111.

[45]*The Oil and Gas Journal*, Feb. 6, 1978, p. 36, estimates a maximum of 60 billion barrels to be in the proved, economically recoverable category. At the 1977 production rate, that would represent a 14-year supply. At any rate, Soviet concern with the proved reserve–production ratio has been apparent for some time. The decision significantly to step up exploration was announced by N. K. Baibakov, chairman of Gosplan.

[46]*The Oil and Gas Journal*, Feb. 6, 1978, p. 36.

[47]*Pravda*, Jan. 27, 1978, p. 2.

[48]*Pravda*, Feb. 28, 1978, p. 2; Aug. 10, 1977, p. 2; Aug. 24, 1977, p. 2; and V. V. Fedynskiy, "Exploration geophysics in the USSR on the 60th anniversary of the October Revolution," *Vestnik Moskovskogo Universiteta, Geologiya*, No. 4, 1977, pp. 123-124. During the 1960s, in the rich petroliferous province of West Siberia, large reserves in stratigraphic traps were found by accident, in the sense that exploration had been directed entirely toward anticlinal structures. The high frequency of finding stratigraphic traps when they were not even being sought suggests that important reserves are yet to be found in such nonstructural traps. Obviously, when the major anticlines in shallower depths have been drilled, the Russians cannot rely on such accidental finds any longer. *The Oil and Gas Journal*, May 8, 1978, p. 328.

[49]*Narodnoye khozyaystvo SSSR za 60 let*, p. 205; *Ekonomicheskaya Gazeta*, No. 6, February 1978, p. 5; No. 31, July 1978, p. 5 and No. 31, July 1977, p. 5.

[50]*Soviet Geography*, April 1978, pp. 273-277 and *The Oil and Gas Journal*, Dec. 26, 1977, pp. 70-72 and July 3, 1978, p. 27.

[51]*Sovetskaya Tatariya*, Nov. 7, 1976, p. 3 and Nov. 27, 1976, p. 2.

[52]*Pravda*, June 5, 1978, p. 2; *The Oil and Gas Journal*, Dec. 26, 1977, pp. 70 and 71 and May 15, 1978, p. 46. The yields of older wells in West Siberia have actually declined more rapidly than was the case in the Volga fields. *Ekonomika neftyanoy promyshlennosti*, 1976, No. 8, pp. 23–24.

[53]*Referativnyy sbornik. Ekonomika promyshlennosti*, 1977, No. 7, V-18.

[54]*Geologiya, bureniye i razrabotka gazovykh mestorozhdeniy*, 1977, No. 4, p. 3.

[55]The recovery rate of gas is higher than that of oil, varying in the USSR from 80 to 90% of explored reserves of a deposit. *Gazovaya Promyshlennost'*, 1975, No. 8, p. 4. Since the West Siberian fields have low reservoir pressure, recovery rates at these deposits may be appreciably lower. This is conceded by Soviet authorities, including the Minister of the Gas Industry himself. *Gazovaya Promyshlennost'*, 1976, No. 11, p. 6 and *Pravda*, Feb. 26, 1977, p. 2.

[56]S. A. Orudzhev, *Gazovaya promyshlennost po puti progressa* (Moscow: Nedra, 1976), p. 12; Yu. Bokserman. "Ways of enhancing the cost-effectiveness of fuel transport," *Planovoye khozyaystvo*, 1975, No. 2, p. 21. Half of all Soviet reserves are found in six supergiant fields, of which four are located in northwest Siberia, one in Central Asia and one in Orenburg Oblast of the European USSR. An additional supergiant is shaping up on the Yamal Peninsula of northwest Siberia. The Urengoy field in West Siberia alone concentrates one-fifth of all explored reserves and is by far the largest gas field in the world discovered to date.

[57]R. D. Margulov et al., op. cit. (1976), p. 39.

[58]*Neftyanaya i gazovaya promyshlennost'*, 1977, No. 1, p. 2.

[59]*Gazovaya Promyshlennost'*, 1977, No. 9, pp. 33–34.

[60]*Gazovaya Promyshlennost'*, 1977, No. 1, p. 2.

[61]*Ekonomicheskaya Gazeta*, No. 6, February 1976, p. 2.

[62]*Gazovaya Promyshlennost'*, 1976, No. 11, p. 2, and *Business Week*, Oct. 17, 1977, p. 52.

[63]The 1,500 kilometer, 56-inch gas pipeline was advancing at a rate of 5 km a day instead of the necessary 15 km or, in another source, 20 to 22 km. (With the onset of spring thaw, construction work must come to a halt; only part of the year is utilizable.) As of February 1978, one of the key construction agencies had welded into place less than 100 km of its assigned 568 km of the trunk line. Only a little over 10% of the pipe was insulated and placed into the trench and 300 km of the route was not even cleared. *Pravda*, Feb. 15, 1978, p. 2 and *Ekonomicheskaya Gazeta*, No. 9, February 1978, pp. 1–2.

[64]*Izvestiya*, Nov. 30, 1976, p. 1.

[65]Hydroelectric plants in 1976 produced 135.7 billion KWH of electricity; nuclear stations in 1975 produced 20.2 billion KWH. *Narodnoye khozyaystvo SSSR za 60 let*, p. 201 and *Energetika . . .* , op. cit., p. 111. Computed at the heat equivalent of power (960 kilocalories per KWH), therefore, these sources added less than 20 million tons of standard fuel to the Soviet fuel balance. Computed at the heat rate of central electric stations (2,359 kilocalories per KWH, i.e., the amount of fuel these plants need to generate 1 KWH), they added about 53 million tons of standard fuel to the fuel balance in 1976. In that year, aggregate fuel consumption from centralized sources in the USSR reached 1,465 million tons of standard fuel. *Narodnoye khozyaystvo SSSR za 60 let*, pp. 83 and 203.

[66]The share of hydraulic power is declining and will drop by the end of the century to less than one-tenth of all electricity, while the contribution of nuclear plants, still small, will grow markedly, to perhaps 40–50% of electric power output. However, since less than one-fourth of gross energy supply today is converted to electricity (a share unlikely to double before the 21st century), fossil fuels, responsible for 96% of all energy used by the Soviet economy at present, should furnish over three-fourths of the total supply even by the year 2000.

[67]*The Daily Telegraph* (London), November 1977.

Chapter 10

ADMINISTRATION AND PLANNING OF THE ENERGY INDUSTRIES

The institutional environment of the Soviet economy is overwhelmingly an administratively directed one. In such an environment, to use Robert Campbell's words, interrelated production processes are coordinated not primarily through lateral communication, but basically by the administrative, command approach. This means not through negotiation and mutual accommodation via market mechanisms between the controllers of such processes, but by the "issuing of direct orders by some superior organization to each lower level overseer involved in the coordination of a particular variable."[1] While both market economies and Soviet-type economies employ both methods, one strongly predominates in each of them. In modern societies, the fuel-energy industries are distinguished by huge investment requirements, scale economies and long payoff periods, combined with comparatively few and locationally restricted resources, a relatively narrow output mix and range of technical coefficients. This results in the more pervasive and, generally speaking, more efficient application of the administrative method than is possible in most other industries. At the same time, the strategic significance of energy leads to a strong degree of government involvement in the energy sector everywhere.

MINISTERIAL STRUCTURE OF THE ENERGY SYSTEM

There exists no single energy authority in the USSR, no overarching policy-making or coordinating energy agency either in the party or in the government apparatus. Still, administrative centralization of the Soviet energy industries has reached an altogether vaster scale than found in the West. The Soviet Ministry of the Coal Industry, for example, is responsible for 99% of all coal output[2] and employs a total labor force of some 2 million (including supervisors, engineers, apprentices, and auxiliary workers), over 600,000 of them working underground.[3] Understandably, one finds some overlap between the Ministry of the Petroleum Industry and the Ministry of the Gas Industry in the production of hydrocarbons. In 1975, the oil agency provided close to a fifth of all natural gas (both associated and nonassociated gases in roughly equal proportion), while the gas agency was responsible for 1.5% of liquid hydrocarbons, mainly gas condensate output.[4] Between them, the two ministries were responsible essentially for all oil and gas withdrawn from Soviet reservoirs. Similarly, in the mid-1970s, the Ministry of Electric Power and Electrification generated 93% of all electricity in the USSR.[5]

Altogether a dozen ministries, a fifth of the Soviet total, are either directly energy-related or transact by far the larger part of their total services for or with the products of the energy industries.[6] To these the State Committee for Utilization of Atomic Energy should be added. Five ministries specialize directly in the output, processing and conversion of primary energy sources.* Two ministries of machine-building and one of construction are directly and specifically geared to supply and serve the energy industries with equipment and facilities,** and two more, those for geology and the railroads, contribute a larger part of their total services to the energy sector than to any other group of industries.[7] Finally, the Ministry of Foreign Trade realizes more than a third of total export earnings and more than half of its hard-currency earnings from the sale of fuels and electricity and imports large quantities of equipment for the energy industries.[8]

Despite the creation of the specialized Ministry for Construction of Petroleum and Gas Industry Enterprises, requirements for such projects are said to be so great that the ministries of both the oil and gas industries have construction departments under their own jurisdiction. The Petroleum Ministry's construction department handles almost a quarter of all building tasks for this industry, while the specialized construction ministry handles three-fifths. The remainder is performed by still other ministries.[9] Most likely the traditional unreliability concerning quality, specification and timing of such work performed in an administrative management framework (long a bane of the Soviet economy), mistrust and lack of clear accountability for the work done, were instrumental in such organizational duplication.

Poor cooperation between ministries has long plagued the Soviet system and resulted in a minimum degree of lateral communication among enterprises

*Ministry of the Gas Industry, Ministry of the Petroleum Industry, Ministry of the Petroleum Refining and Petrochemical Industry, Ministry of the Coal Industry and Ministry of Power and Electrification.

**Ministry of Chemical and Petroleum Machine-Building, Ministry of Power Machine-Building and Ministry of Construction of Petroleum and Gas Industry Enterprises.

subordinated to different ministries. Complaints about the problem are endemic in the Soviet press. Recent experience in the initial phase of the South Yakutian territorial production complex in Siberia may serve as an example. The complex, centered on the new town of Neryungri, is the latest Soviet coking-coal base (well north of the inhabited belt), to be developed primarily for export jointly by the Ministry of Coal Industry and the Ministry of Electric Power. Both agencies planned their own supply bases separately–their own agricultural complexes, their own plants for the manufacture of prefab housing modules, their own railroad stations, garages, warehouses, even roads. The Coal Ministry refused to accept responsibility for housing and services for workers not directly related to coal mining, and calculated that out of the projected quarter of a million residents only 100,000 belonged to the coal industry. The Coal Ministry wanted the first housing project built near the mines; the Power Ministry wanted it next to a proposed power plant. The heat-supply designs for the town of Neryungri alone had five versions, each worked out strictly within a ministerial context, ignoring the interests of other agencies. With two strong ministries evenly matched, neither could prevail. After much delay, the plans were turned over to the State Committee for Construction Affairs, the Soviet Government's umbrella agency for all construction, to come up with a mutually acceptable version.[10] Other sources, having reviewed similar problems with respect to the Tyumen' oil and gas fields, state: "A lack of coordination and disunity of actions at various ministries and agencies is the greatest hindrance to the most effective and planned development at our chief fuels and energy base."[11]

Equipment suppliers of different ministries are slow to deliver new gear or alter old designs to fit specific local conditions. For example, detachable core-barrels (for drilling), gasket pumps, unitized plungers were all developed some years ago, the first in 1973, but not a single one of these reached drillers and producers in the field by the early part of 1977.[12] A high-level authority in a major electric power institute complained that steam generators shipped 15 years ago to Siberian power plants to use Kansk-Achinsk lignite have had to be rebuilt at costs sometimes exceeding the original outlays to suit the physical and chemical composition of that difficult fuel. Fifteen years later the supplying factory had still made no alteration in that basic design.[13]

Poor communication and insufficient coordination may also be rife among enterprises of the same ministry, and between enterprises and their ministry, if the vertical chain of command is too long. Until the middle 1970s, five-six stage hierarchies between ministries and enterprises were not unusual and four-stage hierarchies very common.[14] Traditionally, middle-level management has been represented by the *glavk*, some of which (*glavki*) may be in charge of specialized branches of the industry, certain territorial groupings of enterprises, research institutions, financial activities or some combination of these. In addition, trusts may exist in some industries, as at one time in the gas industry.

In the 1970s, the Soviet economy has again been undergoing extensive administrative reorganization, with the aim of shortening the chain of command and to promote specialization and efficiency among enterprises. The latter are being grouped into production associations, or regional corporations, with one director and internal system of accounting. They are being subordinated to the appropriate ministries either directly or through the intermediary of industrial associations (national corporations), which ultimately may end up to be not too different from the former *glavki*. In the energy sector, the administrative hierarchy of the coal industry was the most drastically simplified. Today large consolidated regional corporations, except those in the Ukraine, are subordinated directly to the Coal Ministry in Moscow. The Ukraine, being a critically important producer, has its own republican coal ministry, acting as an intermediate authority between its regional corporations and the central Coal Ministry in Moscow. The administrative hierarchies in the petroleum, oil-refining and gas industries were also simplified. However, because of the many sub-branches from exploration and drilling to transport, storage and petrochemicals and because of the technological complexity and dynamism of these industries, their organizational structure remains appreciably more complicated than that for coal. Broadly speaking, the lowest level consists of the appropriate enterprises and/or regional corporation (for drilling, transport, production, etc.), linked to the central ministry either directly or through the intermediaries of national industrial corporations for the given sub-branch. As in the case of coal, the major benefit stemming from the pivotal new role of the regional agencies is thought to be greater awareness of geographic differences and problems, so crucial in primary industries because of inherent spatial variations in resources.[15]

Whether bottles of novel shape but hardly fresh materials will really improve the wine remains to be seen. Past penchant for administrative reorganizations seldom wrought significant changes in efficiency. Soviet sources, complaining about lack of coordination and delays in the completion of oil wells, about drilling and producing enterprises working at cross-purposes, sound a familiar note, but all the simplifications intended by the reorganization may not, of course, have been implemented yet.[16] Labor productivity in the oil and gas industries, though not in coal mining, has been rising rapidly, more likely because of a rapid growth of capital investment than because of administrative changes. At any rate, intra-ministerial reorganizations have done little to improve cooperation between the Ministry of Geology and those for gas and oil with respect to prospecting and exploration. Uncoordinated work by these ministries, each having its own interest in mind, reportedly continues, having a detrimental effect on hydrocarbon reserves and their distribution.[17]

The primary production units of the Soviet energy industries, the enterprises and, where they exist, the regional corporations, are roughly similar to such fundamental blocks comprising the energy industries in other countries. In Campbell's words, "a more or less universal matrix of technology generates a system of basic production units not unlike their counterparts in the United States." In addition, these analogues to capitalist firms also are distinguished by a certain independence, control and are responsible for their own performance and finances and, in general, meet expenditures out of their revenues.[18] Because of this measure of independence, output and assortment decisions are not made solely at the ministerial level. Rather, such decisions are the outcome of a constant interaction between higher and lower level agencies. In the oil and gas industries, for example, aggregate output is the outcome of composite choices concerning the development of new petroleum provinces, the finding and development of new fields in already producing regions, intensification measures, the drilling of additional wells and/or the abandoning of old wells. The first of these is essentially the responsibility of the ministry, the second of the regional corporation, the third of the local oil-field administration.[19]

Relative scarcity of inputs in producing a desired output mix implies economic choices at any and all levels of decision-making. In the Soviet context, the necessary economic guideposts, prices, rent payments, charges for capital, etc., essential to produce rough congruence among the objectives of decision-makers at different levels, are not generated by market forces but fixed centrally and not only for the purpose of guiding economic choices. Nor is simple profit maximization the main success indicator for an enterprise even today. Campbell has shown how the setting of rent charges per ton or 1,000 cubic meters of output instead of in a lump sum complicates rational output decisions. It may induce oil-field administrations to abandon wells and forgo further intensification efforts at old reservoirs before the marginal cost of such endeavor approximates marginal costs on the external frontier, i.e., bringing the least advantageous oil and gas sources into production to cover demand.[20] In a somewhat different vein, a Soviet source describes how, in the supply and delivery of oil-field pipe, performance indicators, prices, profits and the whole planning system are based not on length but on weight (presumably by assortments given in diameters). Thus, demand may not be met even though the plan for delivery is technically being fulfilled by shipping heavier pipe with thicker walls than necessary.[21]

ENERGY POLICY CONFLICTS: THE CURRENT DEBATE

The five energy ministries (coal, gas, crude oil, petroleum refining and electric power), the Ministry of Geology and the specialized machine-building and construction ministries in the energy field are unquestionably among the largest and most powerful in the entire Soviet economy. They are responsible not only for current production but also for the feasible alternatives of future development. In their dozens of research institutes, they conduct most of the energy-related research and development (particularly applied research and pilot projects) and do a large part of the modeling for energy planning and strategy. The selection and coordination of alternatives and their incorporation into the five-year plans are done by Gosplan, the main planning agency, which itself includes high ranking members of the more important ministries and, apparently, all of the energy ministries.

Paul Cook has expressed what Soviet officials themselves have intimated: "In this vast, cumbersome bureaucracy, battles range on a scale which puts to shame the infighting found in the relatively minuscule governments in capitalist countries. The current leaders appear well aware of the interrelationships between the many problems besetting the Soviet economy."[22] The potential and far-reaching impact of major decisions concerning any of the fuel-energy industries, both within the strongly interconnected energy system and on the economy at large, is surely obvious to them. As the literature also makes clear, the leadership is equally aware of the need for the long view, well beyond the span of the five-year plans, in the formulation of a more or less comprehensive energy policy, given the lead times and capital requirements in the energy sector. Already in the early 1970s, Soviet planners were doing preparatory work on a 15-year plan for 1976–90, and the fuel-energy economy was to form a cornerstone of this program. Of the seven basic, though very general, long-term objectives announced by the Council for the Study of Productive Forces, the preplanning research agency of Gosplan, two relate to fuel and hydroelectricity production and one to the expansion of energy-intensive branches in resource-rich Soviet Asia. Yet the approval of the 15-year program has been seriously delayed, and no comprehensive energy policy is available at the time of writing even in a rough skeletal form. The program itself has been redesignated a general outline of development until 1990, with no starting date.[23]

Although admittedly sketchy, the available evidence suggests that Soviet planning agencies have lately been engaged in a continuing crucial energy debate, but that no firm decision has been taken on long-term energy policy. The controversy and lobbying among the appropriate ministries, planning and research institutions and regional interest groups appear more intense than at any time in the post-Stalin years. The crux of the debate concerns the adequacy, role, feasible and rational speed of development of hydrocarbon resources compared with those of coal, particularly Siberian lignite. Such controversy, of course, is not peculiar to the USSR and, together with the role of nuclear power, nonconventional energy sources and conservation, is also focal to the search for an energy policy elsewhere, not least in the United States. However, the larger size, the comparatively unprospected and environmentally harsh character of most of the Soviet land mass, combined with a relative backwardness of technology, all lead to a quite different perception of the energy problem in the USSR than in Western countries. Even in the case of oil, where the proved reserve situation is the least satisfactory, the problem is not seen as an absolute shortage of available resources or as a problem of import dependency. Rather, the "energy problem" in the Soviet Union is viewed basically as a question of supply and construction bottlenecks, investment requirements and lead times, retarding expansion. Equally, it is regarded as a question of uncertainty concerning the best path of development for domestic and export needs in the face of increasingly difficult and costly technical-economic choices on the basis of incomplete data. Soviet planners know that the exigencies facing the country a decade hence must be anticipated and confronted today or very soon. But they also know that once heavy commitment of capital and other resources to a certain line of expansion is under way, the decisions will not be easy to reverse, uneconomic though they may turn out to be.

The alignment of forces is complex and defies easy categorization. To a degree, it manifests ministerial, institutional and regional loyalties and long-term commitment to given research fields. At the same time, evidence of deliberate caution regarding the potential of one's field and the speed of future development of given energy sources also exists. Not surprisingly, the cautionary and aggressive, the pessimistic and optimistic dichotomy is most clearly apparent among scientists and officials connected with petroleum geology, research and production, hydrocarbons being notoriously chancy and exhaustible resources, difficult to plan for. In general, applied geologists, oil and gas producers have proved to be consistently more diffident about Soviet petroleum potentials than research geologists and geophysicists. This is explicitly stated by A. A. Trofimuk, director of the Siberian Institute of Geology and Geophysics and deputy director of the Siberian Division of the Academy of Sciences[24] but can also be supported by a few additional sources.

Trofimuk himself is, perhaps, the most sanguine among Soviet officials and scientists concerning the country's petroleum potential and is totally convinced that even by the end of the century a mere fraction of Soviet petroleum reserves will have been expended. He firmly believes that the concept of oil as a relatively scarce resource is a mistaken notion that will harm the Soviet economy. He calls for further increase in the share of both oil and gas to enrich the country and produce a new and higher level of technology.[25] The equally optimistic First Secretary of Tyumen' Oblast's Party Committee* criticizes the timidity and earlier low estimates concerning West Siberia's oil production potential. He reveals the "seat of the pants" methodology of the official targets, then cheerily proceeds to supply his similarly unsubstantiated but highly optimistic figure.[26]

In the case of gas, where confidence is based on vast, proved reserves, supporters of accelerated development argue that comparisons of returns on incremental capital between natural gas and coal have become academic. In contrast to earlier periods, the incremental need for fuel has grown so great that it simply cannot be met without the pivotal role and sharply increased contribution of gas in the fuel balance. Comparable additions to energy output from coal are impossible to achieve.[27] Nor is there much sense, according to these pro-gas forces, in restricting the use of gas in electric stations and other boilers. Soviet resources are so vast that without further rise in the use of gas for the generation of electricity, it will not be possible to consume the output. The environmental and economic benefits from natural gas are so great that to turn back to coal instead would be a great mistake. The Tyumen' Oblast chief declares as irrational any crash program to develop, process and transport (or transmit by wire) Siberian lignites for energy-deficit regions in the European USSR. In his view, this would only divert scarce capital from the economically and ecologically more effective gas and oil industries in West Siberia.[28]

Others are more cautious. Scientists and officials more closely connected with and responsible for the actual production of hydrocarbons appear circumspect and prefer to hedge their bets. While research geologists stress the vast hydrocarbon potential and party whips push for higher output goals, operating

*Tyumen' Oblast covers most of the West Siberian lowland and contains the overwhelming portion of its gas and oil resources.

officials seem to prefer slower rates of growth. Not only would that improve plan fulfillment, but it would enhance ultimate recovery and reduce field damage and waste of which producers lately have become accutely aware. The cautious outlook of the Minister of the Gas Industry and, even more, of the chief of the Scientific Council for Production Problems of Petroleum Fields in the Academy of Sciences, both of whom want to economize on the use of hydrocarbons; the undisguised concern of the late oil minister, Shashin, and the chief of a major West Siberian oil-producing administration, all can be documented.[29]

Equally important is the concern of scientists in the domain of gas consumption. These scholars have misgivings about the current pattern of gas utilization and the direction of development with respect to efficiency. Their concern is not based on perception of a shortage, but on calculations about the low economic gains that can be derived from the vast quantities of gas burned as boiler fuel, when far greater benefits could be reaped in smelting, glass-making, petrochemical synthesis and the like.[30] These calculations would thus favor the substitution of more abundant gas for scarcer oil in such technological fields rather than in power stations and, more importantly, would argue for the reduction of both gas and oil in the boiler market. The latter implies a corresponding rise in the requirement for coal, but an inquiry into that course is not in the domain of scholars concerned with the most efficient use for natural gas.

The forces that press for a more vigorous commitment to coal and for a more significant role of this fuel in the energy balance seem to be found entirely in research institutes and, as a rule, not under the jurisdiction of the coal industry. Industry administrators appear strangely silent, refraining from prognostication about the development of the coal industry. Aside from one statement by the minister himself, intimating a 1990 output short of one billion tons, there does not appear to be a single forecast, appraisal or recommendation made public by the industry concerning its prospects. This cannot be from lack of adequate information. A Soviet source describes a massive study about the development of the coal industry until 2000. The work, consisting of a total of 109 volumes, was completed in the first half of the 1970s by the Central Scientific and Economic Institute for Coal (TsNIEIUgol') and the various mine-design institutes. It approaches the problem both from the demand and from the supply side, details the current state of the industry and presents alternative paths of development, giving consideration to technological change as well.[31] Nothing of that study has been made public; nor have there been published references to its conclusions and results by officials in charge of production and planning.

By contrast, vocal supporters of a strongly coal-oriented energy strategy appear among researchers associated with electric power institutes, sections of Gosplan for regional development, and energy institutes and sections of the Academy of Sciences, particularly its Siberian branch. Z. F. Chukhanov, Ya. Mazover, and N. V. Mel'nikov and the late A. Ye. Probst are among the leading figures who have pressed for greater reliance on coal, but half a dozen younger men at Siberian research centers might also be listed.[32] These are all respected scientists and scholars, but removed from concrete responsibilities of production and plan fulfillment.

The pro-coal forces are aware that the possibilities of increasing solid fuel output in the European USSR are limited and, as in the case of hydrocarbons, expansion must take place overwhelmingly east of the Urals (Siberia and Kazakhstan). Because of larger labor and social overhead requirements and locationally associated production linkages, a coal-oriented energy strategy is more conspicuously, unequivocally an eastern (primarily Siberian) development strategy than one that would continue to favor oil and gas. This is particularly true since the cheapest, rapidly expandable and vast lignite resources of Siberia, mostly in the Kansk-Achinsk Basin, are nontransportable beyond short distances without radical processing. Even widespread local use is difficult without enrichment. The need for huge enrichment plants, doubling capital and labor requirements,[33] would greatly increase backward linkages and the employment multiplier for the region despite the low capital and labor intensity in surface mining. The alternative solution of "coal by wire" projects, with several 6.4 million KW power plants and extra high voltage power lines, is technologically still unfeasible and would be even more capital-intensive, though probably less labor-intensive. The long-term commitment of investment and manpower resources for the production and processing of Siberian lignite on such an immense scale is highly controversial, given the many huge projects already under way east of the Urals.[34] The growing capital and manpower shortages and high labor turnover, especially in Siberia, combined with other economic and demographic problems, also make it doubtful. Some specialists acknowledge that the full-scale utilization of eastern coal resources is beyond the strength of the Soviet state alone. They would like to interest Western countries in technological cooperation to turn these resources into transportable synthetic fuel oil, high-calorie gas, methanol and other derivatives of coal processing both for the Soviet and Western markets.[35]

The issue of the massive development of Siberian coal resources for the national market clearly goes well beyond being only an energy policy question and choice. It becomes intertwined with problems of the territorial dimensions and structure of the economy as a whole. Because of their large investment requirements and size, all energy and other resource projects share this characteristic to a degree. Hydrocarbons, however, generate most of their territorial production linkages around downstream facilities (e.g., refineries) in established market areas, and the smaller, more mobile labor force engaged in their extraction also has a smaller regional impact. The territorial structure of an entire economy is subject to tremendous inertia and can change only slowly, but the local and regional consequences (economic and social) of individual projects can be considerable, just as local and regional conditions have a strong influence on the viability of the project. The problem of regional proportions, development and policy thus impinge on and interact with issues of energy policy, strategy and prospects.

INTERACTION OF REGIONAL DEVELOPMENT AND ENERGY POLICIES

On both the macro-scale, concerning the spatial structure of the country as a whole, and the smaller areal scale, Soviet planners have long been more explicitly involved with issues of regional development and policy than decision-makers in market economies. First, because of immense size and uneven distribution of population and resources, the variables of geography play a more influential if not decisive role in economic development for the USSR than in nearly all other sovereign nations. Second, the Soviet state is multinational, with ethnic groups of different cultural background inherited from the colonial conquests of Imperial Russia. The questions of regional growth and levels of economic development, investment and project allocation cannot be isolated from the ethnic issue. Finally, the regional and ethnic problem, its historic role and resolution are imbued with a deeply ideological meaning. It is associated with Lenin's contribution to the theory of economic relations along territorial, ethnic lines. Under capitalism, according to this view, the unevenness of economic development engenders not only class antagonism and struggle, but equally sharp inequalities, exploitation and eventual confrontation between "metropolis" and "colony", developed and underdeveloped areas and their inhabitants, both on a world scale and within multinational empires. In the evolution of this theory, much of the stigma of the metropolis–colony dichotomy was even bestowed, though with considerable mollification, on the core–periphery division between developed regions of complex economic structure and sparsely inhabited, pioneering areas with nearly exclusive roles of resource exploitation and raw material supplies.

These ideological concerns have, in Jensen's words, imparted over time a high degree of consistency to Soviet regional development policy at the aspiration level.[36] The basic principles of Soviet spatial policy contained a mixture of economic, economic-political (or social) and squarely socio-political desiderata. They aimed at rapid industrialization as much as at the economically, socially and strategically more effective geographic distribution of economic activity.[37] The "ambiguities, not to mention contradictions, in the formulations of these locational principles have allowed for a wide range of 'theoretically' based choices in selecting from among alternative courses of action. This circumstance has provided some operational flexibility in dealing with changing geographic situations so far as regional policy is concerned. Perhaps the major influence of ideology in this regard has been its tendency to minimize locational and environmental constraints. It has (also) highlighted social objectives relative to purely economic factors,"[38] though lately the concerns for social objectives can be shown to have diminished considerably.

Because the expansion of energy supplies is crucial to industrialization, energy-policy decisions have been intertwined with regional developmental issues from the beginning of the Soviet era. Lenin's dictum about Communism being Soviet power plus the electrification of the entire country was unquestionably accepted and considered to have geographic as well as sectoral and social dimensions. The famous Goelro Plan (1920) for electric power development over the subsequent 10 to 15 years was concurrently a plan for regional development on the basis of electrification and heavy-machine industries. The contours of regions were to be mainly determined by the location of central electric stations and the shape and carrying capacity of high-voltage grids.[39] The construction of a modern fuel and electric power base for each backward ethnic group or underdeveloped region has been regarded as a major vector of economic development.[40] The share of electric power generation, together with that of machine-building and, later, the chemical industries, has also been viewed as an important indicator of the economic level achieved by republics and economic regions.[41]

In the past, the theme of equalizing development among ethnic areas has tended to be emphasized or at least acknowledged in statements about regional policy. More recently, that goal has disappeared from among the explicitly declared objectives. Indeed, according to Leonid I. Brezhnev, the Soviet leader, that problem has essentially been solved and planners must now concentrate on the economic problems and the cost-effectiveness of the national economy as a whole. "The focus of attention has shifted to stress the development of about a dozen specific regional complexes, in particular, and the priority development of the eastern regions in general," without consideration of particular ethnic areas.[42] All but one of the major regional economic development complexes listed among the meager information made public about the long-range development plan until 1990 are located in the Asian USSR. Similarly a dozen of the 15 or so geographically more narrowly defined "territorial-production complexes" whose development is under way or in an incipient stage are situated east of the Urals and the Caspian Sea. At least 10 of these are in Siberia and adjoining northern Kazakhstan, in that resource-rich but harsh and sparsely inhabited land, clearly part of the Great Russian cultural realm.[43] To provide the scientific and technological underpinning for that eastern thrust, the Siberian Division of the Academy of Sciences is in the process of drawing up a long-term "Siberian Program", consisting of 24 subprograms, for the comprehensive development of the region's natural resources.[44]

In the Soviet view, territorial-production complexes (TPC's) are formed by the locational association of technological linkages generated by a specific production cycle. In the original formulation of the concept, the stress on the technological form of the linkages resulted in the near-complete neglect of economic aspects and, consequently, also of the important role played by supporting services and infrastructure unless they were an integral part of the production cycle itself. More recently, some scholars have come to admit the significant and more than passive role of such supporting activities and argue for a broader definition. As might be expected in a centrally planned economy, where spontaneous growth impulses are stifled, the Soviet literature on these areal complexes also shows no concern with the propulsive, spreading mechanism of development via the system of industrial linkages.[45] Soviet scholars, however, focus attention on the nature of the principal, region-forming or specialization sectors of such complexes versus the associated industries and services and those that meet local needs. The principal or region-forming industries are said "to determine the place of a region in the national and international division of labor" in a manner strongly reminiscent of Western export-base theory, though with centrally set production goals and planners' priorities taking the place of national demand with sufficiently high income elasticities. These industries are said to be those "that make the most efficient use of natural and economic conditions for the production of a commodity [or commodities] for the national and international markets."[46]

All but one or two of the areal complexes now emerging in the Asian USSR are associated with the energy production cycle. Fuel and electric power production and energy-intensive enterprises in close proximity (electrometallurgical, electrochemical, pulp and paper) form the backbone of these complexes. The others are based on the extraction and processing of copper, other metals and phosphoric minerals, which are also heavy power consumers but whose location is oriented to raw-material rather than power sources. Where oil and gas dominate the production complex, as in Tyumen' Oblast (West Siberian lowland), the preponderance of the energy industries is particularly striking. As already noted, the mobility of these fuels compared to other energy forms generally results in the agglomeration of associated consuming industries at some distance from the often isolated and harsh regions of hydrocarbons extraction. In 1973–75, some two-thirds of all industrial fixed assets and almost one-half of industrial output in Tyumen' Oblast were accounted for by oil and gas extraction and the electric power capacity serving it.[47] Nor is it likely that this share will diminish significantly in the future with the further development of the production complex.[48]

By contrast, local consuming industries inevitably form a crucial part of the complexes associated with the hydroelectric projects and lignite development in the Yenisey-Angara-Baikal zone of southern Siberia and with the surface mining of low quality coal in northeast Kazakhstan. The long-distance transmission and transport of these energy sources are technologically far more problematic and costly. Consequently, the close geographic association of vertically linked industries is imperative. In addition, the more favorable, if still harsh, natural environment, a better developed infrastructure and larger population are giving rise to a greater variety of supportive and more peripherally linked activities than in the primeval swamps of the West Siberian plain. Throughout the 1960s, less than a third of all capital investment in the industrial sector of East Siberia went into the production of energy; the rest flowed into energy-consuming and supporting industries. At the beginning of the 1970s, energy production as such contributed only one-seventh to total industrial output compared to almost one-half in Tyumen' Oblast in 1975.[49] No quantitative data are available concerning the individual production complexes in the Yenisey-Baikal zone, but qualitative descriptions and schematic presentations indicate that they are destined to be more broadly developed and multifaceted than the complexes of Tyumen' Oblast.[50]

The scale and complexity of regional development in pioneering areas, remote from established centers of economic activity, have long been vexing and debated issues in Soviet literature. Linked to these issues are considerations regarding the speed of development, the nature and location of the supply bases and the choice of transport modes. Obviously, the more remote, unpopulated and environmentally rigorous a pioneering province may be, the more difficult and expensive it becomes to strive for a complex, multifaceted economic structure and the less likely is it to succeed. Similarly, the more pressing the nation's need for a given resource or product of a remote region may be, the more likely is a crash attempt to make it available, a strategy that, by definition, cannot result in balanced development and may preclude it even for the long run.

Thanks to its better climate, considerable agricultural land and the Trans-Siberian railway, the southern belt of Siberia is both more habitable, and, especially west of Lake Baikal, more accessible to the established centers of economic activity than regions farther north. In diminishing degree from west to east, the Ural-Baikal zone has already become an integral part of the vital economic triangle of the Soviet Union, a part of the nation's ecumene. Vast though the natural riches of this zone may be, oil and gas so far have not been among them and the resources present have tended to be more expensive and difficult to transport and/or require local use and processing. Since World War II, these resources, with the exception of grain from the Virgin Lands, also did not seem to appear quite so vital and indispensable to planners as hydrocarbons from the wilderness of the West Siberian northland. For all these reasons, the development of the south Siberian belt in recent decades has been proceeding in a relatively measured though deliberate fashion, largely free of crash programs and with some attention given to a degree of balance among economic activities, the creation of regional industries and the long-term needs of a permanent population.

The drastic slowdown in the growth of the labor force, the inability of planners to adequately control and channel migration and a clear shift of emphasis in Soviet planning toward specialization and regional complementarity are all affecting development plans for southern Siberia, especially for its more distant regions east of the Kuznetsk Basin. Mass in-migration is no longer envisaged, labor-intensive industries are now played down, automation and labor-saving technology are stressed. In Jensen's words, there is "a growing acceptance of a classic core–periphery relationship between West and East, with Siberia supplying raw materials and energy to the more developed European regions" for fabrication and further processing.[51] The areal separation of fuel-energy and raw material versus labor-intensive processing-fabricating activities, however, is not a simple matter in many groups of industries and some compromise will continue to be necessary. In addition, the momentum of previous investments in a broad range of activities, an existing, if rather skeletal, urban network and the presence of nearby China, will all combine to make southern Siberia, even east of the Kuzbas, something more than purely a raw-material, energy and power supplying colony of the European USSR.

It is otherwise with the development of the central and northern zones of Siberia. Here the change in Soviet attitudes, strategies and methods has been clearly marked. Robert North of the University of British Columbia, who has examined the problem in depth, has found "that both the conditions for remote area development and accepted criteria for evaluating them are coming closer to those in northern North America." The Soviet view is gradually shifting toward the prevalent American opinion, which is unenthusiastic toward complex Arctic development with large towns, though, perhaps, the analogy should not be taken too far.[52] In northwest Siberia, where the richest resources of Soviet oil and gas are now located, the pressure to produce hydrocarbons for the national and export markets has become so great and the size of these reserves so large that a crash development for these resources alone was decided to be both justified and clearly necessary whatever resources may or may not be exploited later on. At the same time, the rapid development of hydrocarbon resources was found to be possible with a smaller labor force than was thought to be necessary earlier. The current strategy stresses the need to minimize manpower requirements. For the northern part of the West Siberian plain, with its giant deposits of natural gas, the "tour-of-duty" method now holds great interest, whereby employees are flown in from southern base cities to makeshift settlements for a predetermined period and then returned for rest and recreation before their next tour. However, even farther south along the middle Ob', with larger settlements and more diverse development, serious shortcomings in the construction of housing and social infrastructure have resulted in an increasing emphasis on a highly capital-intensive, automated and labor-saving technology.[53]

North's excellent study was published when the crash development of the more accessible oil fields of Tyumen' Oblast was already under way but the exploitation of the more northerly gas fields had barely begun. Similarly, the debate over complex, diversified versus narrow, exploitative development was still very vocal. Since the time of that study, the swiftest possible development of the north Tyumen' gas fields has also become a national priority. The performance of the Soviet energy sector in the 10th five-year plan, and for some years beyond 1980 as well, clearly hinges on hydrocarbons from Tyumen' Oblast, and on natural gas at least as much as on oil. At the same time, as the supply of manpower continues to tighten, the shift away from the strategy of complex development of remote areas appears to have accelerated. A Soviet source reveals that the traditional method of providing permanent residence for all employees and their families in the oil and gas industries, supporting industries and service activities would require a minimum increase of 600,000 in the urban population of Tyumen' Oblast by 1980.[54] Though this approach would still fall short of complex development since it places almost exclusive emphasis on only two related resources, such a strategy is proving clearly unfeasible. The author suggests that the expedition method be added to the tour-of-duty method on a significant scale. The expedition technique entails flying in workers from old oil and gas regions in the European USSR to work temporarily in the oil and gas fields of West Siberia. The method, already being applied, reduces the influx into the base cities of Tyumen' Oblast, where the infrastructure and social services are already strained beyond capacity.[55] The more diversified development strategy with well-serviced, permanent cities near the new gas and oil fields still has a few defenders and may still be regarded as a Soviet ideal.[56] It is clear, however, that demographic and geographic realities and the urgency of national energy needs have overtaken Soviet planners. As these strains increase, the likelihood that ideology and central planning will continue to sustain a developmental strategy for the Soviet North fundamentally different from that followed in the northlands of the West is diminishing every year.

So far neither the champions of complex development nor those of the narrow resource-exploitative strategies could answer the question of how best to implement regional objectives in the prevailing framework of vertical, ministerial administration and planning. The current emphasis on the territorial-production complexes has done nothing to solve that old problem. Indeed, it may have made the issue more acute, for these entities serve to emphasize regional objectives without providing any mechanism for coordination between ministerial and territorial planning. By providing clearly defined foci for regional objectives, the complexes furnish a stage for inter-industry and inter-agency conflicts, as is shown by the experience with the South Yakutian complex, described earlier in this chapter.

Criticism concerning the lack of effective regional planning has always been widespread but lately complaints "have reached a crescendo. Regional planning still seems to amount mainly to the adding up of the relevant parts of plans drawn up by the various ministries," whose desires have priority.[57] And, as shown, the implementation of these regional plans frequently amounts to some ad hoc resolution of conflict among the participating ministries or their departments. The Soviet press has repeatedly carried proposals for the creation of supradepartmental agencies for the integrated management of territorial-production complexes. However, the late Mikhail Pervukhin, chief of the Department of Territorial Planning and the Location of Productive Forces at Gosplan, did not think the idea to be feasible. Given the branch management of the economy, for enterprises and construction trusts to receive directives from two superiors would wreck havoc with the economy.[58] In all likelihood, "Soviet regional development over the next decade will proceed much as now, with much ad hoc planning and implementation, with many projects started but completed only after inordinate delays and skyrocketing costs, with inconsistent scheduling and completion of interdependent projects, with trials and tribulations, but with some spectacular successes and with some forward motion on many fronts."[59]

NOTES

[1] Robert W. Campbell, *The Economics of Soviet Oil and Gas* (Baltimore: Johns Hopkins Press, 1968), p. 25.

[2] G. G. Khorev, *Ekonomika otkrytoy dobychi uglya* (Moscow: Nedra, 1974), p. 5 and *Narodnoye khozyaystvo SSSR za 60 let* (Moscow: Statistika, 1977), p. 206. Data refer to 1972, but there does not seem to have been any change in the degree of concentration since.

[3] *Ekonomicheskaya Gazeta*, No. 34, August 1976, p. 13 and *Ekonomika ugol'noy promyshlennosti*, 1976, No. 9, p. 34.

[4] *Ekonomika neftyanoy promyshlennosti*, 1976, No. 7, p. 47–48.

[5] *Ekonomicheskaya Gazeta*, No. 3, January 1977, p. 2.

[6] This paragraph is based primarily on the brief description of Soviet government structure and accompanying detailed chart in Paul K. Cook, "The political setting," in U.S. Congress, Joint Economic Committee, *Soviet Economy in a New Perspective* (Washington: Government Printing Office, 1976), pp. 12–14.

[7] For example, 59% of all geophysical work in the USSR today is carried on for

petroleum and natural gas and four-fifths of such work is done by the Ministry of Geology and most of the rest by the Ministry of the Oil Industry. V. V. Fedynskiy, in *Vestnik Moskovskogo Universiteta. Geologiya*, 1977, No. 5, pp. 119–120. As to the railroads, almost a third of their total freight turnover, as measured in ton-kilometers, and the same share of their total freight, as measured in tons, are accounted for by coal, coke, petroleum products and crude oil. In the mid-1960s, these shares were somewhat higher still. *Narodnoye khozyaystvo SSSR za 60 let*, pp. 392–393.

[8] *Vneshnyaya torgovlya SSSR v 1976 godu* (Moscow: Statistika, 1977), pp. 18–19.

[9] P. S. Sapozhnikov and G. D. Sokolov, *Ekonomika i organizatsiya stroitel'stva v neftyanoy i gazovoy promyshlennosti* (Moscow: Nedra, 1976), pp. 42–43.

[10] *Sotsialisticheskaya Industriya*, March 12, 1977, p. 2.

[11] *Pravda*, February 27, 1977, p. 2; May 7, 1978, p. 3 and June 5, 1978, p. 2. At various times there has been talk at Gosplan USSR about forming an authoritative, interdepartmental body to coordinate the overall development of natural resources in such crucial areas as the Ob' Basin.

[12] *Pravda*, Feb. 7, 1977, p. 2 and *Bakinskiy Rabochiy*, April 7, 1977, p. 2.

[13] *Sotsialisticheskaya Industriya*, Nov. 19, 1977, p. 2.

[14] If the ministries themselves are so called union-republic ministries, such as those for coal, petroleum refining, electric power and geology, rather than all-union ones, represented by the others in the energy sector, two stages of hierarchy exist already at the ministerial level. All-union ministries are located only in Moscow and supervise production throughout the country without republican-level intermediaries. Union-republic ministries have a central headquarters in Moscow and subordinate ministries in one or more of the republics. See example of the coal industry further in the text. In addition, there are republic ministries, which administer industries of purely local significance.

[15] Alice C. Gorlin, "Industrial reorganization," in U.S. Congress, Joint Economic Committee, *Soviet Economy in a New Perspective* (Washington: Government Printing Office, 1976), pp. 162–188 and Brenton M. Barr, "The changing impact of industrial management and decision-making on the locational behavior of the Soviet firm," in F. E. Ian Hamilton, *Spatial Perspectives on Industrial Organization and Decision-Making* (London-New York: John Wiley, 1974), pp. 411–446, particularly pp. 432–433.

[16] *Sotsialisticheskaya Industriya*, Feb. 16, 1977, p. 1, and May 25, 1977, p. 2.

[17] *Geologiya nefti i gaza*, 1977, No. 4, pp. 5–9 and *Gazovaya Promyshlennost'*, 1977, No. 5, p. 8.

[18] Campbell, op. cit. (1968), pp. 25–26.

[19] Robert W. Campbell, "Price, rent and decision-making: the economic reform in Soviet oil and gas production," *Jahrbuch der Wirtschaft Osteuropas*, Band 2, 1971, p. 294.

[20] Ibid., pp. 293–298.

[21] *Material'no-tekhnicheskoye snabzheniye*, May 1977, No. 5, pp. 35–36.

[22] Paul K. Cook, op. cit., pp. 13–14.

[23] *Soviet Geography*, November 1977, pp. 699–700.

[24] Trofimuk is cited in I. Ognev, "Attainment of a discovery," *Ekonomika i organizatsiya promyshlennogo proizvodstva*, 1976, No. 4, pp. 174–179. The entire discussion by the journalist Ognev (Ibid., pp. 154–180) is fascinating and reveals much of the deep cleavage of opinion among experts concerning future fuel policy and of the zigzags of such a policy in the past. In fairness to Trofimuk, it must be pointed out that he, too, has long been concerned about the increasing disproportion between prospecting and exploratory work and developmental drilling, especially in Siberia. *Sotsialisticheskaya Industriya*, Sept. 13, 1974, p. 2.

[25] Ibid., pp. 170–172 and 174–179.

[26]G. P. Bogomyakov, "The Tyumen' complex and its future," *Ekonomika i organizatsiya promyshlennogo proizvodstva*, 1976, No. 5, pp. 9–11.

[27]G. Z. Khaskin, I. Ya. Furman and V. Ya. Gandkin, *Oznovnyye fondy gazovoy promyshlennosti* (Moscow: Nedra, 1975), p. 129.

[28]Bogomyakov, op. cit., pp. 5–11.

[29]*Gazovaya Promyshlennost'*, 1976, No. 11, p. 6; Ognev, op. cit., p. 171; R. I. Kuzovatkin, "The front and the rear of Samotlor," *Ekonomika i organizatsiya promyshlennogo proizvodstva*, No. 6, 1976, pp. 80–85; *Petroleum Economist*, March 1973, p. 80 and January 1975.

[30]Ye. N. Il'yina and L. D. Utkina, "A cost-effectiveness scale of natural gas use," *Gazovaya Promyshlennost'*, 1978, No. 6.

[31]A. S. Stugarev et al., *Prognozirovaniye razvitiya ugol'noy promyshlennosti* (Moscow: Nedra, 1976), especially pp. 64–73.

[32]Z. F. Chukhanov, "Scientific-technological problems of development of the fuels and energy complex of the USSR," Akademiya Nauk SSR, *Vestnik*, 1976, No. 9, pp. 104–119; Ya. Mazover, "Prospects of the Kansk-Achinsk Basin," *Planovoye Khozyaystvo*, 1975, No. 6, pp. 65–73, and "What kind of a complex is it to be," *Sotsialisticheskaya Industriya*, Aug. 13, 1975, p. 2; *Izvestiya*, July 9, 1975, p. 3; A. Ye. Probst and Ya. A. Mazover, eds. *Razvitiye i razmeshcheniye toplivnoy promyshlennosti* (Moscow: Nedra, 1975), especially pp. 137–138; A. Krichko and L. Semenov, op. cit., pp. 96–98; G. A. Andreyev et al., "Peculiarities of the external linkages of fuel production bases," in *Voprosy vliyaniya razvitiya energetiki na drugiye otrasli narodnogo khozyaystva* (Irkutsk: Energetics Institute, 1975), pp. 29–45. In late 1976, Academician N. V. Mel'nikov insisted in an interview with the present author that Soviet energy production over the next 10 to 15 years must expand more rapidly than a 5% annual rate and that the share of coal in the fuel balance must begin to rise again. This would imply a rate for coal above that of aggregate energy output, an incredibly high rate of expansion.

[33]*Sotsialisticheskaya Industriya*, Nov. 19, 1977, p. 2, and S. K. Karyakin, "An investigation of the mineral composition of Kansk-Achinsk coals in connection with their use in power generation," unpublished candidate's dissertation (Tomsk: Polytechnic Institute, 1975), especially pp. 5, 9, 105 and 129–135. Andreyev et al., op. cit., pp. 30–31.

[34]The 9th five-year plan (1971–75) devoted 29% of total investment to the Asian USSR; the 10th five-year plan allocated almost 30%. These represent 145 billion and 198 billion rubles respectively, provided the investment targets of the 10th year plan are fulfilled. G. I. Granik, *Razmeshcheniye proizvoditel'nykh sil v SSSR* (Moscow: Ekonomika, 1977), p. 33. See also Note 43 and text.

[35]Moscow Narodny Bank, *Press Bulletin*, Oct. 13, 1976, pp. 3–4.

[36]Robert G. Jensen, "Soviet regional development policy and the 10th five-year plan," *Soviet Geography*, March 1978, p. 197.

[37]For an elaboration of these basic principles see F. E. Ian Hamilton, "Spatial dimensions of Soviet economic decision-making," in V. N. Bandera and Z. L. Melnyk, eds., *The Soviet Economy in Regional Prespective* (New York: Praeger, 1973), pp. 237–240 and F. E. Ian Hamilton, "The location of industry in east-central and southeast Europe," in George W. Hoffman, *Eastern Europe: Essays in Geographical Problems* (New York: Praeger, 1971), pp. 184–198; I. S. Koropeckyj, "The development of Soviet location theory before the second world war," *Soviet Studies*, July and October, 1967.

[38]Jensen, op. cit., p. 197.

[39]A. Lavrishchev, *Economic Geography of the USSR* (Moscow: Progress Publishers, 1969), p. 161 and V. Pokshishevsky, *Geography of the Soviet Union* (Moscow: Progress Publishers, 1974), p. 44.

[40] See, for example, K. M. Kim, *Sovershenstvovaniye struktury toplivno-energeticheskogo balansa Sredney Azii* (Tashkent: FAN, 1973), passim; V. A. Shelest, *Regional'nyye energoekonomicheskiye problemy SSSR* (Moscow: Nauka, 1975), passim; V. A. Shelest et al., *Problemy razvitiya i razmeshcheniya elektroenergetiki v Sredney Azii* (Moscow: Nauka, 1964) and A. Ye. Probst, ed., op. cit. (1968), passim.

[41] Feygin, ed., *Zakonomernosti i faktory razvitiya ekonomicheskikh rayonov SSSR* (Moscow: Nauka, 1965), pp. 74–81 and L. N. Telepko, *Urovni ekonomicheskogo razvitiya rayonov SSSR* (Moscow: Ekonomika, 1971), pp. 162–168.

[42] Gertrude E. Schroeder, "Soviet regional development policies in perspective," paper prepared for the 1978 NATO Colloquium, *The USSR in the 1980s: Economic Growth and the Role of Foreign Trade*, Brussels, 1978. The author also cites a long theoretical article in *Istoriya SSSR*, 1976, No. 3, pp. 3–21, which "develops the thesis that there are no longer any backward regions and the reduction of such gaps as do exist is to be subordinated to the common goal of 'building the material-technical basis for communism'." Schroeder, op. cit. (1978), Note 2.

[43] Ibid. See also A. T. Khrushchev et al., *Novyye promyshlennyye kompleksy SSSR* (Moscow: Prosveshcheniye, 1973) and *Adresa Desyatoy Pyatiletki* (Moscow: Molodaya Gvardiya, 1976) and Victor L. Mote, "Predictions and realities in the development of the Soviet Far East," Association of American Geographers, Discussion Paper No. 3, Project on *Soviet Natural Resources in the World Economy* (Syracuse University, 1978), pp. 29–42.

[44] As reported by Academician A. Trofimuk in *Izvestiya*, March 31, 1978, p. 2.

[45] For an excellent analysis of the subject, including a comparison with "growth pole" theory, see George A. Huzinec, "Some initial comparisons of Soviet and Western regional development models," *Soviet Geography*, October 1976, pp. 552–565. The author also presents a carefully selected bibliography of the most relevant Soviet writings.

[46] I. V. Nikol'skiy, "The role of economic sectors in the formation of regional economic complexes," *Soviet Geography*, January 1972, p. 17. See also the cogent discussion on the TPC's in Akademiya nauk SSSR, Sibirskoye otdeleniye, Institut ekonomiki, *Ekonomicheskiye problemy razvitiya Sibiri* (Novosibirsk, 1974), pp. 142–170.

[47] Tyumenskiy Industrial'nyy Institut, Nauchnyye trudy, Vypusk 43, *Kompleksnoye planirovaniye na promyshlennykh predpriyatiyakh* (Tyumen', 1975), p. 76 and Bogomyakov, op. cit., p. 7.

[48] The development of refining and petrochemical capacities (and that of the wood-processing industry with almost no technological links to hydrocarbons) at the southern limit of the West Siberian plain should increase the importance of associated (and other) industries in the future. By that time, however, dry natural gas output from the northern half of the plain will have accelerated very greatly. Most of this gas does and will continue to leave West Siberia immediately upon extraction.

[49] Akademiya nauk SSSR, Sibirskoye otdeleniye, Institut ekonomiki, *Mezhotraslevyye svyazi i narodnokhozyaystvennyye proportsii Vostochnoy Sibiri i Dal'nego Vostoka* (Novosibirsk: Nauka, 1974), p. 185. Bogomyakov, op. cit., p. 7 and Akademiya nauk SSSR, Sibirskoye otdeleniye, *Ekonomicheskiye problemy. . .* (1974), p. 38.

[50] *Ekonomicheskiye problemy. . .* (1975). See Note 46, pp. 144–147 and I. I. Belousov, *Osnovy ucheniya ob ekonomicheskom rayonirovanii* (Moscow: MGU, 1976) inset following page 168.

[51] R. Jensen, op. cit., p. 199.

[52] Robert N. North, "Soviet northern development: the case of NW Siberia," *Soviet Studies*, October 1972, pp. 171–199; reference on pages 198–199.

[53] Ibid., p. 197.

[54]A. Khaytun, "Socio-economic problems of development of the country's oil and gas regions," *Planovoye khozyaystvo*, 1977, No. 9, pp. 88–95.

[55]Ibid., pp. 186–189 and 197–198. The expedition method has now been adopted in the West Siberian oil industry on a large scale. Drillers flown in from the Tatar and Bashkir ASSR's, Kuybyshev and Saratov oblasts are planned to sink 700,000 meters of wells in Tyumen' Oblast in 1978. This amounts to one-tenth or more of all the planned meterage of drilling in the Tyumen' oil industry for that year. *Pravda*, June 5, 1978, p. 2.

[56]Such as V. Perevedentsev, the noted demographer and sociologist, heard at a public lecture in late 1976.

[57]Gertrude Schroeder, op. cit. (1978).

[58]*Literaturnaya Gazeta*, No. 7 (Feb. 18), 1976, p. 11.

[59]Gertrude Schroeder, op. cit. (1978).

Chapter 11

CONCLUSION

Economic development and industrialization have proven compatable with a wide range of political and social organizations. In very different societies, showing few signs of rapprochement, the exigencies of modern technology and industrial management have led to convergent efforts and similar techniques. In the energy sector, such structural influences, resulting in universal trends, problems and efforts, seem particularly strong. The relatively narrow range of products, restricted technical coefficients, the nature of fuel deposits and the like, affect and constrain decision-makers in a similar fashion everywhere.

SOVIET CONGRUENCE WITH WORLD TRENDS

Past and present trends in the Soviet energy economy have shown a general congruence with those of other countries at similar stages of development. The heavy industrial drive of the 1930s and the first postwar decade were associated with a huge expansion of coal-mining capacity, thermal and hydropower generation and transmission. Though far greater in intensity, this coal and steel-oriented drive was not unlike that in the United States before the 1920s. It is true that Soviet planners were slow in appreciating the advantage of hydrocarbons and petrochemicals, that excessive fear of investment risks, structural conservatism, but also military needs prolonged the dominance of coal and iron. The vanishing petroleum reserves of the Caucasus were not replaced by the new Volga-Urals finds until the middle 1950s, though exploration had obviously accelerated somewhat earlier. Most other parts of the world, however, were equally coal-dominated until the 1950s and, when it came, the Soviet shift to hydrocarbons was purposeful and rapid. Until recently, petroleum received most of the emphasis and the natural gas industry suffered from immaturity, but again the Soviet experience is hardly unique. In refinery activities and the use of petroleum, the structural needs of the economy prevailed over the philosophical dislike of burning oil for stationary uses, particularly under boilers. Refinery output in the Soviet Union came to resemble that in Western Europe much more than that in the United States, and for similar reasons. The inadequacy and high cost of coal resources in the European USSR, the country's economic heart, resulted in a huge demand for heavy distillates as industrial and power station fuel.

The increasing global dependence on oil products throughout the 1960s, combined with the tightening of easily available supplies outside the Middle East, culminated in the dramatic actions of OPEC and the radical transformation of the world petroleum market. During the same period in the USSR, a corresponding rapid growth in the economy's dependence on easily accessible hydrocarbons, combined with rising East European needs, also resulted in mounting pressure on supplies. Absolute increases of petroleum output during the second half of the 1960s failed to grow and declined significantly for gas. Yearly rates of increase dropped sharply for both fuels. The USSR extricated itself from that supply squeeze by the crash development of West Siberia's oil and, somewhat later, that region's natural gas resources (Chapter 3). However, the cost and enormous efforts required to make these fuels available for the industrial core of the country were also foreseen a decade ago, though the lead times needed to develop the vast reserves of gas were seriously underestimated.

Soviet energy planners, therefore, entered the 1970s with a more cautious attitude towards their hydrocarbons riches. In particular, a perception of the increasing value and relative scarcity of oil, when compared to other fuels, was beginning to be shared by a growing number of specialists. Publications of the early 1970s already contained projections of a stabilized and later decreasing share in the contribution of oil and a greater role for gas. The contribution of coal was expected to decline further, but, according to most experts, less rapidly than before, implying an acceleration in its rate of growth, particularly after 1975. The construction of large hydroelectric stations was to continue with renewed vigor. Finally, a decision was made to push ahead with the expansion of nuclear capacity in the European provinces, where a steep rise in the fuel deficit was expected. Significantly also, the USSR served notice on the Comecon countries that it would impose a ceiling on its oil deliveries to Eastern Europe.

It may be said, therefore, that the radical transformation of the world energy market found the USSR more or less in step. The new global perception of petroleum as a relatively scarce, increasingly valuable, economically and strategically critical resource was paralleled, even anticipated, in the Soviet Union. The quadrupling of world oil prices and the Arab embargo against the West in 1973 dramatically reinforced already present but still tentative ideas about the change in the direction of Soviet energy development. High-ranking officials began to call for curtailment of the use of petroleum (and even gas) under boilers, the accelerated development of surface coal deposits, nuclear power and the long heralded extra high voltage lines to bring Siberian power to the energy-hungry European USSR. Hydrocarbons were to be devoted increasingly to technological uses or exported to earn hard currency, though this crucial motivation was seldom explicitly stated in published literature. Concurrently with all these, the Soviet press began to stress the theme of energy conservation in all areas of the economy.

Development since the late 1960s show that Soviet planners underestimated the difficulties encountered in the new energy era. In that respect, too, they have not been far out of step with the world at large. They have underestimated the speed with which the output of hydrocarbons from remote and/or inaccessible areas and strata can be expanded and delivered to centers of demand. They have been overconfident about the growth of oil reserves and, outside West Siberia, about the growth of gas resources as well. Finally, they have underrated the cost and lead time required to lessen the dependence on petroleum by restructuring the energy balance toward a greater role of coal and nuclear power.

Since the late 1960s, the expected contribution of gas from northwest Siberia has been successively scaled down and the full development of these enormous fields pushed further into the future. Plans to replace exhausted oil and gas reserves west of the Urals, and even in Central Asia, by tapping deeper strata and offshore deposits, have remained underfulfilled. In 1970–75, for example, 63% of all exploratory drilling was allocated to the European USSR with disappointing results.[1] West Siberian oil resources, still remote but more accessible and more transportable than West Siberian gas, were indeed crash-developed to compensate for both of these shortfalls. Such crash response and adjustment, however, only underscore earlier overoptimism and miscalculations. Attempts to accelerate coal production so far have not borne fruit. Halfway through the 10th five-year plan (1976–80), the coal industry has fallen far short of its target and the long-term problems it faces in accelerating output are proving severe (Chapters 4 and 9). Finally, Soviet planners are also finding that nuclear power is no panacea, that its growth rate, while rapid, is slower than expected and its contribution to the energy balance will remain modest for the rest of the century.

Nor are Soviet planners finding it easier than Western decision-makers to effect a substitution of more abundant fuels for scarcer ones, at least in the short and medium term. The physical structure and design of fixed equipment, transport facilities and modes constrain both groups the same way. In particular, the shift toward greater use of coal in the Soviet Union is proving to be even more difficult than in the United States (Chapter 8). Conservation so far is bringing only minor relief, despite the centralized chain of command that should make implementation easier than in Western countries.[2] The sectors where demand can be limited by fiat without hurting production targets, namely households, private transportation and some services, are small consumers. Soviet managers are insensitive to price and, beyond some tightening of allocation norms, no ready instrument exists to save energy in the industrial and agricultural sectors.

Soviet energy demand is, on balance, structurally determined. It can be reduced only through shifts in the industrial structure, a vast effort of retooling, and/or far-reaching changes in production functions. There is little evidence of any systemic shift since World War II toward a less energy-intensive industrial structure in the USSR. Indeed, net (i.e., useful) energy consumed per unit value of national income rose rather than declined for the last two decades, even when Soviet rates for GNP are accepted. Positive changes in energy-related production functions have been confined largely to technical improvements effected by the shift from coal to hydrocarbon fuels and advances in conventional power-generating technology. Such improvements in efficiency are coming to an end and, for the next decade, are expected to be marginal (Chapter 2). Radically new technology, such as MHD (magnetohydrodynamics), cannot be expected to have an impact much before the end of the century. In the past, the Soviet system has not shown itself particularly good at rapid adjustments in production functions and technical coefficients over the broad front of the whole economy. At best, only slow and modest shifts in the structure and coefficients of energy use may be expected in the future.

In contrast to the erratic, capricious record of the Western nations, Soviet leaders deserve moderately high marks for the steady, purposeful and economically rational planning of their energy riches. They have shown awareness of relative scarcities and opportunity costs both on the world market and at home. Technological progress, industrial growth and modernization have evidently led to developments in the Soviet energy economy roughly analogous to world trends. And Soviet planners also have not been spared many of the problems and dilemmas that confront Western leaders. In particular, world concern over the growth of petroleum demand, supplies and costs found the USSR roughly in phase. Although Soviet energy reserves are vast, they are not cheap. Their exploitation requires long lead times and an ever larger share of the nation's capital and manpower resources, material products and research effort. In addition, proved reserves of oil appear unsatisfactory, and prospects for that pivotal commodity seem more worrisome than for other fuels. It is quite possible that the admittedly great potential will not be transformed into exploitable capacity in time to avoid serious consequences for the entire economic alliance of the Soviet bloc.

Having stressed that general developments in the Soviet energy economy have conformed to world trends, that the system shared certain universal problems, that the perceptions of Moscow planners concerning the role of different fuels are roughly congruent with global views, one must also emphasize that the "energy problem" for the Soviet Union assumes substantially different contours than for the Western nations, including the United States. While it is conceivable that the difficulties facing the Soviet energy economy in the 1980s will be acute, they will express themselves quite differently than in Western countries. The options available in the USSR are also broader; responses to problems can be made from a wider range of choices. These choices are not necessarily clear-cut. Most carry risks and require an increasing share of the nation's economic resources. For these reasons, major decisions of energy policy will not be any easier to make than in the West and may actually be more difficult.

SPECIFIC FEATURES OF THE SOVIET "ENERGY PROBLEM"

Compared to all industrial nations, and to most other nations as well, the USSR is in an enviable position concerning the magnitude and range of its energy resources. In both total and individual energy sources, the Soviet Union is endowed with fuel reserves and hydroelectric potential much larger than any Western nation. With the probable exception of Canada, this holds true on a per capita basis as well. Despite the energy-intensive nature of Soviet industrialization, the USSR has also exploited and depleted its potential less extensively than the Western countries, whose national territories are smaller and whose industrialization began earlier.

In this "resource vault", Soviet gas reserves assume a particularly critical role. Hydrocarbons are the most valued, most sought-after sources of primary energy today; they are also the most risky and exhaustible fuels, for which reserve additions often do not materialize and which are therefore difficult to plan for. The Soviet Union's massive reserves of gas in the proved and indicated categories thus represent a solid, central pillar of long-term energy policy, missing in West Europe and North America. The frequently touted large gas resources remaining in the United States as well as in Canada are less tangible; only a fraction of them may ever be transformed into the proved, recoverable category even with much higher prices. While, in the USSR, location (northwest Siberia), distance and transport bottlenecks impose severe limitations on the annual increments of gas output, these supply increments are far larger than anywhere else in the world. The size of proved reserves and steady improvements in transport technology will guarantee that they will remain so for the rest of the century (Chapter 3). An expanding market in West Europe insures a long-term and increasing role for natural gas in the Soviet Union's export plans. The growing importance of gas cannot entirely compensate for possible shortfalls in the output of other fuels, particularly oil. Natural gas, however, is the ace in Soviet energy plans and provides a critical cushion for the uncertainties faced by planners with respect to other sources of supply.

The sheer size and favorable composition of its energy riches thus sets the Soviet Union apart from its Western counterparts, even the United States. While other developed and most developing countries rely heavily on energy imports, mostly in the form of oil, the USSR is the third largest exporter of petroleum and also of total energy. The Western nations spend billions on imported petroleum and have become vulnerable to any interruption of flow. The USSR earns large amounts of hard currency from the sale of hydrocarbons and its supplies are strategically secure. The Western world faces the prospect of still greater import dependency; the USSR may face an erosion of its export capabilities but will almost certainly remain self-sufficient and retain its strategic independence (Chapters 8 and 9).

Unfavorable developments on the energy scene are affecting the relationship among members of both the Western and Eastern alliances, but, again, in a different fashion. The Soviet Union is the primary and direct supplier of hydrocarbons for the rest of the Comecon. It provides a vital and, at current world prices, irreplaceable economic prop for the entire Eastern bloc, though, of course, at a political price. Soviet domestic requirements and East European needs are in direct conflict. The economic burden on the USSR to maintain stability among its partners is substantial and obvious. In the West, except for some coking coal, the United States is a negligible source of fuel for West Europe and Japan. Conflicts of energy policy involve competition for OPEC supplies, with the resulting impact on prices, balances of payment and world monetary stability. The steadily growing quantity of American oil imports, already by far the largest in the world, is a threat to the United States' allies, whose domestic resources are more limited and whose efforts at conservation have been more noteworthy. With current trends continued, the burden and strain on the Western alliance will intensify. While Moscow's relationship with its Comecon partners is hardly benign, Soviet decisions in the sphere of energy show a significant awareness of international consequences and the impact on East Europe in particular. The glacial pace of United States moves toward a coherent energy policy attested to a surprising lack of such perception.

As shown, the Soviet "energy problem" is rather different and, in some ways, the mirror image of the one confronting the West. It is not a problem of import dependence but of continued export capabilities; not a problem of competing with allies for petroleum on the world market but of supplying Comecon partners in the face of rising domestic demand and the need for hard-currency earnings. And it is less an issue of precarious hydrocarbon potential yet to be found or of reserves unrecoverable at current prices. Uncertainty does exist concerning petroleum resources (though prospects are brighter than in any member state of the Organization of Economic Cooperation and Development), but enormous proved reserves of gas go a long way to compensate for it and make long-term planning possible. The Soviet "energy problem" is basically that of location and distance, of a crippling lack of spatial congruence between demand and resources, actual as well as potential.

In facing such a version of the "energy problem", Soviet planners operate in a political-institutional environment different from the one found in Western countries. They also operate with a technology factor that makes their position quite distinct and presents them with a political dilemma peculiar to closed societies. On balance, the first probably expedites the development of energy potential, the second retards expansion and is likely to do so for a number of years.

While the administrative organization of the Soviet energy economy, though more centralized, has much in common with the corporate structure of its Western counterparts, the Soviet system operates quite free from popular pressures and public concern. The policy struggles present in the USSR take place within the ministerial arrangement, among the administrative bodies responsible for energy development (Chapter 10). Public awareness and articulation of social and environmental issues related to energy development are subdued. When they appear, they often do so after the fact, after plant completion, when remedial actions may be taken. The need for environmental impact statements appears to be less stringent in the Soviet Union than in the

United States. Still less are there injunctions sought by citizen groups and granted by courts against planned and ongoing projects.

The thorough Soviet commitment to nuclear energy and the development of the breeder reactor have spurred no public debate, not even mild questioning. Characteristically, Soviet nuclear plants have had no outer protective shell and they have been put on stream in roughly half the time required for such projects in the United States in recent years (Chapter 6). Safety features are designed essentially for normal operation and "little credence is given to the possibility of a loss-of-coolant accident." There seems to be a slow change in attitude toward a greater appreciation of safety (the Novovoronezhskiy-5 1000 MW reactor does have a containment structure) and toward the ecological problem of the whole fuel cycle.[3] Still, it is safe to say that what constrains the pace of expansion in nuclear power generation is not public or even scientific ambivalence or controversy, but simply bottlenecks in equipment capacity and possibly skilled labor.

Nor does public attitude toward energy producers and the visceral issues of prices, subsidies and proper profit margins have any influence on resource development and policy in the institutional environment of the USSR. Prices and profits are basically accounting devices, not used as direct instruments of investment and labor allocation. Over the past decade, energy prices have tended to express relative scarcities and marginal utility tolerably well and in most ways have been more sound economically than those in the United States (Chapter 8). But apart from retail prices for the small household-communal sector, where they have an impact on conservation, they do not affect energy policy or the direction and proportion of resource development.

If, for better or worse, the nearly complete absence of the public voice in the respective Soviet institutions makes energy development more expeditious than in the West, a relative backwardness of technology restrains expansion considerably, though unequally in different energy sectors. As in the past, this relative backwardness continues to offer Soviet planners opportunities for technology transfer. Because of the rising cost and magnitude of such transfer, however, the latter presents the leadership with increasingly hard choices, which they may be unable or unwilling to make. As in the past, large-scale technology imports will have to be paid for mostly by the sale of natural resources, among which hydrocarbons will continue to be pre-eminent. Coal, particularly from eastern basins, where nearly all increments must originate, is not an exportable commodity on a large scale. Nor can nonfuel minerals and forest products even remotely approach oil and gas in their hard-currency earning potential.

There is general agreement that among the USSR's energy industries, the electric power branch is technologically the most advanced. Compared to their counterparts in leading Western nations, especially the United States, the Soviet oil, gas and coal industries are further behind. Because of the huge reserve cushion, the leadership has been willing to mortgage large quantities of natural gas in long-term compensatory agreements for Western pipe and equipment to speed up the production of this fuel. Under such contracts, gas exports in the 1980s may surge still more. For several years, however, hard-currency earnings from natural gas cannot hope to equal those derived from the sale of petroleum, let alone provide both increased earnings and compensate for possible declines in oil exports. In contrast to gas, the USSR so far has been reluctant to make long-term commitments on petroleum deliveries. The protracted negotiation and prevarication with Japan in the first half of the 1970s over the sale of West Siberian oil (now apparently a dead issue) has already proved that. Since that time, the problems and uncertainties facing the oil industry have increased and so has Soviet reluctance to mortgage this resource in long-term contracts.

The most significant development specific to the Soviet energy situation, therefore, is yet to come. Barring some extraordinary luck (i.e., the discovery of one or two supergiant oil fields in reasonably accessible provinces), Soviet energy policy is approaching a divide. Whichever road the leadership is likely to take, the choice will imply profound consequences for the Soviet economy as a whole and probably for the entire Comecon system.

Of all industrial countries, the USSR alone has the option again to turn inward and cut itself off from world trade if determined to do so. The country does have the technical ability to develop its riches alone or with some Comecon help, albeit at a slower pace than could be achieved by a massive surge of technology infusion from abroad. Thanks in particular to its mammoth gas reserves already on the shelf and the easy substitutability of gas for oil in most stationary uses, the USSR could weather the peaking, probably even a downturn, of petroleum production. Nor should it be forgotten that, given the size of Soviet sedimentary basins and the relatively low level of exploration, the oil industry is unlikely to have yet passed its mid-point. The Russians can continue to discover and produce more oil without foreign technology even if at a much slower pace than before. Even with a downturn, the collapse of the industry is not imminent. With the elimination of oil exports to the West, the Soviet Union can still continue to earn enough hard currency to avoid a disaster in case the harvest fails. It will also continue to receive pipe and equipment for the gas industry under barter contracts already in effect and likely to be enlarged. Most other hard-currency imports, however, would have to be cut. In the longer run, the processing and transport and therefore massive use of Siberian lignite should also become possible since Soviet research on this problem is quite advanced.

The cost of this option in economic retrenchment and slower technological growth is likely to be high. If oil production declines, it could also require cutting East Europe loose from Soviet petroleum. The political consequences of this action could prove unacceptable, though one Western analyst suggests that as far as Soviet leverage is concerned, the increased supply of gas, nuclear reactors, enriched uranium, reprocessing and electric power may become an adequate substitute for control of oil deliveries.[4] At any rate, costly though this scenario may prove, the option of autarky in the energy field exists. Energy independence for the USSR is a feasible alternative, something no Western country (with the possible exception of Canada) can contemplate. And as shall be pointed out later, rather than making a conscious choice, the Soviet leadership may back into it gradually, step by step.

The alternative scenario, as Campbell noted recently, would be a plunge into economic and technological cooperation with the West to break the energy bottleneck and especially to turn the oil situation around again. But technical assistance from abroad would have to be applied "across a broad spectrum in all the related technological fields–exploration, production, refining, transport and utilization, not just in certain key areas as heretofore." And if this scenario is to work, decisions on it will have to be taken soon, because of the long lead times needed before such measures can bear fruit.[5] Like all actions in the oil business that involve going after yet-to-be discovered reserves that may not materialize, such a plunge is an obvious gamble. However, according to most experts, Soviet potential resources are promising; the odds should favor the gambler. Concurrently with this endeavor, present efforts to attract foreign technology into the gas industry would have to be intensified and closer cooperation sought in the development of nuclear energy as well.

In a recent communication, Professor Campbell goes further. He suggests that the refutal by experience of the optimistic investment and cost expectations for West Siberian oil and gas, on the one hand, and recent quantitative evidence about the low relative productivity of gas in power stations, on the other, may generate greater efforts to solve the problems preventing the upsurge of the coal industry. And here, too, the bottleneck may have to be opened by increased trade. Soviet need for energy technology may shift accordingly, with greater emphasis on coal.

A conscious choice for either of these scenarios would demand boldness and, in the case of drastic retrenchment, even ruthlessness, which at present the aging Soviet leadership does not seem to possess. To throw the door open to Western assistance across the whole spectrum of the energy field, to make such broad participation financially secure and economically attractive against opportunities elsewhere, would demand considerable institutional modifications in the Soviet system. Attempts at technology transfer over such a broad spectrum, particularly with the coal industry added, would also intensify institutional rivalry, complicate and slow decision-making through the vast, cumbersome Soviet bureaucracy. Still more importantly, it would require a political climate of sufficient confidence and trust that is absent today and for which prospects are dim.

On the other hand, the repercussions of economic retrenchment and retreat into autarky are likely to be more disturbing than at any time since World War II. First, as Goldman argued, integration with the world economy has now reached a level where the sheer volume of trade results in a momentum that is increasingly costly to break. This applies not only to the rising role of imports and the infusion of modern technology but the growing impetus to export as well. To cut either off could be disruptive to the system.[6] Second, for the first time since the war, the Soviet economy is entering a period of a sustained squeeze on all three factors of production, land, labor and capital. The closing of the agricultural frontier (no more virgin lands) and depletion of most high-grade and/or easily accessible mineral reserves constrain the supply of this factor. The growth of the labor force has drastically slowed and except among non-European ethnic groups will be essentially nil through 1990. Largely because the supply constraints on these two factors produce pressures for a higher level of capitalization, automation and intensification, but also because of concurrent claims from the military, space, foreign affairs and even the consumer, competing demands on capital resources have intensified. With such a multifold squeeze that will last at least through the 1980s, the high additional cost of turning inward again cannot be contemplated with equanimity.

The writer does not believe that the aging bureaucrats of the Kremlin have the imagination, the feeling of security and the flexibility to make a whole-hearted commitment for open economic cooperation. Nor is the political climate of East-West relations likely to promote such a course in the near future. On the other hand, the ruthlessness and relentless drive to make it alone also seem to be absent. Whether the next generation of leaders will possess either of these qualities is a moot point. Times are not sufficiently desperate or out of joint to produce either another Peter the Great or a Stalin. At any rate, given the Soviet political system, it should take some time for a new hierarchy and set of power relations to get established after the present leaders pass from the scene.

Soviet failure to adopt the long-heralded 15-year plan (1976–90), to publish it even in skeletal form and, by some indications, even to complete it in sufficient detail, may well be a straw in the wind. As the protracted and so far inconclusive dispute between pro-hydrocarbon and pro-coal forces (Chapter 10) also seems to show in a more restricted subfield, Soviet planners appear cognizant of the shape of impending problems. There are no signs, however, that they are willing to face the consequences of a retreat on the energy front, which would bring with it a retrenchment toward autarky through the entire economy. But they are no more ready for the leap into a much more open economy and for its unsettling effects. As in the United States, where the choices are more limited, procrastination seems the order of the day.

The writer therefore doubts that in the foreseeable future the decision for either alternative will be made. But this should not mean that the Soviet energy economy, and by implication the whole economy, can stay on its present course. It seems that the leadership is gambling that dramatic new oil finds will break the tightening bottleneck. In the unlikely but not impossible event that at least one readily exploitable supergiant, together with smaller discoveries, is found before this decade closes, the bottleneck may indeed be broken again. The energy economy can be held more or less on course. If that fails, inaction will back the Soviet leaders gradually toward greater autarky. Economic isolation need not be complete (it never is), or even reach the situation of the 1950s. But if oil exports beyond the Comecon indeed falter, and the debt-service ratio* rises much beyond the recent level (23% in 1976), both less favorable Western terms and Soviet concern itself will begin to limit trade with hard-currency countries.[7] Step by step, the freedom of action will be reduced, economic responses to world market developments increasingly circumscribed. In the

*Debt-service ratio = hard-currency interest payment plus debt repayment divided by exports to the developed West.

opinion of this analyst, such a gradual retrenchment is more likely than a conscious choice of autarky. It is also more probable than the robust Western participation in Soviet resource development called for by the opposite scenario.

Rational energy planning involves the balancing of economic and strategic interests, short-term and long-term gains and concerns in all societies for their own stability. The stability of the Soviet system and the power of those who run it are plainly the paramount goal of Moscow policy-makers. However, economic strength and well-being, as determined by the leadership, are perceived as a fulcrum of system stability. The evidence is undisputable that in the USSR today the rationality of energy decisions embodies economic realities at home and abroad. As one Western scholar put it, "where 'political' ends figure in energy decisions, Soviet decision-makers are well aware of the opportunity costs incurred and appear ready to give up the political ends if the economic opportunity costs go too high."[8]

On the whole, the Soviet energy system today is managed with competence and rationality. For many years now it has also been less buffeted and warped by political pressures and interests than its counterpart in the United States. Its responses to higher energy prices have been in the right direction, have been expeditious and consistent, compliments that cannot be bestowed on the actions of most Western governments. Yet, in the longer haul, this writer perceives and is inclined to emphasize the constraints to continued development and adjustments. The main constraints in the USSR are not the results of political pressures or even strategic priorities, but those of modern technology and the physical environment.

There is merit in the economists' notion that resources are a function of human appraisal and not substances on the shelves of cataloguing clerks, that resource adequacy is more a matter of policy than of physical occurrence and that, as far as sheer presence is concerned, the world is far from running out of energy or any other substance. In a state that holds sovereignty over one-sixth of the earth's surface, the continued presence of vast amounts of energy and other minerals cannot be doubted. The writer, however, is wary of the mechanical formality often displayed in economic analysis. He doubts the ease with which substitution is supposed to take place and supply and demand, if unregulated or simulated by computers, balance each other along a smooth curve. He prefers to stress the high degree of lumpiness of production factors (of physical capital, financial and institutional arrangements, the increasingly strict physical and chemical specifications of energy and raw material inputs), the growing size and lengthening lead times of projects, together with the imperfect knowledge with which decisions about undiscovered reserves must be made. All these obstruct, delay and sometimes downright confound the adjustments called for by economic forces both on the supply and on the demand side. And, finally, as a geographer, the writer is acutely aware of the constraining roles of distance and physical environment so critical in the USSR. They serve to increase the lumpiness of production factors but impede economic growth and adjustment in other ways as well. This assessment of the problems and prospects of the USSR energy economy, therefore, is relatively somber. The vast Soviet resource base and energy potential are recognized and described, the acumen and rationality of Soviet energy planners are acknowledged, Soviet determination and achievements are given due credit. But in the realm of human endeavor, the shadow that falls between idea and reality may be long indeed.

NOTES

[1]Yu. I. Maksimov and Z. P. Tsimdina, *Optimizatsiya razvitiya i razmeshcheniya neftegazovoy promyshlennosti* (Novosibirsk: Nauka, 1977), pp. 58–59.

[2]The 1978 annual plan notes that the five-year targets for saving electricity, fuels and metals are not being met. Val Zabijaka, *Summary of the USSR Annual Plan for 1978.* (Washington: U.S. Department of Commerce. Overseas Business Reports, July 1978), p. 2.

[3]Robert W. Campbell, *Soviet Energy R&D: Goals, Planning and Organizations* (Santa Monica: RAND Corporation, R-2253-DOE, May 1978), pp. 32–35.

[4]Robert W. Campbell, "Implications for the Soviet economy of Soviet energy prospects," mimeographed paper, United States Department of State, September 1977, p. 16.

[5]Ibid., pp. 13–15.

[6]Marshall I. Goldman, "Autarchy or integration–the USSR and the world economy," in U.S. Congress, Joint Economic Committee, *Soviet Economy in a New Perspective* (Washington: Government Printing Office, 1976), pp. 81–96.

[7]Herbert S. Levine and David L. Bond, "Soviet responses to hard-currency problems" (Stanford Research Institute, Washington Office, 1977).

[8]Arthur W. Wright, "The Soviet Union in world energy markets," in Edward W. Erickson and Leonard Waverman eds., *The Energy Question: An International Failure of Policy*, Vol. 1 (Toronto: University of Toronto Press, 1974), p. 85.

INDEX